T0291140

Nanoscale Computing

Nanoscale Computing

The Journey Beyond CMOS with Nanomagnetic Logic

Dr. Santhosh Sivasubramani
Indian Institute of Technology Hyderabad
Hyderabad, India

WILEY

Published by John Wiley & Sons, Inc., Hoboken, New Jersey.
Published simultaneously in Canada.

For general information on our other products and services or for technical support, please contact our Customer Care Department within the United States at (800) 762-2974, outside the United States at (317) 572-3993 or fax (317) 572-4002.

Wiley also publishes its books in a variety of electronic formats. Some content that appears in print may not be available in electronic formats. For more information about Wiley products, visit our web site at www.wiley.com.

Library of Congress Cataloging-in-Publication Data:

Names: Sivasubramani, Santhosh, author.
Title: Nanoscale computing : the journey beyond CMOS with nanomagnetic logic / Santhosh Sivasubramani.
Description: Hoboken, New Jersey : Wiley, [2025] | Includes index.
Identifiers: LCCN 2024043576 (print) | LCCN 2024043577 (ebook) | ISBN 9781394263554 (hardback) | ISBN 9781394263578 (adobe pdf) | ISBN 9781394263561 (epub)
Subjects: LCSH: Nanotechnology. | Semiconductors.
Classification: LCC T174.7 .S5455 2025 (print) | LCC T174.7 (ebook) | DDC 620/.5–dc23/eng/20241118
LC record available at https://lccn.loc.gov/2024043576
LC ebook record available at https://lccn.loc.gov/2024043577

Cover Design: Wiley
Cover Image: © Ali Kahfi/Getty Images

Set in 9.5/12.5pt STIXTwoText by Straive, Chennai, India

Dedicated to my

Grand Mother Paapa R
Grand Father Rajendran M
Mother Lakshmi R
Family &
Professor Amit Acharyya

Contents

About the Author

Dr Santhosh Sivasubramani received his master's and doctoral degrees from Indian Institute of Technology (IIT) Hyderabad, India. His PhD thesis work is on "Rebooting Computing: Nanomagnetic Logic based Computing Architecture Design Methodology" funded by Redpine Signals Inc., PhD Fellowship in the Department of Electrical Engineering. His MTech thesis work is on "Simulation and Experimental Investigation on Electronic Transport and Magnetic Properties of Graphene for its Applications in Nanomagnetic Computing." His bachelor's
thesis work is on "Design Scheme of Maintaining Power System and Protecting Nuclear Reactor by Intelligent Robot Employed with Fuzzy Logic."

Dr Santhosh also possess experience as an INUP Visiting Scholar at the Centre of Excellence in NanoElectronics, IIT Bombay, and as a Project Assistant – DEITY project on "IOT for Smarter Healthcare" Center for Cyber Physical Systems & IoT, Department of Electrical Engineering, IIT Hyderabad. He is also the Founder of RSL Quantum. Santhosh also have gained post-PhD/industrial experience as the Institute Post-Doctoral Research Fellow/Research Associate in the Advanced Embedded Systems and IC Design Laboratory, Department of EE, IIT Hyderabad, and as Research Scientist in the Nanomagnetics Research Lab, Redpine Signals India Pvt Ltd where he was involved in the generation of the IP portfolio (25 INTL. Patents Filed/4 US Patent Granted/2 US Pre-Grant Publication/3 IN Patents Granted/6 IN Pre-Grant Publication). He is also a freelance science journalist who reports for India Science Wire a DST initiative. Santhosh also served as the General Chair of IEEE Summer School on Nano 2022, 2023, 2024. Santhosh is also the key member/organizer/convener of the 1st/2nd/3rd/4th IEEE Hyderabad Nano Day, IEEE NTC World Nanotechnology Marathon 2021, IEEE NTC R10 YP DL/Webinar Series, IEEE NTC YP R10 Panel Discussion, IEEE NTC HSC

Euro-Neuro/QuNa show (2021–2025), and volunteered for DST/AICTE – TEQIP, GIAN, SPARC, FDP, Sci-Tech Councils and on the academic/research – governing board of institutions/NGO organizations across India. SS is the Chair – IEC/IEEE 62659™ Standard for Nanomanufacturing – Large Scale Manufacturing for Nanoelectronics. SS is also the Rebooting Computing Architect.

Santhosh is also the Senior Member of IEEE and currently serve in various administrative voluntary roles:

- Secretary – IEEE Standards Association – Nanotechnology Council (2021–Present);
- Pro-Com/Advisory Com/Technical Com-Member 2021/22/23/24 AtC-AtG IEEE Magnetics Conference;
- Advisory Committee – DST Star – National level Faculty Development Programme on Nano;
- IEEE NTC YP Region 10 (Asia-Pacific) Representative (January 2021–Present);
- Founding Chair – IEEE NTC Hyderabad Section Chapter (2021–Present);
- Elected Secretary – IITH Alumni Association Governing Body (2021–2023);
- IEEE NTC Technical Committee – TC12 Nanomagnetics/TC1 Nanorobotics and Nanomanufacturing;
- Ad-Com Member IEEE Nanotechnology Council;
- IEEE Communications Society Data Storage Technical Committee Member (2020–Present);
- IEEE Computer Society Task Force on Rebooting Computing Member (2018–Present);
- IEEE NTC Young Professionals India Representative (2020);
- Reviewer for various IEEE, IOP, APS, RS, Elsevier Journals, Magazines and Transactions;
- Awardee – IEEE MGA Young Professionals Achievement Award 2023;
- IEEE STEM Champion Awardee 2024–2025; IEEE Transmitter Impact Creator-PV Initiative – 2023–2025;
- IEEE STEM Inspire Grant Awardee – TryNano – Pre Univ School Girls Program – NanoQuest 2024;
- IEEE SMC EDI Systems Engineering Education Initiative Grant Awardee 2024–2025.

Awards/Academic Recognitions: (a) Awarded with the Certificate of Appreciation for Research in Electrical Engineering, IIT Hyderabad. (b) Article "10.1109/TNANO.2018.2874206" has been featured in the top 10 list of articles, in the IEEE TNANO. (c) Selected for and participated in a fully sponsored on-site research visit to IMEC Belgium. (d) Nominee – INAE innovative student projects award (MTech Thesis). (e) Elected as Doctoral Council Representative, Student Gymkhana, IIT Hyderabad. (f) Recipient of Silver Medal from the

Educational Minister, Government of Tamil Nadu for securing lead score in SSLC in the Educational District. (g) Article "10.1088/1361-6528/ab295a" has been featured in the INTERNATIONAL ROADMAP FOR DEVICES AND SYSTEMS 2020/2021/2022/2023 EDITION BEYOND CMOS (IRDS™: Beyond CMOS) international focus group report. (h) Article "10.1038/s41598-020-63360-6" has been featured in the Top 100 in Materials Science 2020 by Nature Publishing Group SN Insights. (i) Participated and won first place in the South Indian Robo Championship League. (j) Selected for participation in DAE, IGCAR sponsored Nuclear Awareness Festival. (k) Associate Editor – IEEE International Conference on Nano 2022. (l) Track Chair 4 & 10 – IEEE INDICON 2023. (m) Session Chair – IEEE APCCAS. (n) IEEE NANO SDC 2024/2025 – Organizer – NANO Int Conf.

Preface

The continuous advancement of technology is driven by the need to process information more efficiently. The traditional transistor-based computing; Complementary metal-oxide-semiconductor (CMOS), powering the digital revolution, encounters a fundamental challenge at the atomic scale. As transistors shrink beyond nanometers, challenges such as leakage currents, heat dissipation, and miniaturization limitations alarm computing power growth. *Nanoscale Computing: The Journey Beyond CMOS with Nanomagnetic Logic* envisions the future of information processing under the umbrella of "Rebooting Computing." This book is a fundamental textbook and an active guide, inviting readers to explore the historical perspective of 75 years of Transistor and 100 years of Spin. It leads to a domain where electrons step back, and spin becomes the key of logic. For students, whether undergraduate or graduate, this book helps them understand nanoscale computing from multiple view points. It explains basic ideas and shows where current technology has some problems, which might lead to new and creative solutions. Readers learn about nanomagnetic logic (NML), where tiny magnets are used for computing, which could make devices more powerful and efficient. For tech experts and researchers, the book serves as a conversation partner and fellow explorer. Together, they explore novel architectures beyond CMOS limitations, unveiling material design fundamentals crucial for next-generation logic. The exploration extends to edge AI, where nanomagnetic logic promises efficient interactions with the world. For the first of its kind, this book demonstrates transformative power of interdisciplinary collaboration, bridging theoretical concepts with practical applications. It showcases how nanoscale computing and nanomagnetic logic intersect with AI, presenting case studies in various fields, from medical diagnostics to environmental sensing. Acknowledging challenges, the book discusses hurdles like reliability, scalability, and integration with existing technologies. This discussion paves the way for future research and innovation, speculative laws on NML, ensuring the journey beyond CMOS extends beyond the book's final page. The author highlights, *"Performing AI computing on the edge with approximate nanomagnetic logic deployed on the magnetic ICs is an envisaged vision towards energy efficient sustainable futuristic computation. It aim towards UNSDGs-carbon net zero, directly impacting climate change mitigation strategies also lead by the IEEE as a forerunner."* This fundamental textbook is an invitation to be part of a Rebooting Computing revolution. It calls students to shape the future, encourages researchers to push boundaries, and challenges tech experts to turn imagination into reality. Turn the page, and let the journey into the future of computing begin. This book has been carefully crafted to meet the current demands of individuals interested in learning about technology and implementing it in their lives. Its contents have been meticulously developed over a period of five years, with special emphasis on new educational models aligned with UN SDG goals. For the first time, a fundamental textbook has been introduced with this approach, which will keep undergraduate students and readers engaged and thoughtfully provoked. This represents an experiment! With excitement and anticipation.

Hyderabad, 2024   *Santhosh Sivasubramani*

Acknowledgements

This book has undergone several stages, including ideation, process development, content curation, proofreading, flow/coherence checks, figure population, and fact-checking. It's important to note that this was not a simple task, and this fundamental textbook has been carefully crafted to serve its purpose for undergraduates, graduates, and advanced readers alike. This five-year effort would not have been possible without the support of many individuals who have contributed directly and indirectly. This space is dedicated to acknowledging all of them and also to people who motivated me and supported throughout. My special thanks to Mr. Chandrasekaran Lakshmanan, former Head at VHN, India, Mr. Nagarajan, and Mr. Selvan Arputhraj for motivating me to enter the wonder-world of research. I convey my warm regards to Dr. S. Thabasu Kannan, former Director of PCET, India, Mr. Krishnamoorthy, Mr. Bakrutheen, Mr. JayaMurugan, and Mrs. Arul mozhi, who guided me. I thank Mrs. Jayalakshmi, who deserves a special appreciation. I am indebted to my PhD advisors Professors Amit Acharyya and Swati Ghosh Acharyya for their continuous support. Mere words cannot express my gratitude to Dr.Amit who mentored me and shaped me as an independent meticulous researcher/educationist and also had an impact on me in my personal life. This is my pleasure to acknowledge Mrs. Sanghamitra Debroy who helped me in all aspects of experimental research. Special thanks to Dr. Chandrajit Pal. I also acknowledge INUP, Centre of Excellence in Nanoelectronics at the IIT, Bombay, Dr. Nageshwari, and Ms. Gayathri Vaidhya. I acknowledge the support from IEEE Nanotechnology Council (NTC), IEEE NTC TC-12 Nanomagnetics, IEEE NTC Standards, IEEE NTC YP, IEEE P62659, IEEE EDS/YP, IEEE Standards, IEEE Young Professionals, IEEE Education Society, IEEE Magnetics Society, IEEE AtC-AtG, IEEE Communications Society Data Storage TC, IEEE Computer Society – Task Force on Rebooting Computing, IEEE MGA, IEEE Hyderabad Section, IEEE India Council, IEEE Region 10, and IIT Hyderabad Alumni Association. I extend my sincerest thanks to Ms Redpine Signals Inc., for the support. Special mention to Dr. Venkat Mattela CEO Ceremorphic Inc., I thank Mr. N Venkatesh, Silicon

Labs, for his support. I thank and acknowledge the support and guidance from Professor and Chair of Department of ECE, UC Davis, Dr. M. Saif Islam. My special thanks to Dr. Vanlin Sathya and Dr. A.R. Aravinth Kumar. Special mention to Mr. Arun Ramamoorthy, Dr. Karthick Thangavelu. Special acknowledgment to my colleagues in AESICD Laboratory. Special mention to Mr. Rangesh, and Mr. N. Syed Ghouse. Special thanks to Mrs. Malini Raguraman, and Mr. Yedu Kondalu. I extend my sincerest thanks to Dr. Sparsh Mittal, Dr. Sushmee Badhulika, and Dr Rishad Shafik. Special thanks and appreciation to Mr. Kishore C. and Ms. Suma M. for their continuous support. I acknowledge my students and interns who have collectively contributed a lot. Finally, I offer my warmest thanks to my mom Lakshmi R., without whose support and inspiration it would not have been possible for me to devote five years of time, alongside my other primary commitments, to bring this book out. My wholehearted thanks to my family members – Mr. Rajendran M., Mrs. Paapa R., Mrs. Mariyammal R., Mr. Murugananthan A.G, Pavithra, Gayathri, Mr. Singaravel R., Mrs. Ponna S., Shwetha, Pradeep, Mr. Prabhakaran R., Mrs. Jothimani P., Sabarivasan, Karthik, Mrs. Meena K., Mr. Kumaresan B., Poornima, Harshini, Mrs. Priya R., Mr. Ramesh E., Sushil, and Dharoon.

Dr. Santhosh Sivasubramani

Acronyms

AD	Architectural Design
AFM	Atomic Force Microscopy
AI	Artificical Intelligence
ASIC	Application-Specific Integrated Chip
BC	Beyond CMOS
CMOS	Complementary Metal-Oxide Semiconductor
CNT	Carbon Nano Tube
CPU	Central Processing Unit
DIBL	Drain-Induced Barrier Lowering
DVFS	Dynamic Voltage and Frequency Scaling
DW	Domain Wall
GPU	Graphics Processing Unit
IEC	Interlayer Exchange Coupling
IIOT	Industrial Internet of Things
IOT	Internet of Things
IRDS	The International Roadmap for Devices and Systems
LLG	Landau–Lifshitz–Gilbert
MD	Material Design
MLG	Majority Logic Gate
MRAM	Magnetic Random Access Memory
MTJ	Magnetic Tunnel Junction
NC	Nanoscale Computing
NML	Nanomagnetic Logic
NVM	Nonvolatile Memory
PCM	Phase-Change Memory
PCP	Performance Comparison Parameters
QC	Quantum Computing
QD	Quantum Dot
RC	Rebooting Computing

SCE	Short-Channel Effect
SOT	Spin–Orbit Torque
STM	Scanning Tunnelling Microscopy
STT	Spin-Transfer Torque
SW	Spin Wave
TMR	Tunneling Magnetoresistance

Introduction

This book is being written with the purpose of addressing a diverse audience, including graduate and undergraduate students, as well as tech experts and researchers. This book offers an exciting exploration of cutting-edge technology that goes beyond traditional computing methods. It gets into the nanoscale computing and introduces the concept of nanomagnetic logic, which has the potential to revolutionize how we process information.

Detailed Rundown of Subjects Treated:
1) **Nanoscale Computing Basics:** The book starts with an accessible introduction to nanoscale computing, explaining the fundamental principles and significance of operating at such tiny scales, under the umbrella of "Rebooting Computing."
2) **Limitations of CMOS:** Readers will learn about the challenges faced by traditional computing methods and the challenges in achieving higher performance and efficiency.
3) **Nanomagnetic Logic:** This is where the excitement builds up. Nanomagnetic logic is introduced as a game-changing concept that relies on the magnetic properties of nanoscale materials to process information in innovative ways.
4) **Beyond CMOS Architectures:** The book ventures deep into various architectural designs that go beyond CMOS, showcasing how nanomagnetic logic can be integrated into next-generation computing systems.
5) **Edge AI Devices:** The practical applications come into focus as the book explores how nanomagnetic logic can advance edge AI devices. This includes discussions on AI-driven technologies that can operate closer to the data source, enabling faster and more efficient processing.

This book stands out from competing titles because of its unique emphasis on the intersection of nanomagnetic logic with AI and architecture for edge devices. It bridges the gap between theoretical concepts and real-world applications, making it suitable for a broader range of readers, including both beginners and seasoned tech professionals. One of the book's highlights is its emphasis on the applications of nanomagnetic logic in edge AI devices. It explores the integration of nanomagnetic logic in AI-driven technologies and provides case studies demonstrating its advantages. Additionally, hybrid computing systems that combine nanomagnetic logic with CMOS are explored, presenting a promising direction for energy-efficient computing. The challenges and future directions of nanoscale computing are also addressed, including reliability, scalability, and integration with existing technologies. The book concludes with a glimpse into the potential applications of nanomagnetic logic in emerging technologies, such as neuromorphic computing, IoT, bio-nanoelectronics, and environmental sensing leading to rebooting computing.

Chapters' arrangement is as follows:

Chapter 1 introduces the foundational concepts of nanoscale computing, tracing its historical trajectory and practical significance. It begins by examining the connections between nanoscale computing and the rebooting computing initiative, emphasizing the need for a re-evaluation of traditional computing methods. The chapter then explores the limitations of CMOS technology and introduces nanomagnetic logic as a potential solution. It also discusses the role of edge AI devices in driving advancements in nanoscale computing and addresses architectural considerations and material design strategies. Lastly, the chapter outlines the objectives of the book and the key points it aims to cover.

Chapter 2 examines the practical constraints of traditional CMOS technology, discussing challenges including scaling limits, power consumption, and performance bottlenecks. It explores their implications on computing systems, including reduced processing power and energy efficiency concerns. The chapter also addresses technological and economic challenges, emphasizing the importance of adaptability and innovation. Additionally, it introduces nanomagnetic logic as a potential solution, outlining its advantages over CMOS. Educational emphasis is placed on delivering accessible technical content suitable for undergraduate audiences.

Chapter 3 provides a foundational understanding of nanomagnetic logic, starting with its basic principles and historical context. It explores essential elements such as coupling mechanisms, material selection, and spin dynamics crucial for designing logic gates. Signal processing aspects, generation, detection, and propagation, are examined alongside energy considerations and efficiency strategies. The chapter prioritizes accessible explanations for undergraduates, supplemented with practical examples and applications to enhance understanding. It concludes with a chapter-end quiz aimed at reinforcing comprehension and facilitating student engagement in exploring the intricacies of nanomagnetic logic.

Chapter 4 gets into nanomagnetic logic architectures, providing a comprehensive examination of their evolution and design principles. It starts by outlining the importance of architectural considerations and progresses to explore major architectures for both combinational and sequential logics. The chapter analyzes fundamental aspects such as magnetic implementation, quantum considerations, and benchmarking concepts. It further investigates parallel and pipelined architectures, focusing on enhancing computational throughput. Additionally, reconfigurable architectures are discussed, including dynamic reconfiguration concepts and their applications. The chapter concludes with a summary of key architectures and offers a chapter-end quiz to reinforce understanding.

Chapter 5 introduces material design for nanoscale computing. It emphasizes material selection's importance. The chapter explores magnetic and nonmagnetic materials. It discusses their unique characteristics and applications. Additionally, it covers challenges and optimization strategies. The integration of multiferroic and spintronic materials is analyzed. Case studies in nanomagnetic logic computing are examined. The chapter concludes with future directions and a quiz.

Chapter 6 introduces nanoscale computing at the edge. It explores edge computing and defines its characteristics, and objectives. The chapter examines the intersection of nanoscale computing and edge AI, highlighting the role of nanomagnetic logic. It discusses AI integration in edge devices and its applications in IoT, healthcare, and robotics. Additionally, it offers a tutorial on developing edge AI applications, emphasizing hands-on learning. The chapter concludes by summarizing key concepts in edge AI, preparing for further discussions on hybrid computing in subsequent chapters.

Chapter 7 introduces hybrid computing systems and emerging applications. It defines hybrid computing systems, tracing their historical evolution and milestones. The chapter explores emerging applications driving hybrid systems, emphasizing the integration of nanomagnetic logic with other technologies. It delves into nanomagnetic–CMOS hybrid architectures, discussing benefits, performance enhancements, and applications. Additionally, it examines neuromorphic hybrid systems, focusing on advancements in cognitive computing and real-time learning. The chapter also explores emerging applications in industry, particularly in industrial automation, smart manufacturing, robotics, transportation, and agriculture. Lastly, it emphasizes demystifying hybrid computing concepts for educational purposes, preparing students for future industry challenges.

Chapter 8 introduces challenges, conclusions, a roadmap, and future perspectives in nanoscale computing. It outlines technical, educational, and policy challenges, emphasizing reliability, error rates, and scalability solutions. Additionally, it addresses the environmental impact and integration with other technologies like quantum computing, IoT, and edge AI. The chapter presents a roadmap for nanoscale computing technologies, discussing characteristics and conclusions. It explores research opportunities, inviting feedback and student involvement. Finally, it speculates on the mathematical trajectory plots concerning nanomagnetic logic, aiming to identify gaps in research and inspire future exploration in the field.

In conclusion, "Nanoscale Computing: The Journey Beyond CMOS with Nanomagnetic Logic: A Fundamental Textbook Advancing Edge AI Devices through Architecture and Material Design" under the umbrella of "Rebooting Computing" provides a comprehensive understanding of nanomagnetic logic and its applications, catering to a diverse audience of readers interested in cutting-edge technology.

About the Companion Website

This book is accompanied by a companion website:

www.wiley.com/go/sivasubramani/nanoscalecomputing1

This website includes:
- Appendix A: End of Course Assessment Q&As - Chapter Wise
- Appendix B: Fundamental - Micro-magnetic Simulation Tools
- Appendix C: Architecture - Micro-magnetic Simulation Codes
- Appendix D: Material Modeling Simulation Tools and Codes

1

Introduction to Nanoscale Computing

Specific, Measurable, Achievable, Relevant, and Time-bound (SMART) learning objectives and goals for Chapter 1 are to:

- Understand the historical development of nanoscale computing (NC) and its synergies with the rebooting computing (RC) initiative.
- Define and explain the significance and applications of NC.
- Comprehend the challenges posed by traditional complementary metal-oxide-semiconductor (CMOS) technology.
- Familiarize the emergence of nanomagnetic logic (NML) as an alternative to CMOS.

NC is an emerging field which processes information at tiny scales. This is setting up the trend in technological innovations, which couldn't have been possible earlier in information processing. NC is a multi-interdisciplinary field which changes the way we perceive and process information. It deals with the design and development of nanometer-scale components which can perform complex high-performance computations on the go. Novel material exploration, including carbon nanotube (CNT), graphene, and nanodots alongside novel fabrication techniques, is one of the key elements of NC. These material explorations could potentially replace or complement the silicon-based components used in traditional computing devices with their remarkable thermal and electric properties. This comprehends to the bigger goal of designing smaller, energy-efficient devices for RC. NC is also home to quantum computing, which performs computation by exploiting the principles of quantum mechanics. Qubits exist in dual states simultaneously, enabling quantum computers to process information beyond the capabilities of traditional computing. Superconducting circuits, trapped ions, and topological qubits are key nanoscale approaches for developing qubits to perform computationally intensive tasks. Though NC possesses enormous potential in designing and developing next-generation computing platforms for edge artificial

Nanoscale Computing: The Journey Beyond CMOS with Nanomagnetic Logic,
First Edition. Santhosh Sivasubramani.

intelligence (AI) devices, significant challenges are associated with overcoming them, ensuring energy-efficient operations without degradation in performance.

This chapter will talk about the brief overview in Section 1.1, the evolution Beyond CMOS in Section 1.2, edge AI devices as a driving force in Section 1.3, the architectural and material design in Section 1.4 and concludes the chapter detailing the scope of the book alongside the themes and the interactive assignment for encouraging readers in Sections 1.5 and 1.6. This ensures readers to actively participate in the new initiative of redefining educational models, serving as a motivation for undergraduate students and also a potential learning via fun exercise for advanced readers.

1.1 Overview of Nanoscale Computing

This section introduces the readers to the overview of NC. It briefly talks about the synergies in NC and rebooting computing (RC) in Section 1.1.1, defines the NC in Section 1.1.2 and historical development and the significance/applications in Section 1.1.3. Now let's read through the synergies and intertwined perspective.

1.1.1 Synergies in Nanoscale Computing and the Rebooting Computing Initiative

RC is an IEEE initiative to rethink the design and build of computers (cf. Figure 1.1). Von Neumann's architecture, introduced in the 1940s, is the basis for traditional computers, and it has started implications due to scaling concerns arising. RC aims to design and develop "from soup to nuts" – holistically in how computing is perceived, mitigating the challenges posed by the aforementioned.

NC and RC are intertwined areas focused on overcoming the limitations posed by traditional computing technologies. An extensive synergistic complement list of each other's goals and vision is as follows:

- Energy Efficiency
- New Computing Paradigms

Figure 1.1 IEEE rebooting computing
File:IEEE Rebooting Computing TaskForce Logo.png
https://commons.wikimedia.org/wiki/File:IEEE_
Rebooting_Computing_TaskForce:Logo.png. Source:
IEEE Logo.

- Scalability
- Emerging Technologies
- Performance Enhancement

Integrating nanoscale components into RC architectures paves the path toward advancing technologies.

1.1.1.1 Redefining Computing: Exploring the Interconnection

One significant aspect of this synergy is the exploration of (a) alternative materials and (b) fabrication techniques. RC encourages integrating these innovations into a broader framework, emphasizing the need for more efficient and powerful computing. RC recognizes the potential of quantum computing and deploys strategies for research and development. At the nanoscale, quantum effects bring in uncertainties and computational errors. RC acknowledges the need for resilient computing that works with fault tolerance. Thus, NC and RC collectively contribute to developing novel information-processing technologies.

The International Roadmap for Devices and Systems (IRDS) rightly articulates the emphasis on the Beyond CMOS computing paradigm and its corresponding application pulls, as depicted in the Figure 1.2. The device's specification lies in the range of ∽100 up to 1 nm. To realize Beyond CMOS, the major three pillars of focus are as follows:

- Emerging Architectures
- Emerging Materials
- Emerging Devices/Process

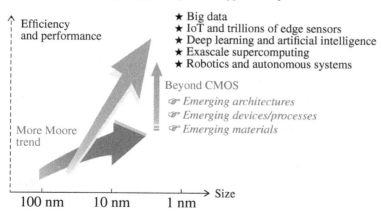

Novel computing paradigms and application pulls

Efficiency and performance

★ Big data
★ IoT and trillions of edge sensors
★ Deep learning and artificial intelligence
★ Exascale supercomputing
★ Robotics and autonomous systems

Beyond CMOS

☞ *Emerging architectures*
☞ *Emerging devices/processes*
☞ *Emerging materials*

More Moore trend

100 nm   10 nm   1 nm   Size

Figure 1.2 IRDS perspective on Beyond CMOS – next-generation computing paradigms File:Beyond CMOS IRDS.jpg https://commons.wikimedia.org/wiki/File:Beyond_CMOS_IRDS.jpg. Source: Errolhunt/Wikimedia Commons/CC BY 4.0.

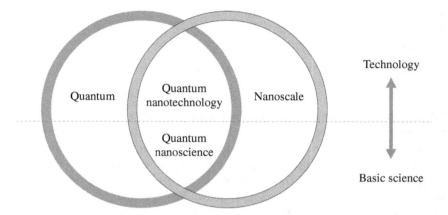

Figure 1.3 Intersection of quantum and nanoscale technologies File:Technology applications of quantum nanoscience.jpg https://commons.wikimedia.org/wiki/File: Technology_applications_of_quantum_nanoscience.jpg. Source: Rickinasia/Wikimedia Commons/CC BY-SA 4.0.

This book talks about all of these three focus pillars throughout. As depicted in Figure 1.3, the intersection of quantum, an individual future direction initiative from IEEE, with the NC, emerges as a critical area of consideration and development touching base on basic science and technology nodes. This book covers a detailed run-through on quantum computing (QC) in later advanced chapters reflecting its goals and visions aligned toward RC. In the subsequent section, let's understand more details on the NC (cf. Section 1.1.2).

1.1.2 Understanding Nanoscale Computing

In this section, we will get into more detail about understanding the NC. This section comprises the following subsections: Section 1.1.2.1 defines the scale, Section 1.1.2.2 contextualizes nanoscale technologies, Section 1.1.2.3 underscores its relevance to modern computing, and Section 1.1.2.4 presents a real-time analogy — for the clarity of the readers.

1.1.2.1 Defining the Scale

The three major factors determining the definition of scale for NC are as follows:

- Size
- Quantum effects
- Fabrication methods

The fundamental three measurable parameters emerging as a key component in the NC paradigm are as follows:

- Control of information
- Different properties from its macroscopic counterparts
- Reliability and efficiency

As aforesaid, NC refers to manipulating and controlling information at the nanometer scale, involving structures and devices operating on the order of nanometers. For understanding, one nanometer equals one billionth of a meter cf. Figure 1.4. At this minuscule size, the behavior of materials and devices differs

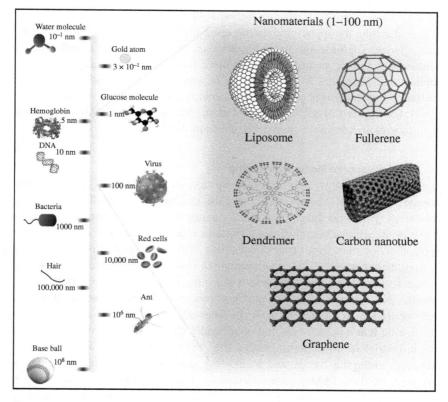

Figure 1.4 Understanding the size of nano-scale definition [from glucose molecules at 1 nm to virus at 100 nm, nanomaterials cover a wide range of sizes. Take a look at this informative diagram] File:Comparison of nanomaterials sizes.jpg https://commons .wikimedia.org/wiki/File:Comparison_of_nanomaterials_sizes.jpg. Sureshbup/Wikimedia Commons/CC BY-SA 3.0.

significantly from their macroscopic counterparts, leading to unique opportunities and challenges in computing. Quantum effects introduce superposition and entanglement, differing from the classical behaviors observed in macroscopic systems. This necessitates a paradigm shift in how we conceptualize and engineer. Additionally, the fabrication techniques employed in NC play a pivotal role. Conventional manufacturing methods are often inadequate for creating structures at such minute dimensions. Therefore, it paved the way for exploring innovative self-assembly techniques and bottom-up manufacturing. These methods involve precise control; by comprehending and advancing these fabrication techniques, scientists aim to build reliable and efficient NC components, further emphasizing the interdisciplinary nature of this field.

1.1.2.2 Contextualizing Nanoscale Technologies

A multidisciplinary approach drawing upon the principles from physics, materials science, engineering, and computer science is required to contextualize nanoscale technologies and creates a holistic framework for understanding and applications. Emphasis is placed on the requirement of collaborative efforts to overcome the challenges of thermal and energy considerations. As components shrink, (cf. Figure 1.5) efficient thermal management and energy-efficient design technologies are critical to ensure reliable operation.

1.1.2.3 Relevance to Modern Computing

The prime focal point with relevance to modern computing is revisiting data storage. Traditional storage mediums face challenges in handling the rising volume of generated data. NC offers a paradigm shift by exploring innovative data storage approaches, ensuring faster and more powerful storage nodes. The application of nanoscale technologies also promises exponentially increasing computational speed. This acceleration is vital in today's scenarios, where complex calculations and simulations are integral in scientific research and AI systems. NC systems exhibit a significant reduction in energy requirements for their operation in alignment with the ongoing environmental considerations, positioning it as a sustainable and eco-friendly alternative. Global academicians and industrial fellows also emphasize the compatibility of nanoscale technologies with emerging trends and technology predictions. The advent of Large Language Models (LLMs), edge computing, cloud computing, and the Internet of Things (IoT) necessitates powerful computing systems that are inherently adaptable to diverse and dynamic environments. Nanoscale components, with their versatility and scalability, seamlessly align with these evolving computing architectures, as pointed out by the IEEE International Roadmap for Devices and Systems.

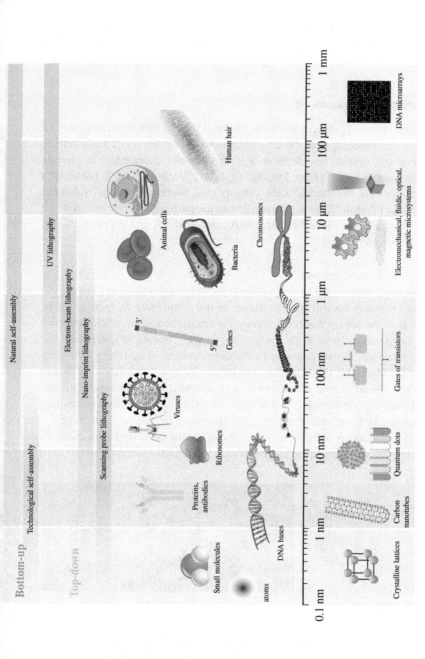

Figure 1.5 Biological and technological scales compared File:Biological and technological scales compared File:Biological and technological scales compared-en.svg https://commons .wikimedia.org/wiki/File:Biological_and_technological_scales_compared-en.svg. Source: Guillaume Paumier/Wikimedia Commons/CC BY 2.5.

1.1.2.4 Analogy

Think of the nanoscale as the world of ants. If traditional computing is like building structures for humans, NC is like creating complex ant colonies. The scale is tiny, but the complexity and organization are interesting (cf. Figure 1.6).

1.1.3 Historical Development

In this section, we will get into more detail about the historical development of the NC. This section comprises the following subsections: Section 1.1.3.1 defines the evolutionary milestones, Section 1.1.3.2 presents pioneering nanoscale technologies and key advances leading to NML, Section 1.1.3.3 presents a real-time analogy for the clarity of the readers, and Section 1.1.3.4 provides the significance/application. This sets the readers on the pathway of progression in the NC, unveiling the chronicle of breakthroughs, innovations, and transformations that help shape this field.

1.1.3.1 Evolutionary Milestones

Physicist Richard Feynman's 1959 speech is the starting point for exploring NC. He outlined his vision for influencing matter at the atomic and molecular levels. This gave rise to the idea of nanotechnology. He envisioned nanotechnology even before the term was coined. With the famous saying, "Seeing is Believing", the revolution in navigating NC began with the development of scanning tunneling microscopy (STM) in the 1980s, which made it possible for scientists to view and work with individual atoms. An increase in investment, with governments and industry players contributing billions to improve chip production and technological capabilities, is a significant ongoing evolutionary milestone. Uncertainties regarding the ongoing geopolitical tensions also affect the semiconductor industry globally alongside the expanding use cases of AI and LLMs, driving up demand

World of ants   World of NC
(a)   (b)

Figure 1.6 (a, b) Real-time analogy to compare and understand the world of ants and the NC. Source: (a) lirtlon/Adobe Stock Photos. (b) catalin/Adobe Stock Photos.

for specialized, high-performance CPUs and GPUs. Concerns about security and sustainability, including environmental issues and the growing significance of cybersecurity, are crucial in determining the path taken. Developments in brain–computer interfaces, quantum computing, and NML pave the way for a revolution in the next-generation computing industry and fundamentally alter how human–machine relationships are perceived, targeting Industry 4.0, 5.0, and Beyond (cf. Table 1.1).

1.1.3.2 Pioneering Nanoscale Technologies

- The design and development of CNT and the atomic force microscopy (AFM) opened up new avenues for NC. Due to their remarkable qualities, CNTs started to be considered for use in constructing nanoscale transistors.
- The demonstration of CNT forecasted nanoscale materials to be tiny and more efficient than their microscopic graphite counterparts. This signaled a change from abstract concepts to real-world developments.

Key Advances Leading to Nanomagnetic Logic:

- The investigation of nanomagnetic logic is one of the major developments in NC. With its unique nanoscale magnetic characteristics, this technology node offers an alternative to conventional semiconductor-based logic.
- NML offers higher computational density and less power consumption than its counterpart, showcasing promise for enhanced device performance and deployment.

1.1.3.3 Analogy

Consider the development of smartphones as a walk-through time analogous to NC (cf. Figure 1.7). From bigger, simple cell phones to advanced smartphones with embedded features. Comparably, NC has progressed from simple architectures to extremely sophisticated systems.

1.1.3.4 Significance and Applications

Considering the viable significance of NC and its real-time applications, sensors emerge as a transformative healthcare measure, enabling focused and accurate medical diagnosis. NC has its consequences for energy storage, opening the door to developing batteries with higher efficiency and PCP. These highlight NC's concrete advantages and its potential to solve urgent global issues. NC emerges as a technological paradigm change that answers the problems encountered by conventional computing. The use-cases, historical evolution, and relevance of NC highlight the same. The field of computing is set for rapid developments, leading

Table 1.1 Key Evolution in the History of Semiconductors

Year	Key evolution
1821	Thermoelectric effect observed in semiconducting metals
1874	Point-contact semiconductor rectification discovered
1894	Use of crystals to detect radio waves
1926	Patent for a field-effect semiconductor device
1931	Alan Wilson's "The Theory of Electronic Semiconductors"
1940	Discovery of the p–n junction and photovoltaic effects
1947	Development of the bipolar point-contact transistor
1951	Invention of the static induction transistor
1954	Design of the first silicon junction transistor
1956	Creation of prototype silicon devices
1958	Invention of the microcircuit featuring active components
1959	Invention of the planar manufacturing process
1959	Creation of the first MOSFET
1963	Development of CMOS fabrication process
1964	Introduction of the first widely used analog IC
1965	Gordon Moore introduces Moore's Law
1965	Manufacturing of semiconductor read-only-memory chips
1967	Report on a floating gate MOSFET
1968	Development of silicon-gate technology for ICs
1971	Release of the Intel 4004 microprocessor
1971	Coining of the term "Silicon Valley"
1971	Introduction of erasable, programmable read-only memory
1974	Samsung's entry into the semiconductor business
1977	Release of the Apple II personal computer
1978	Development of programmable array logic devices
1980	Creation of Flash Memory
1984	Demonstration of the first double-gate MOSFET
1984	Foundation of ASML, specializing in photolithography
1985	Creation of Qualcomm as a semiconductor and tech company
1986	Japan becomes the biggest supplier in the global market
1987	Founding of TSMC in Taiwan

(Continued)

Table 1.1 (Continued)

Year	Key Evolution
1993	Foundation of Nvidia, specializing in graphics processing
1997	First meeting of the World Semiconductor Council
2000	Establishment of SMIC in Shanghai
2007	iPhone revolutionizes smartphones
2016	Google introduces Tensor Processing Unit for ML
2020	Global chip shortage leading to semiconductor reevaluation
Jan 2021	TSMC announces $100 billion investment for semiconductor expansion.
Aug 2021	SMIC invests $8.87 billion for self-sufficiency in chip production.
Oct 2021	Google unveils Tensor Processing Unit (TPU) v4 for machine learning.
Dec 2021	India launched the Semicon India Programme – India's Semiconductor Mission (ISM) for self-reliant sustainability
Feb 2022	Intel announces plans for two new chip factories
Apr 2022	Samsung confirms $45 billion investment to expand logic chip production.
Jun 2022	US–China trade war continues to impact semiconductor industry.
Sep 2022	NVIDIA acquires Arm for $40 billion.
Nov 2022	International Chips and Science Act, providing $52 billion for semiconductor research.
Jan 2023	TSMC announces development of 3nm chip manufacturing process.
Mar 2023	India unveils $70 billion semiconductor incentive program.
Jun 2023	First quantum computers with commercial applications begin to emerge.
Sep 2023	EU Chips Act proposed, to invest €43 billion in semiconductor research.
Dec 2023	Cybersecurity issues related to chip design and manufacturing gain prominence.
2024 and Beyond	Continued global focus on supply chain diversification and domestic chip production.
2024 and Beyond	Emergence of specialized chips for AI, AR/VR, and autonomous vehicles.
2024 and Beyond	Growing momentum in sustainability efforts for eco-friendly green chip manufacturing.

Source: Compiled by the author from various sources including: Computer History Museum.

Figure 1.7 Real-time analogy to compare and understand the evolution of NC File:Mobile Phone Evolution 1992–2014.jpg https://commons .wikimedia.org/wiki/File: Mobile_Phone_Evolution_ 1992_-_2014.jpg. Source: Jojhnjoy / Wikimedia / Public Domain.

to the expansion of what can be accomplished with the continuous investigation and utilization of the unique characteristics of matter at the nanoscale.

The key significant application-oriented keywords for further exploration are: Quantum Leap; Beyond Binary; Medical Marvels; Energy Efficiency; Interdisciplinary Innovation; Practical Progress.

1.2 Evolution Beyond CMOS

This section introduces the readers to the evolution Beyond CMOS. It briefly talks about the dual historical perspective of the 75 Years of Transistor and the 100 Years of Spin in Section 1.2.1, defines the traditional challenges in Section 1.2.2, and elaborately introduces the emergence of NML in Section 1.2.3. Now, let's walk through the evolution and the emergence.

1.2.1 The 75 Years of Transistor and 100 Years of Spin: A Dual Historical Perspective

Over the decades, the transistor and spin have significantly impacted the development of technology. For the past 75 years, silicon-based CMOS technology has been at the forefront of electronic device development. Constant evolution and progress in transistor performance and energy efficiency have been the driving parameters since the first planar-gate MOSFET demonstration in the late 1950s. This has revolutionized the world of electronics we use today in this new post-COVID-19 digital era (Figures 1.8, 1.9, and 1.10).

In parallel, research on understanding spin-particles' intrinsic angular momentum has been evolving. However, spintronics-based devices have just gained traction. The Datta-Das SPINFET, which used the spin of electrons to modify and control information in a transistor, marks a significant advancement. Compared to traditional CMOS technology, this achieves low-power and high-speed operation. Appelbaum and Monsma have proposed the transit time spin field-effect transistor in Silicon.

The relationship between spin and transistors is notable in the history of technology. For the past 75 years, the transistor, which can control and amplify electrical signals, has ruled contemporary computing. Utilizing the inherent qualities of particles, spin has created new avenues for information processing and storage. The combination of the transistor and spin allows advanced development in computing and communication. This emerging study area can completely change our thoughts on the fundamental building blocks of next-generation computing. The intertwining perspective of 75 years of transistors and 100 years of spin highlights advancing technology's continuous evolution and integration.

(a)   (b)

Figure 1.8 (a,b) 75 Years of transistors and the 100 year of spin – dual historical perspective. a) File:Bardeen Shockley Brattain 1948.JPG https://commons.wikimedia.org/wiki/File:Bardeen_Shockley_Brattain_1948.JPG Source: (a) AT&T; photographer: Jack St / Wikimedia / Public Domain. b) File:Wolfgang Pauli.jpg https://commons.wikimedia.org/wiki/File:Wolfgang_Pauli.jpg (b) CERN Document Server/CC BY-4.0.

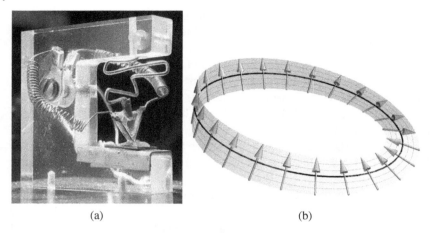

(a)   (b)

Figure 1.9 75 Years of transistors and the 100 year of spin – dual historical perspective – prototype representations. a) File:Nachbau desersten Transistors.jpg https://commons.wikimedia.org/wiki/File:Nachbau_des_ersten_Transistors.jpg Source: (a) Stahlkocher at German Wikipedia/Wikimedia/CC BY SA 3.0. b) File:Spinor on the circle.png https://commons.wikimedia.org/wiki/File:Spinor_on_the_circle.png Source: (b) Slawekb at English Wikipedia, CC BY-SA 3.0 https://creativecommons.org/licenses/by-sa/3.0, via Wikimedia Commons.

(a)   (b)

Figure 1.10 (a,b) 100 Years of spin — historical perspective — hand in hand theory and experimentation. Source: a)File:SternGerlach2.jpg https://commons.wikimedia.org/wiki/File:SternGerlach2.jpg (a) Peng/Wikimedia/CC BY SA 3.0. b) ElectronSpinLeiden2017.jpg https://commons.wikimedia.org/wiki/File:ElectronSpinLeiden2017.jpg Source: (b) Vysotsky/Wikimedia/CC BY SA 4.0.

1.2.2 Challenges of Traditional CMOS Technology

CMOS technology has been the backbone of the semiconductor industry, enabling the development of efficient electronic devices. However, as we go toward more complex applications and push the limits of miniaturization, significant issues result. This section will examine the obstacles researchers and engineers confront around three significant challenges: scaling, power dissipation, and performance. Subsequently, Section 1.2.2.1 details scaling limitations, Section 1.2.2.2 discusses power dissipation issues, and Section 1.2.2.3 provides a detailed brief on performance bottlenecks.

1.2.2.1 Scaling Limitations

The focus on the decreasing transistor size has propelled the industry forward. However, we face severe hurdles as we go closer to nanoscale levels. Quantum phenomena and fundamental physical limits traditional CMOS scaling. Quantum tunneling effects become more evident at such small scales, leading to increased leakage currents and decreased dependability. The three key parameters are as follows:

- Quantum effects and variability
- Gate oxide thickness and tunneling currents
- Innovations beyond traditional CMOS

One critical issue is that transistor channel length approaches the de Broglie wavelength of electrons. Electron flow regulation becomes less predictable, resulting in variable transistor behavior and increased sensitivity to manufacturing fluctuations. Statistical variances are amplified by a reduction in the number of channel electrons, generating uncertainty in transistor properties. Overcoming these scalability constraints necessitates innovative materials and structures, instigating researchers to explore beyond traditional CMOS paradigms.

1.2.2.2 Power Dissipation Issues

With the growing demand for energy-efficient electronic devices, power dissipation becomes a crucial challenge. Power density increases as transistor sizes reduce, posing heat dissipation and energy consumption difficulties. Leakage current is a significant contributor to power dissipation. Subthreshold and gate leakage become increasingly prominent as transistor size shrinks in typical CMOS. Subthreshold leakage happens when a transistor conducts current despite being off-state. This phenomenon is amplified in nanoscale transistors, considerably impacting the system's total power efficiency. The three key parameters are as follows:

- Leakage currents and subthreshold conduction
- Gate leakage and quantum tunneling
- Power gating and dynamic voltage and frequency scaling (DVFS)

Furthermore, the gate oxide thickness reduction to overcome scaling constraints impacts overall power consumption. In order to address power dissipation, novel transistor topologies, low-power design methodologies, and improved materials are gaining momentum. Power gating and DVFS are used for the purpose of having its critiques.

1.2.2.3 Performance Bottlenecks

The interconnect delay is a key bottleneck, considering the signal time to cross the metal interconnects. As transistor size shrinks, parasitic capacitance and connection resistance play a crucial role. This causes increased signal delay, resulting in a decrease in overall performance. On the other hand, high-performance computing, quantum computing readout devices, and 5G communication systems require faster signal propagation. The key parameters are as follows:

- Interconnect delays and signal propagation
- Variability in transistor characteristics
- Emerging technologies to overcome performance bottlenecks

Variability in transistor properties introduced by manufacturing processes and environmental circumstances is another bottleneck. The impact of unpredictability on circuit performance complicates obtaining consistent and reliable operation for critical accuracy-sensitive applications.

1.2.2.4 Analogy

To understand, NC is like upgrading from handwritten letters to emails. It changes how we process information, making it faster and more efficient. The impact is like how emails transformed communication.

1.2.3 Emergence of Nanomagnetic Logic

In this section, we will get into more detail about the emergence of NML (cf Figure 1.11). This section comprises the following subsections: Section 1.2.3.1 defines the principles and fundamentals, Section 1.2.3.2 presents comparative advantages, Section 1.2.3.3 provides role in overcoming CMOS limitations, and Section 1.2.3.4 presents a real-time analogy for the clarity of the readers. This sets the readers on the progression pathway in the NML, unveiling the breakthroughs, innovations, and transformations that help shape this field. The development of NML has drawn interest from academics and business experts in the rapidly evolving field of computing. This groundbreaking techno node offers a viable substitute for conventional CMOS by exploiting the unique characteristics of nanomagnets to carry out logic operations.

1.2.3.1 Principles and Fundamentals

Unlike CMOS's reliance on the electric current flow, NML stores information in the magnetization state of ferromagnetic elements. NML utilizes the intrinsic magnetic moment of these elements, manipulated by an external magnetic field or spin-transfer torque. The basic building block of NML is the single-domain

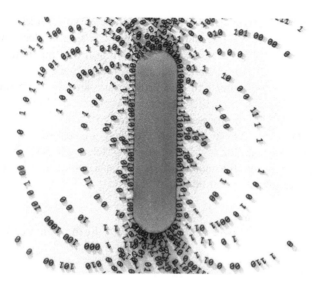

Figure 1.11 Information processing paradigm–conceptual illustration using nanomagnets and logic data flow as if the magnetic flux is formed https://ieeexplore.ieee.org/document/7226612. Source: 10.1109/MSPEC.2015.7226612 IEEE Spectrum (Volume: 52, Issue: 9, September 2015) Page(s): 44–60 Publisher: IEEE; Authors: Wolfgang Porod, Michael Niemier.

nanomagnet, a tiny ferromagnetic island with a well-defined magnetization direction. These nanomagnets can be arranged in various geometries, such as dots, wires, or rings, enabling the implementation of logic gates and circuits. The transitions between logic state 0 and 1 (e.g. parallel or antiparallel magnetization) represent logical operations like NOT, AND, and OR. Various NML devices have been proposed, a few including:

- Spin valves: Two ferromagnetic layers separated by a nonmagnetic spacer, and its relative resistance depends on the magnetization alignment.
- Nanowires: Wires with tailored shapes and magnetic properties used for data transmission and logic operations.
- Skyrmions: Topologically protected magnetic textures with unique properties offering high-density and low-energy consumption.

Magnetic Logic Gates: At its core, NML represents nanoscale magnets as the fundamental entities for constructing logic gates. These gates encompass standalone designs of majority logic, encoding and processing information through manipulating magnetic states. This departure from the charge-based CMOS introduces RC.

Spintronics and Magnetization Reversal: The field of spintronics is fundamental to the operation of NML. Exploring the intrinsic spin of electrons by exploiting the spin degree of freedom plays a pivotal role in the magnetization reversal. Achieving controlled and precise magnetization reversal is essential for logical operations. Spin-transfer and spin-orbit torque are key mechanisms facilitating the controlled switching of magnetic orientations, enabling reliable and energy-efficient information processing.

Nanomagnetic Devices: These nanodevices' size, shape, and magnetic properties influence the scalability, speed, and energy efficiency of NML circuits. Achieving a balance between thermal stability and susceptibility to external magnetic fields ensures robust operation.

1.2.3.2 Comparative Advantages

NML offers several advantages over CMOS, particularly in the miniaturization environment as follows:

- Lower power consumption
- Higher density
- Non-volatility
- Radiation tolerance

Energy Efficiency and Non-volatility:
NML introduces comparative advantages over CMOS in terms of energy efficiency, contrasting to CMOS, which incurs dynamic power consumption for the movement of charge carriers. NML stems from this salient approach of manipulating magnetic states rather than facilitating charge flow. Moreover, NML advocates non-volatility, a distinctive characteristic wherein magnetic states persist even after the power is turned OFF.

Scalability and Beyond Von Neumann Architecture:
The inherent scalability of NML challenges the limits of Moore's Law. The fundamental blocks in NML allow for denser packing of logical elements, enabling higher computational density. Furthermore, NML disrupts the traditional Von Neumann architecture, aiding paths for in-memory computing and non-Boolean logic. This differentiation opens up new computational efficiency and performance avenues via models that bypass established boundaries.

Fault Tolerance and Radiation Resilience:
The absence of charge carriers reduces susceptibility to errors induced by thermal effects and radiation, making it inherently fault tolerant. This characteristic positions NML as a tech prediction for applications in harsh environments, such as space or nuclear facilities, where conventional electronic devices may fail. The increasing need for reliable and resilient computing solutions in these crucial fields aligns with NML's adaptability.

1.2.3.3 Role in Overcoming CMOS Limitations

A paradigm change in computing has occurred with the introduction of NML. A vital first step in realizing its full potential is comprehending its principles, foundations, and comparative benefits toward investigating its function in surpassing CMOS limitations. With regard to next-generation computing technologies, NML is a transformational techno node due to its fault tolerance, scalability, energy efficiency, and compatibility with CMOS. NML has the potential to solve today's technology problems and change the information processing landscape in the future as research and development in the field advance. The progression of NML from principles to applications demonstrates its capacity to transform how we ultimately approach and carry out computational tasks. NML offers the following:

- Continued scaling
- Novel computing architectures
- Reduced environmental impact

However, challenges remain in the domain of:

- Material science and technology exploration
- Integration with CMOS
- Device reliability and testing

Beyond Moore's Law: Traditional CMOS technology confronts scaling limitations dictated by Moore's Law. The unique attributes of NML, including its energy efficiency and scalability, position it in the tech predictions as a potential successor. NML challenges the conventional trajectory of silicon-based computing and opens new avenues for continued advancements.

Heterogeneous Integration: Rather than seeking to replace CMOS entirely, NML is crucial in heterogeneous integration. This approach involves coexisting NML and CMOS technologies on the same chip, complementing each other. Although NML is more adept at handling particular computational workloads, CMOS is still a dependable workhorse for general-purpose computing. This symbiotic relationship maximizes the benefits of both technologies.

Applications in Specialized Computing:

NML finds its niche in specialized computing applications with significant advantages and unique characteristics. Quantum-dot cellular automata, in-memory computing, and neural network acceleration are among the domains where NML shows promise.

1.2.3.4 Analogy

Consider CMOS as a conventional car running on a straight road. NML is like a Formula One car that can pass through twists and turns more efficiently and robustly. It offers advantages in navigating the complex paths of computing challenges toward RC.

1.3 Edge AI Devices: A Driving Force

This section introduces the readers to the edge AI devices as a driving force. It briefly walks you through the role of NC for edge AI in Section 1.3.1 and defines the applications and use cases in Section 1.3.2 and a real-time analogy in Section 1.3.3. Now, let's read through the comprehensive study.

1.3.1 Role of Nanoscale Computing in Edge AI

In this section, we will get into more detail about role of NC in edge AI. AI is developing rapidly, with the rise and spread of Edge AI devices changing our perception and implementation of AI applications including LLMs.

- **Foundational Principles of Edge AI Devices**:
  Edge AI devices differ from conventional centralized AI models by bringing computational power closer to the data source. Unlike traditional setups where data is transmitted to a remote cloud for processing, edge AI devices perform computations locally. This paradigm shift is the basic edge principle, emphasizing decentralized data processing in enhancing efficiency, reducing latency, and addressing bandwidth constraints.
- **Embedded AI and On-Device Processing:**
  Integrating embedded AI capabilities and on-device processing is the fundamental component of edge AI devices. Machine learning algorithms are deployed onto the device to enable real-time analysis and decision-making. This benefits applications that require extremely low latency, such as industrial automation, healthcare monitoring, and autonomous cars.
- **Sensory Integration and Context Awareness**:
  With their sensory integration module, edge AI devices collect data from their surroundings using sensors, cameras, and microphones. AI algorithms and sensory input make these devices context-aware. Edge AI devices are more versatile and efficient as they recognize and react to contextual inputs in image identification, natural language processing, and environmental monitoring.
- **Transformative Impact – IoT and Smart Devices**
  Edge AI devices are essential to achieve the goal of smart, interconnected systems in the emerging IoT. These devices offer efficiency, automation, and responsiveness, becoming ubiquitous in smart homes with security cameras, intelligent thermostats and industrial IoT apps optimizing manufacturing processes.

- **Transformative Impact – Autonomous Systems and Robotics**
  Edge AI is essential in robotics and autonomous systems making split-second decisions. Drones, robotic systems, and vehicles equipped with these are able to navigate dynamic terrain and obstructions, offering adaptability. It paves the way for the development of driverless vehicles and intelligent infrastructure.

- **Key Factors Projecting Adoption – Latency Reduction and Real-Time Processing**
  Real-time processing and lower latency are the main drivers. Edge AI devices' local processing capabilities aid applications requiring quick responses, such as augmented/virtual reality and critical infrastructure monitoring. As an outcome, low reliance on cloud servers and quick decision-making are achieved.

- **Key Factors Projecting Adoption – Privacy and Data Security**
  Adopting Edge AI requires careful consideration of privacy issues and data security. Sensitive information stays within predetermined bounds with local processing, reducing the data transmission risks. This is consistent with user expectations for data privacy and changing legislative frameworks.

- **Key Factors Projecting Adoption – Bandwidth Efficiency**
  By lowering the amount of data sent to central servers, edge AI devices improve bandwidth efficiency, where bandwidth is expensive or scarce, targeted for applications in smart agriculture, 5G/6G networks, and remote geo-sensing.

- **Optimizing Power Consumption:**
  Edge devices, especially those deployed in remote or battery-operated settings, demand optimal power consumption. NC components, designed for energy efficiency, contribute toward extended battery life and reduced environmental impact. This optimization is crucial for applications in environmental monitoring, wearable devices, and smart infrastructure.

- **Integration with AI – Adaptive Learning:**
  NC's adaptability and reconfigurability initiate the design of edge AI devices with adaptive learning capabilities. These systems learn from their surroundings, adapt to changing conditions, and have context awareness. This versatility benefits smart homes and industrial automation applications. The function of NC in edge AI provides convergence in tackling difficulties, improving efficiency, and opening new opportunities for embedded intelligence. Integrating NC into edge-AI devices forecasts powerful computation and energy efficiency via its linkage to the physical environment. As technology improves, the symbiotic link between NC and Edge AI potentially reinvents intelligent, edge-centric computing possibilities.

1.3.2 Applications and Use Cases

In this section, we will get into more detail about the applications and use cases. This section comprises the following subsections: Section 1.3.2.1 presents the real-world implementation, Section 1.3.2.2 enhances edge-AI performance, and Section 1.3.2.3 underscores its contributions to emerging technologies for the clarity of the readers.

In the ever-evolving landscape of technology, edge-AI devices have emerged as a driving force, shaping the way we interact with and harness the power of AI. Here you will explore the diverse applications and use cases of edge-AI devices, unraveling their transformative impact on various industries. From real-world implementations that redefine healthcare, smart cities, and Industry 4.0 to the enhancement of edge-AI performance through on-device inference and customized accelerators, and the contributions to emerging technologies like 5G, blockchain, and quantum computing, this exploration provides a holistic understanding of the pivotal role Edge-AI devices play in our connected world.

1.3.2.1 Real-World Implementations
Healthcare Revolution: Remote Patient Monitoring and Personalized Treatment:

Edge-AI devices revolutionize the healthcare industry by bringing patient care via remote monitoring. Wearable medical technology tracks vital signs, identifies irregularities, and delivers timely information. These improve prevention and diagnostic healthcare, offering personalization plans, early disease detection, and continuous data analysis. Telemedicine platforms facilitate virtual consultations, transcend geographical barriers, and ensure medical access for all.

Smart Cities: IoT-Driven Urban Transformation:

Edge-AI devices ideate the concept of smart cities as a reality. It pushes urban change through traffic management, waste management, and energy-efficient infrastructure. IoT sensors strategically deployed across the city collect data in real-time and input it into edge-AI platforms for analysis. This data-driven strategy leverages informed decision-making, optimizes resource utilization, and reduces environmental impact.

Industry 4.0: The Rise of Intelligent Manufacturing:

At the forefront of Industry 4.0 are changes in manufacturing processes with edge AI. Predictive maintenance, automated quality control, and agile production systems have all been implemented. It examines data from embedded sensors in machinery to forecast equipment breakdowns, optimize manufacturing workflows, and improve overall operational efficiency. This leads to less

AI in Business: What's hot in the latest research?

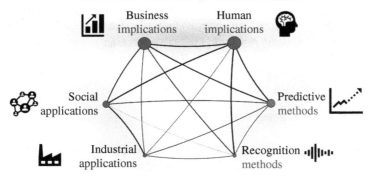

Figure 1.12   The infographic summarizes the results of a literature review, showing the six main topics of business-applied AI research, namely, Business implications, Human implications, Industrial applications, Social applications, Prediction methods, and Recognition methods. The picture shows the network visualization of the topic model, grouped by Implications, Applications, and Methods. Nodes' size is proportional to the relative presence of the topic in the corpus while the width of each edge shows the level of inter-topic distance in our model. File:Research topics in Business-applied Artificial Intelligence.png https://commons.wikimedia.org/wiki/File:Research_topics_in_ Business-applied_Artificial_Intelligence.png. Source: Aboutbigdata/Wikimedia Commons/CC BY-SA 4.0.

downtime and higher product quality, toward a new era of intelligent production (cf. Figure 1.12).

1.3.2.2   Enhancing Edge-AI Performance

- **On-Device Inference: Catalyzing Real-Time Decision-making:**
  On-Device inference is conducted quickly and accurately by the enhanced performance of edge-AI applications. Innovations in hardware and algorithms have facilitated real-time decision-making. Edge-based facial recognition and intelligent cameras benefit from on-device inference.
- **Customized Edge AI Accelerators: Tailoring Hardware for Efficiency:**
  Customized hardware accelerators play a crucial role in catering to the growing demands of edge AI and are instrumental in enhancing performance. These AI workloads-specific designs optimize power consumption and computational efficiency. Tailored hardware accelerators are key for real-world implementations of edge devices, from smartphones to IoT sensors. Equipped with efficient handling of complex AI tasks, contributing toward integrating intelligence into day-to-day life.

- **Federated Learning: Collaborative Intelligence at the Edge:**
  Federated learning has emerged as a fundamental approach to enhancing performance. Models are trained collaboratively in this decentralized learning approach without transmitting raw data. Predictive text suggestions on smartphones and localized voice recognition are the key areas of interest. Diverse datasets are leveraged to improve model accuracy and ensure sensitive data remains on the device, addressing privacy concerns.

1.3.2.3 Contributions to Emerging Technologies

- **5G and Edge Computing Synergy: Enabling Low-Latency Connectivity:**
  Combining 5G networks and edge AI has enabled unparalleled connection breakthroughs. Real-world applications requiring ultra-low latency include augmented and virtual reality (AR/VR) experiences, autonomous vehicles, and remote robotic surgery. Edge AI processes data locally, using 5G networks' high speed and low latency. This advances the development of immersive and responsive technology.
- **Blockchain and Edge AI: Ensuring Trustworthy Decentralized Systems:**
  The combination of blockchain technology with edge AI caters to data integrity and security in decentralized systems. Real-world applications include supply chain security and transparent/traceable healthcare data. Edge AI algorithms process data locally, while blockchain assures that the data is immutable. This synergy improves the stability of decentralized systems.
- **Quantum Computing at the Edge: Redefining Computational Limits:**
  The arrival of QC at the edge marks a fundamental shift in computational capability. Solving complex optimization problems, simulating molecular structures for drug discovery, and improving machine learning algorithms are emerging. This partnership redefines the boundaries of secured and powerful computational capabilities.

Edge-AI devices have applications and use cases in various industries, including healthcare, urban life, and manufacturing. Edge-AI applications perform better when on-device inference, tailored accelerators, and federated learning are integrated seamlessly. Furthermore, contributions to developing technologies, including 5G, blockchain, and quantum computing, highlight the capability of Edge AI devices. With advancements, their significance as a driving force in our interconnected world becomes more apparent.

1.3.3 Analogy

Imagine NC in Edge AI as the brainpower in a superhero suit. It enhances the capabilities of devices, making them more innovative and responsive. Just like how Iron Man's suit enhances his abilities.

1.4 Architecture and Material Design

This section introduces the readers to the architectural and material design. It briefly talks about the architectural considerations in Section 1.4.1, elaborately introduces the real-time analogy in Section 1.4.2, and defines the material selection and design strategies in Section 1.4.3. Now, let's walk through the design selection and considerations.

Exploration toward architecture and material design in NC highlights the transition Beyond Moore's Law to pioneer a new era of computational possibilities and craft atomically thin materials. NC acknowledges the challenges of fabrication precision, material imperfections, and energy efficiency. The pursuit of innovations in NC promises to surpass the limits of classical computing. As researchers navigate the nanoscale frontier, integrating novel architectures and material designs is key to shaping the perception of information processing and technology.

- **The Nanoscale Computing Landscape; Breaking the Traditional Mold:**
  Transitioning from macroscale to nanoscale: A paradigm shift
  Beyond Moore's Law – architectures for post-CMOS era
- **Material Design at the Nanoscale; Crafting the Future of Computing:**
  2D materials – atomically thin wonders
  Topological insulators – facilitating quantum computing
  Nanoscale magnetic materials – spintronics revolution
- **Challenges and Future Directions; Navigating the Nanoscale Frontier:**
  Challenges in nanoscale fabrication and integration
  Material imperfections and reliability concerns
  Energy efficiency and thermal management

1.4.1 Architectural Considerations

In this section, we will get into more detail about understanding architectural considerations of NC. This section comprises the following subsections: Section 1.4.1.1 provides the details of design paradigms, Section 1.4.1.2 presents system integration challenges, and Section 1.4.1.3 explains the aligning with edge AI architectures for the clarity of the readers.

1.4.1.1 Design Paradigms for Nanoscale Computing
Nanoscale Quantum Architectures: A Quantum Leap

QC emerges as a frontrunner in the advent of NC. Quantum bits provide processing capabilities using superposition and entanglement. Nanoscale quantum architectures transform information processing by pushing beyond classical bits.

It leverages quantum states for parallelism and computation at the atomic scale. The issues of qubit stability, error correction, and scalability must be addressed for efficient nanoscale quantum structures.

Spintronics and Memristor-Based Computing: Rethinking Information Processing:

Alternate design paradigms include spintronics, using electron spin for information processing. Memristor-based computing integrates memory and processing. Spintronics presents spin-based transistors for energy-efficient computing. On the other hand, memristors deploy neuromorphic computing by simulating the synaptic plasticity of the human brain. Exploring these paradigms necessitates novel methods for transistor design, memory architectures, and the seamless integration of different materials.

1.4.1.2 System Integration Challenges

Nanoscale Fabrication Precision: Engineering at Atomic Scales:

A key obstacle in NC is the need for precision in fabrication methods. In atomic engineering, breakthroughs in nanolithography, deposition methods, and material synthesis are required. The dependability and reproducibility of nanoscale devices are prominent. Achieving this level of breakthrough in manufacturing technologies requires investigating 2D materials and top-down/bottom-up hybrid methods.

Material Compatibility and Heterogeneous Integration:

Material compatibility and heterogeneity concerns are crucial in integrating nanomaterials into functional computer architectures. 2D topological insulators and materials exhibit distinct electrical characteristics. To ensure their seamless integration into this cohesive computing system, the key focus areas to address and investigate are as follows:

- Material compatibility difficulties.
- Interface effects minimization.
- New methods for integration.

Quantum Error Correction and Fault Tolerance:

Quantum architectures, while promising, are intrinsically subject to mistakes – influences and material flaws. Quantum error correction and fault tolerance techniques must be implemented to achieve scalable quantum computing. To address these, error-correcting codes, quantum gates with enhanced-fidelity, and novel fault-tolerant quantum computing paradigms must be developed.

1.4.1.3 Aligning with Edge-AI Architectures

Energy-efficient computation and processing capabilities are the key strengths of nanoscale architectures for edge AI. Spintronics, QC, and memristor-based devices are ideally suited to the demands of Edge AI. NC is compatible with edge

Figure 1.13 Nano IoT.
Source: MMxeon/Adobe
Stock Photos.

devices' required power limits and can perform complicated computations with minimum energy consumption. It opens paths for intelligent edge applications in healthcare, IoT (cf. Figure 1.13), and autonomous systems. The convergence with edge AI reflects a symbiotic connection in which NC improve the performance. At the same time, edge-AI applications drive the demand for energy-efficient and powerful nanoscale structures. As these technologies interact, they construct the future of computing. It offers a new era in which the constraints of conventional architectures are overcome, and decentralized intelligence is developed. The key areas are as the follows:

- Edge-AI design philosophy: decentralized intelligence
- Enhanced edge-AI performance: nanoscale contributions
- Forging the future of computing

1.4.2 Analogy

NC architecture is like designing a compact, efficient city. It is about organizing components in a way that minimizes travel time (signal transmission) and maximizes functionality, much like planning a smart city.

1.4.3 Material Selection and Design Strategies

In this section, we will learn more about understanding material selection and design strategies. This section comprises the following subsections: Section 1.4.3.1 defines the materials for NML, Section 1.4.3.2 details impact on device performance, and Section 1.4.3.3 bridges architecture and material science. In the NC era, the selection of materials and design strategies play a pivotal role. Let's understand in detail.

1.4.3.1 Materials for Nanomagnetic Logic

Beyond Traditional Semiconductors: Exploring New Frontiers:

Two-dimensional (2D) materials, like graphene and transition metal dichalcogenides, have promising electrical characteristics and mechanical flexibility. Topological insulators offer exciting spintronic potential with their shielded conducting surface states. Nanomagnetic materials, defined by their magnetic moments, emerge as significant players.

Materials for Nanomagnetic Logic: Harnessing Magnetic Moments:

Magnetic nanomaterials can be ferromagnetic, antiferromagnetic, or ferrimagnetic. The material used determines the magnetic state stability, manipulation, and overall performance of NML devices.

Magnetic Tunnel Junctions: A Synergistic Approach:

The tunneling magnetoresistance effect in the magnetic tunnel junction determines the resistance with the relative orientation of the magnetic layers. High magnetoresistance materials: ferromagnetic and antiferromagnetic alloys are essential in developing efficient spintronic devices.

1.4.3.2 Impact on Device Performance

Scalability and Quantum Effects: A Material Dilemma:

Quantum phenomena: tunneling and quantized conductance changes our perception on scalability. With atomically thin structures, 2D materials exhibit quantum confinement effects, redefining electronic properties.

Magnetic Anisotropy and Switching Dynamics:

Tailoring magnetic anisotropy is crucial in determining stable magnetic states and reliable logic operations. Materials having controllable magnetic anisotropy rare-earth transition metal alloys exhibit viable choices for NML. Material properties determine switching dynamics, critical for improving device performance and lowering energy usage.

1.4.3.3 Bridging Architecture and Material Science

Co-design Strategies: Synergizing Architecture and Materials:

Co-design strategies, in which architectural and material considerations are intertwined, emerge as a paradigm for innovation. Tailoring material qualities to suit specific architectural requirements is necessary to obtain the desired device performance.

Material Challenges and Architectural Solutions:

Material science challenges can be harnessed into novel architectural solutions, including defects, stability, and manufacturability. The critical elements of such an architectural framework include an error correction system, adaptive control algorithm, and redundancy scheme.

1.5   Scope of the Book

"NanoScale Computing: The Journey Beyond CMOS with Nanomagnetic Logic" walks readers through the fundamental exploration of NC. This comprehensive and in-depth book examines the transformative potential of nanomagnetic logic and its applications. The book's scope is strategically designed to cater to a diverse audience, ranging from graduate and undergraduate students to researchers and technology professionals.

This book is carefully handcrafted to be generic, increasing its potential, serving as a fundamental textbook, and also covering applied technological research through last few chapters to elaborate on advanced concepts.

1.5.1   Objectives and Goals

The scope unfolds in a structured manner across several chapters to provide a systematic and engaging learning experience. The objectives described in Section 1.5.1.1 are aligned with the desired learning results. The book aims to lay a transparent and learner-centered foundation, to navigate complicated NC systems.

1.5.1.1   Establishing Learning Objectives

Specific, Measurable, Achievable, Relevant, and Time-bound (SMART) learning outcome-based objectives and goals of this book are to:

- Understand: NC Basics for Diverse Applications and Innovations.
- Evaluate: Limitations of CMOS and Explore Challenges in NC Computing.
- Gain: Profound Insights into Fundamentals and Operations of NML.
- Explore: Diverse NML Architectures for Innovative Computing Solutions.
- Comprehend: Material Design Principles Essential for NC Technologies.
- Define: Edge AI Devices and Applications, Unveiling Opportunities and Challenges.
- Examine: Synergies in Hybrid Computing Systems for Enhanced Performance Solutions.
- Address: Challenges and Future Directions, Assessing Reliability, and Scalability Issues.
- Explore: Real-world Applications, from Neuromorphic Computing to Environmental Sensing.
- Conclude: With a Vision for NML Impact and Evolution.

1.5.2   Interconnected Themes and Concepts

The flow of the content is strategically organized maintaining technical clarity and seamless transition between interconnected themes and concepts (cf. Figure 1.14 depicting Interconnected themes and concepts of this book - ISBN – 9781394263554).

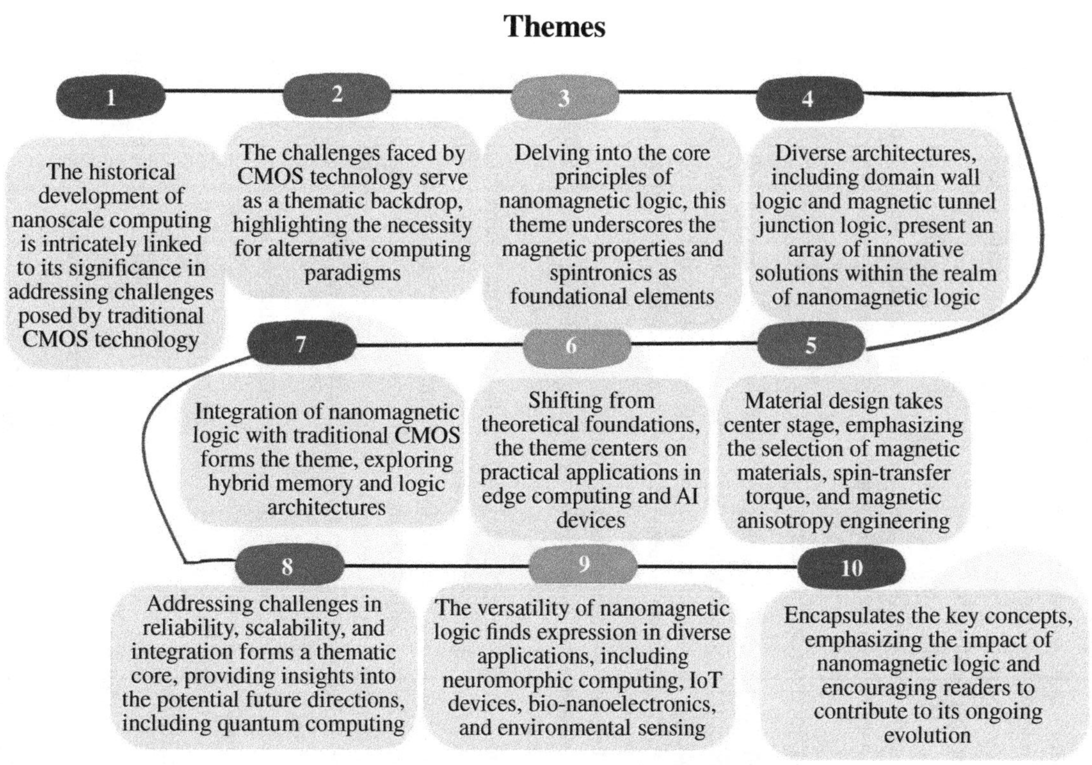

Themes

1

2

3

4

The historical development of nanoscale computing is intricately linked to its significance in addressing challenges posed by traditional CMOS technology

The challenges faced by CMOS technology serve as a thematic backdrop, highlighting the necessity for alternative computing paradigms

Delving into the core principles of nanomagnetic logic, this theme underscores the magnetic properties and spintronics as foundational elements

Diverse architectures, including domain wall logic and magnetic tunnel junction logic, present an array of innovative solutions within the realm of nanomagnetic logic

5

6

7

Material design takes center stage, emphasizing the selection of magnetic materials, spin-transfer torque, and magnetic anisotropy engineering

Shifting from theoretical foundations, the theme centers on practical applications in edge computing and AI devices

Integration of nanomagnetic logic with traditional CMOS forms the theme, exploring hybrid memory and logic architectures

8

9

10

Addressing challenges in reliability, scalability, and integration forms a thematic core, providing insights into the potential future directions, including quantum computing

The versatility of nanomagnetic logic finds expression in diverse applications, including neuromorphic computing, IoT devices, bio-nanoelectronics, and environmental sensing

Encapsulates the key concepts, emphasizing the impact of nanomagnetic logic and encouraging readers to contribute to its ongoing evolution

Figure 1.14 Interconnected themes and concepts of this book – ISBN – 9781394263554.

1.6 Conclusion

Chapter 1, "Introduction to Nanoscale Computing", provides a foundational journey into the NC. It sets the stage for an in-depth exploration. The key points highlighted in this chapter can be summarized as follows.

1.6.1 Summary of Key Points

- Explores the fundamental principles and significance of operating at the nanoscale.
- Emphasizes the synergies with the RC initiative, providing a broader context.
- Presents a dual historical perspective, commemorating 75 years of transistors and 100 years of Spin.
- Examines the challenges faced by traditional CMOS technology, leading to the emergence of NML.
- Defines edge computing and its relevance, establishing a connection between NC and edge AI.
- Showcases applications and use cases, illustrating the practical impact of NML.
- Introduces architectural considerations and material design strategies in the context of NML.
- Utilizes analogies to enhance understanding, laying the groundwork for more technical explorations.

1.6.2 Active Engagement – Educational Model

As you progress through the chapters, actively engage with the themes. Consider analogies, case studies, and prospective applications. Your participation is essential for realizing the entire range of NC potential. Accept curiosity, ask questions, and use your exploration to better grasp this dynamic field. Here is the "learning and fun based activity" for you to participate:

- **Reflect and Share:** What aspect of NC sparked your curiosity the most? Share your thoughts! Tag with the **#9781394263554**
- **Tech Trivia:** Test your knowledge! Can you name a famous scientist or inventor in the field of NC? Tag with the **#9781394263554**
- **Analogies Unleashed:** Draw an analogy between NML and something from everyday life. Let your creativity flow! Tag with the **#9781394263554**
- **Predict the Future:** What potential applications of NML do you foresee in the next decade? Share your predictions! Tag with the **#9781394263554**
- **Meme Challenge:** Create a meme that humorously captures the essence of NC. Let us add a touch of humor to the tech world! Tag with the **#9781394263554**

1.6.3 Chapter End Quiz

An online Book Companion Site is available with this fundamental textbook in the following link: www.wiley.com/go/sivasubramani/nanoscalecomputing1

For Further Reading:

Taha et al. (2022), Zahoor et al. (2023), Goldstein (2005), Wright et al. (2013), Kim et al. (2020), Eshaghian-Wilner (2004), Beckett and Jennings (2002), Teodorov (2011), Passian and Imam (2019), Nathan et al. (2023), Chandrakasan and Brodersen (1995)

References

Paul Beckett and Andrew Jennings. Towards nanocomputer architecture. In *Conferences in Research and Practice in Information Technology Series*, volume 19, pages 141–150, 2002.

Anantha P. Chandrakasan and Robert W. Brodersen. *Low Power Digital CMOS Design*. Springer New York, NY, 1 edition, 1995. doi: 10.1007/978-1-4615-2325-3. Springer Book Archive.

Mary Mehrnoosh Eshaghian-Wilner. The architectural designs of a nanoscale computing model. *Journal of Systemics, Cybernetics, and Informatics*, 2(4), 2004.

Seth Copen Goldstein. The impact of the nanoscale on computing systems. In *ICCAD-2005. IEEE/ACM International Conference on Computer-Aided Design, 2005*, pages 655–661. IEEE, 2005.

MeSuk Kim, ALam Han, TaeYoung Kim, and JongBeom Lim. An intelligent and cost-efficient resource consolidation algorithm in nanoscale computing environments. *Applied Sciences*, 10(18):6494, 2020.

Arokia Nathan, Samar K. Saha, and Ravi M. Todi, editors. *75th Anniversary of the Transistor*. Wiley-IEEE Press, July 2023. ISBN 978-1-394-20246-1.

Ali Passian and Neena Imam. Nanosystems, edge computing, and the next generation computing systems. *Sensors*, 19(18):4048, 2019.

Taha Basheer Taha, Azeez Abdullah Barzinjy, Faiq Hama Seaeed Hussain, and Togzhan Nurtayeva. Nanotechnology and computer science: Trends and advances. *Memories-Materials, Devices, Circuits and Systems*, 2:100011, 2022.

Ciprian Teodorov. *Model-Driven Physical-Design for Future Nanoscale Architectures*. PhD thesis, Université de Bretagne Occidentale (UBO), Brest, 2011.

C David Wright, Peiman Hosseini, and Jorge A Vazquez Diosdado. Beyond von-neumann computing with nanoscale phase-change memory devices. *Advanced Functional Materials*, 23(18):2248–2254, 2013.

Furqan Zahoor, Fawnizu Azmadi Hussin, Usman Bature Isyaku, et al. Resistive random access memory: Introduction to device mechanism, materials and application to neuromorphic computing. *Discover Nano*, 18(1):36, 2023.

2

Limitations of CMOS Technology

Specific, Measurable, Achievable, Relevant, and Time-bound (SMART) learning objectives and goals for this chapter. After reading this chapter, you should be able to:

- Understand – scaling limitations in CMOS technology.
- Evaluate – implications on power and energy efficiency.
- Recognize – technological barriers and adaptability.
- Explore – application of nanomagnetic logic (NML) as an alternative.

This chapter talks about the challenges in traditional CMOS technology in Section 2.1, briefs the implications for Computing systems in Section 2.2, describes the technological and economic challenges, in detail in Section 2.3, presents the Bridging to Nanoscale Computing in Section 2.4, and elaborates the educational emphasis in Section 2.5.

2.1 Challenges in Traditional CMOS Technology

This section talks about the challenges in traditional CMOS technology, it briefs the scaling limitations in Section 2.1.1, describes the power dissipation issues in Section 2.1.2, and presents in detail the performance bottlenecks in Section 2.1.3.

CMOS technology is vital in the semiconductor industry with key technological advancements. On the other hand, the IEEE International Roadmap for Devices and Systems (IRDS) has identified key challenges toward enhancing sustainability and advancement. Pushing the state-of-the-art silicon technology requires;

- Long-term investments
- Collaboration in fundamental research.
- Computational techniques – development.

Nanoscale Computing: The Journey Beyond CMOS with Nanomagnetic Logic,
First Edition. Santhosh Sivasubramani.
© 2025 The Institute of Electrical and Electronics Engineers, Inc. Published 2025 by John Wiley & Sons, Inc.
Companion website: www.wiley.com/go/sivasubramani/nanoscalecomputing1

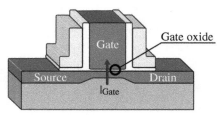

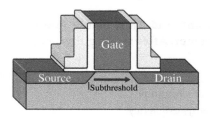

Figure 2.1 Semi-log graph of 125 years of Moore's Law.
File: 51391518506_7453df3ace:o.png https://commons.wikimedia.org/wiki/File:Leakage_Current_(2_models).PNG Source: STosaka/Wikimedia Commons/CC BY-SA 3.0.

Here, we will discuss: high-gate leakage currents, source-to-drain leakage, gate stack reliability, and channel mobility degradation challenges alongside their impacts and effects.

1. **High-Gate Leakage Currents:**
   This issue arises from the thinning of the gate oxide layer (cf. Section 2.1.2 briefing on power dissipation issues). This layer is crucial for insulating the gate efrom the channel. With shrinking, this layer is susceptible to electron tunneling. Electrons overcome the barrier, resulting in leakage currents (cf. Figure 2.1).

   Impact on Power Dissipation and Reliability: It poses significant consequences as power dissipation increases. Energy is lost through these leakage paths. This escalation leads to heat generation concerning device reliability. As temperatures rise, it affects the overall performance and lifespan of CMOS devices.

2. **Source-to-Drain Leakage**: It is directly associated with the short-channel effects (SCEs) (cf. Section 2.1.1) and reduced transistor channel length (cf. Figure 2.2). With shrinking, SCEs arise at their peak, increasing source-to-drain leakage (cf. Figure 2.3). This phenomenon significantly contributes to

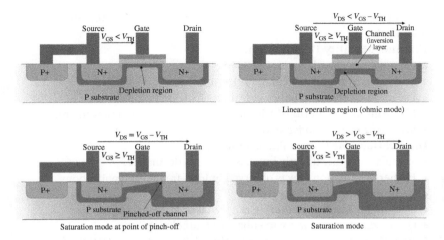

Figure 2.2 Source tied to the body to ensure no body bias. (a) Subthreshold, (b) ohmic mode, (c) active mode at the onset of pinch-off, (d) active mode well into pinch-off channel length modulation evident File: MOSFET functioning https://commons.wikimedia.org/wiki/File: MOSFET_functioning.svg Source: Olivier Deleage and Peter Scott/Wikimedia Commons/CC BY-SA 3.0.

power dissipation. Also, greatly concerns heat management and overall device reliability.

SCEs and Power Dissipation: These effects include lowering of the drain-induced barrier. It also impacts ionization, disrupting the ideal behavior of transistors (cf. Figure 2.3). To mitigate these adverse effects necessitates innovative design strategies.

3. **Gate Stack Reliability Issues**: The gate stack, composed of the gate oxide, gate electrode, and metal interconnects, faces reliability concerns. The quality and integrity of these components influence the performance and lifespan (cf. Figure 2.4).

Factors Influencing Gate Stack Reliability: Gate oxide defects, variations in thickness, and metal interconnect failures affect the reliability of the

Figure 2.3 MOS capacitor on p-type silicon showing inversion and depletion layers. File: MOS Capacitor https://commons.wikimedia.org/wiki/File:MOS_Capacitor.svg Source: Fred the Oyster/Wikimedia Commons/CC BY-SA 3.0.

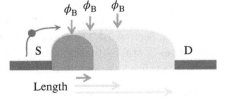

gate stack. Addressing these requires an understanding of materials and its manufacturing processes.

4. **Channel Mobility Degradation**:
   - Channel mobility degrades in CMOS.
   - Scattering by impurities hinders electron flow.
   - Defects within silicon lattice contribute to degradation.
   - Electrons face scattering during source-to-drain channel movement.
   - Scattering events hinder smooth electron flow.

   Implications for Device Performance:
   - Reduced mobility directly impacts device performance.
   - Higher degradation effects observed in high-frequency applications.
   - Rapid electron mobility is crucial.
   - Minimizing electron scattering is a trivial task.
   - Optimization is necessary to enhance channel mobility.

5. **Navigating the Future: Computational Techniques**: CMOS challenges demand strategic solutions. Simulation reveals intricate dynamics of CMOS devices. Modeling and analysis provide essential insights. Researchers harness its capabilities for effective problem-solving.

   Role of Computational Techniques: Computational techniques explore materials, designs, and processes. Simulating CMOS components helps identify optimal configurations. This approach speeds design refinement, enhancing CMOS performance. Traditional CMOS faces challenges in evolving. Balancing power efficiency, reliability, and performance is challenging at nanoscale. Issues like gate leakage currents and channel mobility degradation pose hurdles. (cf. Figure 2.5).

2.1.1 Scaling Limitations

This section talks about the scaling limitations throughout. This section briefs the implications of Moore's Law in Section 2.1.1.1, describes the quantum effects and miniaturization in Section 2.1.1.2, and presents in detail the size reduction challenges in Section 2.1.1.3.

Scaling challenges persist in CMOS technology, hindering miniaturization goals. Electron thermal energy emerges as a fundamental factor limits shrinking. Two-dimensional (2D) electrostatic scale length complicates miniaturization efforts significantly. Tunneling leakage through gate oxide adds to scaling limitations (cf. Section 2.1.2 briefing on power dissipation issues). Overcoming these challenges is crucial for enhancing chip performance. Performance issues arise due to these fundamental limiting factors.

A. **SCEs**: Its impacts as discussed in Section 2.1 play a significant role. DIBL decreases transistor threshold voltage, altering currents (cf. Figure 2.4).

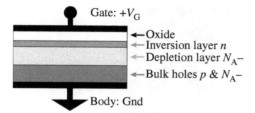

Figure 2.4 As channel length decreases, the barrier ϕ_B to be surmounted by an electron from the source on its way to the drain reduces. File: IEEE Rebooting Computing TaskForce Logo.png https://commons.wikimedia.org/wiki/File:\ignorespacesBarrier_lowering.PNG Source: Brews ohare/Wikimedia Commons/CC BY-SA 4.0.

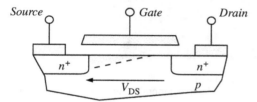

Figure 2.5 Cross section of a metal-oxide-semiconductor field-effect transistor (MOSFET) operating in its saturation region. File: Mosfet saturation https://commons.wikimedia.org/wiki/File:Mosfet_saturation.svg Source: Cyril BUTTAY/Wikimedia Commons/CC BY-SA 3.0.

when the off-state current increases, the on-state current decreases. This phenomenon hinders transistor performance, limiting size reduction progress.

B. **Process Variation**: Process variation, a scaling limitation, results from manufacturing differences. Shrinking transistors increases process variation. It also reduces the yield of functional chips. Consistency amid reducing dimensions are becoming challenge.

C. **Hot Carrier Injection and Power Consumption**: Scaling in CMOS technology faces problems with hot carrier injection and power consumption. Injection impacts the reliability of transistors; power use increases as transistors get smaller.

Scaling CMOS requires careful consideration amidst challenges. SCE, process variation, hot carrier injection, and power consumption limit the scaling. These challenges slow down reduction in the transistor size, illustrating need for the innovative solutions.

2.1.1.1 Implications of Moore's Law

Moore's Law (cf. Figure 2.6), by Gordon Moore in 1965, guided semiconductors, predicting transistor doubling annually. True for decades, shrinking circuits now

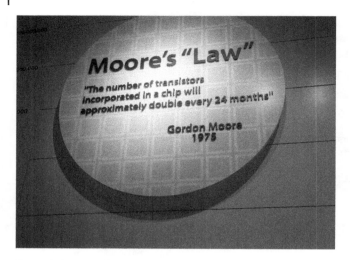

Figure 2.6 Moore's Law – Computer History Museum.
File: Moore's Law – Computer History Museum Source: Moore's Law Section at Computer History Museum/CC BY 2.0

faces technology bounds. This exploration examines into Moore's Law implications, far-reaching technology effects, and challenges requiring alternative approaches (cf. Figure 2.7).

The Evolution of Moore's Law: Initially, Moore's Law fueled computing power growth, transistors doubling yearly. This growth led to a substantial increase in capabilities. Today's circuit is two billion times more powerful than 1960.

Moore's Law Impact: Moore's Law has wide-reaching implications in technology. It goes beyond transistors and microchips, influencing various landscapes. Cloud computing and social media thrive on predicted computing growth, boosting transistor density per square inch. This leads innovation in multiple sectors.

Business Decision-Making: Moore's Law impacts business decisions, influencing strategic investments. The doubling of transistors every two years enables businesses to foresee and also help adapt to changes. Advancements impact market competitiveness, shaping business strategies.

Policy Considerations: Moore's Law extends beyond business, affecting national policies. Policies must align with technological trends to enhance competitiveness and security. Understanding technology trajectories guides innovation, aims to foster economic growth, and enhances national security.

The Call for Alternative Technologies: Moore's Law influenced computing evolution but has limitations identified. Exponential power increase sparks innovation with persistent challenges. Cutting-edge silicon pushes sustained investments, collaborations, and computational techniques. Overcoming these

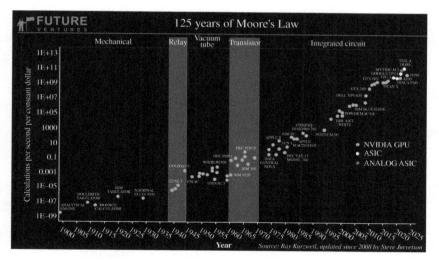

Figure 2.7 Semi-log graph of 125 years of Moore's Law.
File: 51391518506_7453df3ace:o.png https://www.flickr.com/photos/jurvetson/
51391518506/ Source: Steve Jurvetson/Flickr/CC BY 2.0.

challenges at future tech nodes requires affordable solutions. As Moore's Law reaches limits, next-generation computing nodes rely on interdisciplinary research and innovative approaches. It helps sustain required computational power.

2.1.1.2 Quantum Effects and Miniaturization

As we explore miniaturization, quantum effects pose challenges and opportunities.

A. **Electrons to Waves Transition**: Microscopic world reveals electrons' shift from particles to waves. Beyond classical physics, electrons accept their wave nature. This quantum shift challenges traditional transistor foundations. In the macroscopic realm, transistors operate predictably under classical physics. Electrons flow switching its states as instructions. But miniaturizing, below 10 nm, reveals a transformation. The electron's wave nature emerges. As the electron's wavelength matches the transistor's width, quantum effects dominate. Electrons tunnel through barriers, violate classical restrictions. Quantum phenomena result in leakage currents, destabilizing circuits and limiting transistor performance. They challenge CMOS miniaturization.

B. **The Tunnel Barrier and Quantum Tunneling**: Quantum miniaturization leads to significant electron tunneling. Insulating oxide layer serves as potential barrier for electrons. Quantum mechanics grants non zero probability for electron tunneling. In the classical realm, electrons lack energy for barrier traversal. Tunneling occurs even when transistor is in "off" state. Unintended switching jeopardizes circuit stability affecting logic integrity. Leakage current pose challenges, wasting power and generating heat (cf. Figure 2.8).

C. **Wave function Overlap**: Miniaturization confines electrons spatially within the channel. Wave function leads to unintended coupling between transistors. Overlapping wave functions cause electrons to interact in novel ways. False signals arise, impacting circuit functionality adversely. Extreme cases result in spontaneous transistor switching, inducing errors.

D. **Band to band Tunneling**: Shrinking transistors reduce energy separation in silicon substrate. Threat to switching ability and transistor performance. Quantum phenomenon: electrons jump, bypassing gate control, compromising functionality.

E. **Quantum Confinement**: Miniaturization promises density and speed but impacts electron properties. Translates to reduced clock speeds and transistor

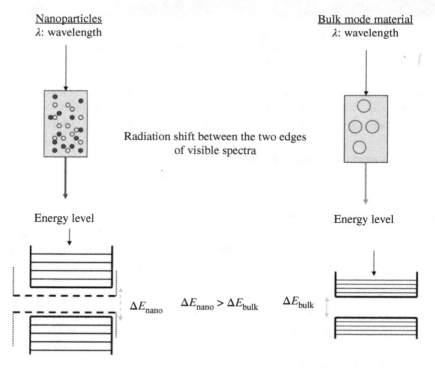

Figure 2.8 Quantum confinement is responsible for the increase of energy difference between energy states and band gap, a phenomenon tightly related to the optical and electronic properties of the materials.
File: Quantum confinement 1 https://commons.wikimedia.org/wiki/File:Quantum_confinement_1.png Source: Yshen8/Wikimedia Commons/CC BY-SA 3.0.

gain. Confined channel geometry reduces electron mobility, slowing device performance.

F. **The Uncertainty Principle: A Fundamental Limit to Control**: Nanoscale magnifies limitation, challenging accurate electron control. Quantum mechanics imposes fundamental control and measurement constraint. Predictable transistor behavior becomes increasingly difficult.

G. **Implications for Chip Design and Performance**: Wave function overlap and band-to-band tunneling demand innovative design techniques. Quantum effects complicate CMOS design, introducing issues like leakage currents (cf. Figure 2.9). Moore's Law faces potential roadblock due to quantum limitations. Novel materials and advanced manufacturing processes mitigate quantum effects. Exploration needed for alternative computing paradigms. Spintronics, neuromorphic computing, and quantum computing (QC) offer glimpses of the future.

2.1.1.3 Size Reduction Challenges

Think about: (a) The challenges of scaling down? (b) Power dissipation and heat generation? (c) Variability and reliability? Challenges arise as devices scale down. Dominant issues include power dissipation and heat (cf. Section 2.1.2 briefing power dissipation issues). Variability in manufacturing becomes a critical bottleneck (cf. Figure 2.10). Scaling promises increased power but encounters problems. Efficient heat dissipation becomes a formidable reliability challenge (cf. Section 2.1.2.2 detailing heat dissipation challenges). Power density surges as

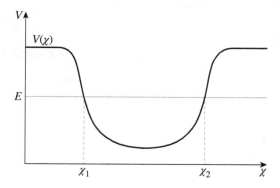

Figure 2.9 A generic potential energy well. In quantum physics, potential energy may escape a potential well without added energy due to the probabilistic characteristics of quantum particles; in these cases, a particle may be imagined to tunnel through the walls of a potential well.
File: Potential energy well https://commons.wikimedia.org/wiki/File:Potential_energy_well.svg Source: Benjamin D. Esham/Wikimedia Commons/Public Domain.

transistors shrink, raising temperatures. Challenges extend to reliability, performance consistency, and device functionality (cf. Figure 2.7). Dopant, process, and statistical variations amplify the challenges.

2.1.2 Power Dissipation Issues

This section talks about the power dissipation issues throughout. This section briefs the dynamic and static power consumption in Section 2.1.2.1, describes the heat dissipation challenges in Section 2.1.2.2, and presents in detail the impact on energy efficiency in Section 2.1.2.3. Traditional CMOS technology faces scaling challenges. Moore's Law guides transistor evolution. Challenges emerge in power dissipation. Exploration reveals intricacies, challenges, and mitigation avenues.

Understanding power dissipation includes the following components:
1. Basic Principles:

- Switching induces heat, energy loss.
- Measured in watts, crucial in nanoscale.

2. Dynamic and Static Power:

- Dynamic: Capacitance, transistor switching.
- Static: Leakage currents in inactive states.
- Addressing both is vital.

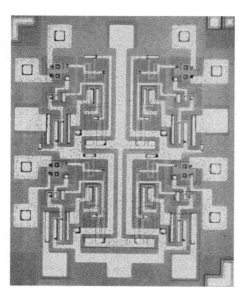

Figure 2.10 NAND gate with four inputs in transistor–transistor logic; seven transistors on the chip, smallest structure width 20 µm (see Radio Fernsehen Elektronik 10/1990, p. 649). File: 4-fach-NAND-C10JPG https://commons.wikimedia.org/wiki/File:4-fach-NAND-C10JPG Source: Dgarte/Wikimedia/CC BY SA 3.0.

3. Challenges in Power Dissipation:

a) **Scaling Dilemmas**:
- Transistor size reduction escalates dynamic power.
- Lowered thresholds, supply voltages increase leakage.

b) **Subthreshold Leakage**:
- Static power intensifies in scaled CMOS.
- Electrons tunnel, even when state "off."
- Managing involves trade-offs.

c) **Gate Oxide Leakage**:
- Reduced sizes, thinner gate oxide.
- Higher tunneling probability, elevated standby power.
- Mitigating leakage is crucial.

4. Mechanisms Governing Power Dissipation:

a) **Switching Losses**: Switching induces dynamic power via capacitive load changes. Transistor speed impacts performance and power trade-offs.

b) **SCEs**: Miniaturization introduces SCE (cf. Section 2.1.1), affecting dynamic and static power. DIBL and subthreshold slope degrade efficiency.

c) **Temperature Dependence**: Power links with temperature, forming a feedback loop causing thermal runaway. Heat affects device characteristics, requiring temperature management for reliability.

5. Mitigation Strategies:

a) **Advanced Transistor Designs**: Designs like FinFETs address power dissipation. Three-dimensional structures control electrostatics, reducing leakage and enhancing efficiency.

b) **Low-Power Design Techniques**: Voltage scaling, clock gating, and power gating optimize dynamic power.

c) **Materials Innovation**: Explore high-k dielectrics for gate oxides, mitigating leakage. Improved insulating properties lower static power dissipation.

Power dissipation challenges demand continuous innovation in semiconductor devices. Holistic approaches combine design, materials, and system-level optimizations for balance. Power dissipation challenges persist in traditional CMOS technology. Understanding dynamic and static power interplay at nanoscale dimensions is crucial.

2.1.2.1 Dynamic and Static Power Consumption

Challenges grow as transistors shrink, impacting power consumption dynamics. Investigating into this, we explore underlying mechanisms, challenges, and mitigation avenues. Here we will discuss dynamic and static power consumption, challenges, their underlying mechanisms, and mitigations.

1. Dynamic Power Consumption: CMOS devices experience energy consumption during switching activities. It is a transient form. It occurs as transistors toggle between logic states 0 and 1. The dynamic power (P_{dyn}) can be expressed as: (cf. Eq. 2.1)

$$P_{dyn} = \frac{1}{2} C_{load} V_{dd}^2 f \qquad (2.1)$$

where C_{load} is the load capacitance, V_{dd} is the supply voltage, and f is the switching frequency.

2. Static Power Consumption: It is the power dissipated when the device is in a static or idle state. It occurs even in the absence of switching activities. It includes subthreshold leakage and gate leakage currents. The static power (P_{static}) can be expressed as:

$$P_{static} = I_{static} \times V_{dd} \qquad (2.2)$$

where I_{static} is the static leakage current. (cf. Section 2.2)

Challenges in Dynamic Power Consumption: (a) Scaling challenges (b) SCE (cf. Section 2.1.1) and (c) Temperature dependence.

Challenges in Static Power Consumption: (a) Subthreshold leakage and (b) Gate oxide leakage.

Mitigation Strategy: 1. Dynamic Power Mitigation:

(a) **Clock Gating**: Disabling clock signals to idle components reduces dynamic power consumption.

(b) **Dynamic Voltage and Frequency Scaling (DVFS)**: Adjusting the supply voltage and frequency based on workload minimizes dynamic power.

Mitigation Strategy: 2. Static Power Mitigation:

(a) **Transistor Sizing**: Optimizing transistor dimensions balance static and dynamic power trade-offs.

(b) **Power Gating**: Completely shutting off power to idle components mitigates static power dissipation.

2.1.2.2 Heat Dissipation Challenges

Transistors shrinking, heat rises, and threatening chip integrity. This joule heating results in a consistent temperature rise, posing a threat to circuit performance, and reliability. Even minimal leakage currents contribute to baseline heat. Rapid on-off cycling adds dynamic heat spikes. The following key components plays a crucial role as detailed below:

1. **Thermal Conduction:**
   Regarding thermal conduction, silicon exhibits excellent electrical conductivity but lags behind in thermal conductivity (cf. Figure 2.11). Heat buildup creates local hot spots, worsening the problem. Limited diffusion hinders heat removal from transistors.

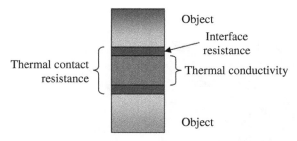

Figure 2.11 Thermal conductivity of thermal interface materials and thermal contact resistance. File: https://commons.wikimedia.org/wiki/File:Difference_between_thermal_conductivity_of_thermal_interface_materials_and_thermal_contact_resistance.png Source: Dtc5341/Wikimedia Commons/Public Domain.

2. **Clocks and Unstable Circuits**:
   High temperatures impact circuit performance directly. Increased thermal agitation introduces noise, compromises data integrity. Also, excessive heat degrades the mobility and lifetime of charge carriers within transistors. It results in performance slowdowns and, ultimately, chip failure. Clock signal delays increase with elevated temperatures introducing errors.
3. **Trading Power for Safety**:
   Thermal management uses clock throttling to prevent overheating. When temperatures rise, clock speed dynamically reduced for stability. Trade-off sacrifices processing power for chip integrity.
4. **Beyond CMOS**:
   Thermal limits of traditional CMOS lead to exploration of alternatives. Technologies like spintronics, neuromorphic computing, and QC are projected to tackle the challenges. These technologies often rely on fundamentally different physical principles, presenting new opportunities for information processing with inherently lower heat dissipation (cf. Section 2.1.2.2 detailing heat dissipation challenges). Exploring graphene toward chip design adds advantages which will be discussed further.
5. **Toward Thermodynamically Aware Design**:
   Addressing heat requires a shift in chip design philosophy (cf. Figure 2.12). Moving beyond a sole focus on transistor density and clock speed. New designs prioritize efficient circuit layout, power gating, and innovative materials. Combining electrical and thermal conductivity crucial for keeping "silicon or beyond-silicon cool."
6. **The Future of Computing: Balancing Power and Performance**:
   Bridging the gap between computational demands and thermal management is a challenge. Continued research in heat dissipation, alternative materials (e.g. graphene and other emerging 2D materials), and energy-efficient paradigms

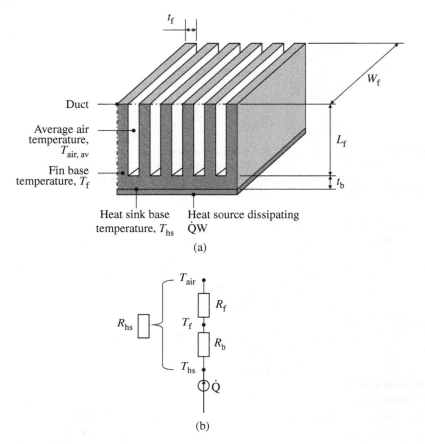

Figure 2.12 Heat sink with thermal resistances used to calculate thermal performance. File: Heat sink thermal resistances https://commons.wikimedia.org/wiki/File:Heat_sink_ thermal_resistances.png. Source: Dtc5341/Wikimedia Commons/Public Domain.

crucial. Achieving a technology node where computing power does not result in excessive heat or "the excessive heat generated could be used to power the chip" is feasible through sustainable measures.

2.1.2.3 Impact on Energy Efficiency

Moore's Law Impact: Shrinking transistors increase computing power. Yet, hidden cost: energy efficiency dramatically declines. The following components (a-l) describes in detail:

a) **Scaling Challenges**: Smaller transistors may seem more power-efficient initially. Yet, as we reach nanoscale, advantages of scaling reduce. Leakage currents, like quantum tunneling, persist even in "off" state. This unintended power drain cancels miniaturization benefits, increasing energy use.

b) **Short Channels and Power Challenges**: Shrinking transistors weaken electrostatic control (cf. Section 2.1.1), causing SCE. SCEs contribute to unintended leakage currents and premature switching. Reducing voltage to mitigate leakage can amplify SCEs, creating a power inefficiency cycle.

c) **Interconnect Challenges**: Denser circuits in a chip require complex metal networks. Inherent resistance in interconnects causes voltage drops during current flow. Signal integrity maintenance demands additional current, increasing chip power consumption.

d) **Clock Distribution Challenges**: Distributing clock signals faces growing difficulty at higher frequencies. Networks with buffers amplify power dissipation, complicating issues. Higher frequencies demand frequent transistor switching, elevating power use.

e) **Static versus Dynamic Power: A Changing Landscape**: Traditionally, static power comes from leakage currents and constant voltage. CMOS power once dominated by static dissipation. Miniaturization and higher clock frequencies introduce dynamic power dominance. Frequent transistor switching now takes the lead in power consumption. This shift poses an optimization challenge, balancing leakage and switching. Reducing leakage may increase switching activity, and vice versa.

f) **Thermal Considerations: Heat Impact on Chips**: Power dissipation leads to heat within the chip. Excessive heat affects performance, reliability, and chip lifespan. Overheating triggers thermal throttling, reducing clock speed for safety. This action counteracts the benefits gained through miniaturization.

g) **Environmental Considerations in Computing**: Modern CMOS chips, with high power consumption, pose environmental challenges. Data centers, powered by numerous processors, consume considerable energy, emitting carbon. Urgent need for energy-efficient computing solutions to ensure sustainability.

h) **Mitigating Energy Inefficiency**: Researchers explore strategies for power optimization. Techniques include high-k gate dielectrics, reducing leakage currents. Clock gating and power gating selectively shut down idle circuits. Exploring alternative device architectures, like FinFETs, shows promise.

i) **Beyond CMOS: Exploring Green Computing**: Traditional CMOS faces energy limitations, leading to alternative paradigm exploration. Spintronics, neuromorphic computing, and QC show potential for significantly lower power consumption. These technologies operate on distinct physical principles, creating opportunities for inherently energy-efficient information processing.

j) **Impact on Performance and Battery Life: 1. Performance-Per-Watt Considerations**: Energy efficiency links directly to performance per watt.

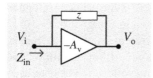

Figure 2.13 An ideal voltage inverting amplifier with an impedance connecting input to output. File: File:Impedance Multiplier. https://commons.wikimedia .org/wiki/File:Impedance_Multiplier.png Source: Constant314/Wikimedia Commons/CC BY-SA 3.0.

Enhancing computational ability while cutting power usage improves the overall ratio. This ratio holds significance in contemporary electronics.

k) **Impact on Performance and Battery Life: 2. Mobile and Internet of Things (IoT) Devices**: Energy efficiency is crucial for smartphones and IoT devices (cf. Figure 2.13). Optimize power-hungry components and use efficient sleep modes. Contribute to extended operational periods by doing so.

l) **Sustainable Computing Ahead**: Tackling energy efficiency issues is a crucial step. Ongoing research will shape a greener era of information processing. Computation must responsibly manage energy for a sustainable future. Pursuit of smaller, faster chips should avoid excessive energy use. "Innovation aims for spin dance that's efficient."

Important Questions for you to think upon: "Can you minimize the need for computing?"

"Can you do less computing, when you have less energy? – Interruptible computing"

"Can you think of an arising situation where, Economy is measured in terms of Energy Efficiency and not in terms of money anymore?"

2.1.3   Performance Bottlenecks

This section talks about the performance bottlenecks throughout. This section briefs the performance challenges in CMOS technology in Section 2.1.3.1. The following key components are crucial to grasp the understanding.

Performance in Electronic Devices: Attributes include processing speed, efficiency, and responsiveness. CMOS tech, with transistors, is a semiconductor industry foundation. Pursuit of increased performance faces challenges, causing the below bottlenecks, understanding performance bottlenecks (UPB) [1–5] is crucial.

UPB: 1. Clock Frequency Limits: Performance relies on clock frequency, determining processor speed. In CMOS, limits arise from signal propagation time in the circuit. As devices shrink, signal distances become comparable, restricting clock frequency.

UPB: 2. Heat Dissipation Challenges: (cf. Section 2.1.2.2 detailing heat dissipation challenges) Enhanced performance can raise power use and heat.

Managing heat is crucial for component reliability and longevity. Traditional CMOS faces challenges in heat dissipation efficiency. Efficiently managing generated heat is vital to overcome thermal bottlenecks.

UPB: 3. Memory Access Latency: Retrieving data from memory affects overall system speed. Latency, the time to fetch data, arises problems in fast-processing conditions. Processing speed surpassing memory speed can amplify this bottleneck.

UPB: 4. Interconnect Limitations: Connecting chip components encounters challenges during downsizing. Resistance, capacitance, and signal crosstalk lead to delays and integrity issues. These issues contribute to data transfer bottlenecks within the interconnects.

UPB: 5. SCEs: (cf. Section 2.1.1) Reducing transistor sizes results in SCEs. These effects involve heightened leakage currents and degraded performance. Variations in threshold voltage from SCEs impact transistor reliability. This, in turn, affects the overall performance of the circuit.

2.1.3.1 Performance Challenges in CMOS Technology

Challenges arise in enhancing processing power. Recognizing bottlenecks is key for effective strategies. Solutions include multicore architectures and beyond-CMOS tech. The semiconductor industry strives for performance advancements. Mitigating bottlenecks ensures CMOS evolution and explores new paradigms. Computing systems rely on swift and efficient information processing. Clock speed, signal propagation, and memory access directly affect performance. Awareness of constraints in these aspects is crucial for device evolution.

Clock Speed Constraints:

1. Clock Speed Basics: Clock speed, measured in Hertz, guides processor instruction execution. It is a core metric for computational speed and efficiency assessment. Industry has seen steady growth, with Moore's Law and advancement in technology.

2. Physical Limitations: As devices shrink, clock speed faces physical limits. Signal traversal time approaches clock period, hindering increase in frequency. This challenge hampers the historical trend of clock speed improvements.

3. Heat Dissipation Challenges: Higher clock speeds increase power consumption, posing heat dissipation challenges. Smaller, denser devices require efficient heat management. Balancing clock speed and heat influences overall device performance.

Signal Delay and Latency:

1. Signal Propagation Challenges: Latency becomes evident as devices scale down. Propagation delays directly influence computing system speed and responsiveness. Distances within a chip, constant and challenging, affect synchronization.

2. Interconnect Limitations: Interconnecting components within a chip poses signal propagation challenges. As devices shrink, interconnect limitations become more evident, requiring attention. Resistance, capacitance, and crosstalk contribute to signal degradation and delays.

Memory Access Limitations:

1. Memory Hierarchy Challenges: Memory access significantly impacts system performance. Latencies in data retrieval can lead to idle processor cycles, impacting efficiency. Hierarchical memory structure poses challenges for access time optimization.

2. Shortcomings in Data Transfer Rates: Processor and memory speeds misalignment limits data transfer. Affecting smooth data flow, resulting in below average-optimal performance. Bridging this gap is vital for unlocking any computing system's potential.

Clock speed, signal delay, and memory access pose hurdles. Industry navigates these complexities seeking efficient solutions. The evolution of speed, signal propagation, and memory access is challenging. Breakthroughs redefine achievable limits, encompassing dynamic progress. The journey sets stage for a future of enhanced capabilities.

2.2 Implications for Computing Systems

This section talks about the implications for computing Systems. This section briefs the impact on processing power in Section 2.2.1 and describes the energy-efficiency concerns in Section 2.2.2.

Ongoing changes in semiconductor tech significantly impact computing systems. Constraints in clock speed, signal delay, and memory access reshape semiconductor design. These changes have notable implications for computing systems' performance and energy efficiency. This exploration (comprising A–C components) examines into the consequences within computing systems, affecting architecture and efficiency.

A. **Performance Dynamics**:

1. Impact of Clock Speed on Performance: Clock speed influences overall system performance in computing. Historically, higher clock speeds led to improved computational capabilities. Expectations from users and industry stakeholders are shaped by this correlation.

2. Transition to Multi-core Architectures: Clock speed limits driving a shift to multi-core systems. Relying less on rising clock speeds, focus turns to parallel processing. Multicore setups allow simultaneous task execution, easing clock speed constraints.

3. Performance Efficiency: Clock speeds face limits, shifting focus on performance per watt (cf. Section 2.1.2.3 detailing impact on energy efficiency). Computing systems navigate the delicate balance for efficiency. Efficiency metrics guide architectural decisions in this era.

B. **Energy Efficiency Imperatives**:

1. Heat Dissipation Challenges Rising power dissipation from higher clock speeds requires strategic solutions. Traditional cooling struggles with thermal equilibrium, urging reevaluation. Crucial for sustainable systems: energy-efficient cooling and advanced thermal management.

2. Green Computing Paradigm: Sustainability links with energy efficiency in the green computing paradigm. Strategies include efficient hardware, optimized algorithms, and renewable energy integration. Computing faces pressure to minimize the environmental impact. The broader goal is to establish eco-friendly computing infrastructures.

3. Energy-Aware Computing Architectures: Computing systems explore energy-aware architectures. DVFS (cf. Section 2.1.2.1 briefing on dynamic and static power consumption) adapts to workload demands. Intelligent power management maximizes efficiency without performance compromise.

C. **Architectural Transformations**:

1. Interconnect Impact on Computing Systems: Signal delay and latency significantly affect system interconnects. As components shrink, interconnect design gains importance. Solutions include basic materials, high-speed tech, and smart routing.

2. Revamped Memory Hierarchy: Memory access constraints spark a rebooting computing system memory hierarchy. Optimized hierarchical memory fits multi core needs, enhancing data access and task execution efficiency. Caches, formerly secondary, now crucial to cut memory access delays.

3. Innovative Computing Models: Traditional CMOS scaling issues encourage novel computing approaches. In-memory computing gains attention by executing computations within memory. Quantum and neuromorphic models explore non classical principles for alternative architectures.

Navigating Computing Challenges: Clock speed limits, signal delays, and memory constraints impact computing. Future trajectories in performance, energy efficiency, architecture, and developmental pathways shape evolution.

2.2.1 Impact on Processing Power

This section talks about the impact on processing power throughout. This section briefs the computational performance threshold in Section 2.2.1.1, describes the challenges in meeting increasing demands in Section 2.2.1.2, and presents in detail the consequences for modern computing devices in Section 2.2.1.3.

The following key components (I–VI) are crucial to grasp the understanding. Increased processing power shapes the semiconductor industry. Explore dynamics of challenges in processing power. Implications on performance, efficiency, and computational foundations emerge.

(I) **The Clock Speed Challenges**: Computing power historically tied to processor clock speed. Moore's Law aimed at doubling transistor density, boosting clock speeds. Linear clock speed growth encounters a bottleneck, limiting processing power.
- Historical Trajectory
- Decreasing Performance
- Shift to Multi core

(II) **Energy Efficiency Imperatives**: Processing power and energy efficiency become intertwined priorities. The industry emphasizes performance-per-watt metrics over sheer computational capability. The challenge is to boost processing power sustainably.
- Performance-Per-Watt Metrics
- Heat Dissipation Challenges
- Green Computing Paradigm

(III) **Architectural Transformations**:
- Interconnect Challenges
- Memory Hierarchy Reimagined
- Innovative Computing Models

(IV) **Future Trajectories**:
- **Quantum Processing**: Exploration into QC seeks enhanced processing power. The shift from classical to quantum processing redefines processing power boundaries. Quantum processors leverage superposition and entanglement for computational gains.
- **Edge Computing Dynamics**: Scaling Challenges impact the evolving edge computing. Processing tasks at network edges highlight constraints of traditional architectures. Need to address processing power requirement in constrained environments.

(V) **Navigating Scaling Challenges**: Scaling Challenges impact processing power significantly. Clock speed challenges urge shifts toward efficiency and parallelism. Alternative computing models require exploration for sustainable progress. Architectural changes are crucial for system design, interconnects, and memory.

(VI) **Industry's Evolution**: The industry faces transformative possibilities amid these challenges. Quantum leaps and edge computing dynamics mark future trajectories. Adaptation, innovation, and relentless pursuit define the quest for computational frontiers.

2.2.1.1 Computational Performance Threshold

Modern computing faces persistent challenges in enhancing performance (cf. Figure 2.14). Societal demands and technological goals drive efficiency. However, devices confront areas hindering performance trajectory. This exploration examines computational performance threshold, highlighting challenges and impacts. The escalating demands (TEDs) are as follows:

TED: A. Emergence of Data-Intensive Applications: Computing changes significantly with data-intensive applications. The need for processing power intensifies due to vast datasets and complex algorithms. Artificial intelligence (AI) (The Rise of large language models [LLMs]), machine learning, and big data analytics demand substantial computational resources.

TED: B. Connectivity and Networking Challenges: Computing devices exist in interconnected systems, not in isolation. 5G technology and the rise of IoT devices increase strain on resources. Demand on performance extends to connectivity and networking challenges.

TED: C. Immersive Technologies and Real-Time Processing: Immersive technologies like virtual reality (VR) and augmented reality (AR) impact computational demands. Meeting expectations, responsive interactions is

Figure 2.14 A laptop PC clock generator, based on the Silego chip. File: Silego clock generator https://commons.wikimedia.org/wiki/File:Silego_clock_generator.JPG Source: AAAndrey A/Wikimedia Commons/Public Domain.

crucial for computational performance. Real-time processing needs for immersive experiences and interactive applications create challenges, "crucial for retail industries moving toward Virtual Commerce platform beyond their state-of-the-ar e-commerce platform experience."

2.2.1.2 Challenges in Meeting Increasing Demands
This includes as follows:

A. Processing Power: Computational performance slows down as traditional scaling faces challenges. Increased transistor counts pose power-density constraints, causing performance degradation. Dennard scaling, enabling power reductions with smaller transistors, reaches limits.

B. Heat Dissipation Challenges: Increasing computational needs result in significant heat generation. Cooling systems struggle with high-performance device heat dissipation. Thermal limitations impact device performance.

C. Memory-Bandwidth Bottlenecks: Modern devices process faster than efficiently retrieving data from memory. The gap between computational and memory access speed hinders performance. Memory-bandwidth limitation significantly impacts overall performance.

2.2.1.3 Consequences for Modern Computing Devices
This includes as follows:

A. Shift Toward Specialized Architectures: General-purpose computing limitations drive a move to specialized architectures. Graphics processing units (GPUs), field - programmable gate arrays (FPGAs), and task-specific accelerators gain attention.

B. Rise of Quantum and Neuromorphic Computing: Exploring quantum and neuromorphic computing challenges classical boundaries. Quantum processors use principles beyond classical physics, while neuromorphic architectures mimic the human brain. These technologies have the potential to reshape computational performance.

C. Adaptive Algorithms and Edge Computing: Adaptive algorithms, adjusting based on resources, become essential. Edge computing, distributing tasks closer to data sources, mitigates CPU bottlenecks.

Modern computing transforms amid increasing demands and performance hindrance. Challenges like scaling, heat dissipation, and memory-bandwidth limitations drive strategic adaptations. Specialized architectures emerge, and quantum and neuromorphic computing are to be explored. The pursuit of computational excellence continues to meet evolving demands in a data-centric world.

2.2.2 Energy-Efficiency Concerns

This section talks about the energy-efficiency concerns throughout. This section briefs the growing importance of low-power Computing in Section 2.2.2.1, describes the environmental and economic considerations in Section 2.2.2.2, and presents in detail the necessity for sustainable technological solutions in Section 2.2.2.3. The following key components are essential to grasp the understanding.

Energy efficiency is crucial in modern computing. As the digital realm grows, challenges in energy efficiency increase. It influences system architecture, design, and operation. This exploration (I–VII) examines various dimensions of energy-efficiency concerns in computing systems.

(I) **The Core Energy-Efficiency Concerns**:

A. Defining Energy Efficiency in Computing Systems: Energy efficiency in computing aims for optimal output with minimal consumption. The pursuit aligns with sustainability, cost-effectiveness, and environmental responsibility. Foundational principles apply to both hardware and software in computing.

B. Significance Across System Hierarchies: Energy efficiency concerns impact various computing layers. Individual components and larger entities need a systemic approach.

(II) **Memory Hierarchy and Energy Efficiency**:

A. Caching Strategies: Memory system energy efficiency links closely to caching methods. Effective cache management reduces main memory access, reducing energy use. Caching design involves balancing energy efficiency and potential latency increases.

B. Memory Access Patterns: Analyzing and optimizing memory access patterns boosts energy efficiency. Predictable patterns optimize cache use, reducing higher-latency memory access. Techniques like pre-fetching and adaptive caching align with energy-efficient task execution.

C. Non-volatile Memory (NVM) Technologies: Exploring NVM enhances energy efficiency understanding. However, overall energy considerations include write endurance, access latency, and integration challenges. Memories like Flash and emerging technologies (Resistive RAM) offer idle state energy advantages.

(III) **Energy Efficiency Concerns in Storage Systems**:

A. Storage Hierarchies and Access Patterns: Storage systems, including HDDs, SSDs, and emerging technologies, presents energy challenge. Optimization of access patterns and balancing storage hierarchies is crucial. Tiered storage architectures align resources with diverse energy efficiency needs.

B. Data Compression and De-duplication: Techniques like data compression and de-duplication aid storage energy efficiency. Reduced data volume results in energy savings, but algorithmic overhead requires scrutiny. Consider trade-offs associated with computational overhead in compression and deduplication.

C. Emerging Storage Technologies: Novel technologies like magnetic recording and phase-change memory (PCM) offer energy-efficient options. Balancing performance and energy considerations is essential in adopting emerging storage technologies.

(IV) **Networking and Communication Energy Efficiency**:

A. Network Infrastructure Challenges: Networking components, like routers and switches, pose energy efficiency challenges. Efforts focus on energy-efficient protocols, routing algorithms, and adaptive communication. Continuous data transmission demands introducing energy overheads.

B. Data Center Network Optimization: Energy-aware scheduling aligns computational tasks with network conditions for enhanced efficiency. Network architecture design significantly affects energy consumption in data centers. Efficient data routing and load balancing contribute to energy savings.

C. Wireless Communication and Mobile Devices: Techniques like advanced modulation and energy-efficient standards contribute to prolonged battery life. Energy efficiency is crucial for wireless communication in mobile devices. Energy expended during data transmission affects battery life.

(V) **Software and Algorithmic Contributions to Energy Efficiency**:

A. Algorithmic Complexity and Execution Efficiency: Software considerations impact energy-efficient computing. Algorithmic complexity and parallelism influence energy consumption. Optimizing for specific hardware and energy-aware algorithms boosts system efficiency.

B. Runtime Management and Task Scheduling: Aligning computational demands with available resources enhances energy efficiency. Dynamic management and scheduling adapt to changing workloads. Techniques like dynamic voltage scaling and intelligent task scheduling minimize energy waste.

C. Energy-Aware Programming Paradigms: Tools for profiling and optimizing energy consumption empower software development. Programming paradigms evolve with energy-aware approaches. Languages and frameworks enable energy-efficient computations.

(VI) **Environmental and Societal Considerations**:

A. Carbon Footprint and Green Computing: Energy use in data centers affects the environment. The carbon footprint emphasizes the need for green computing. Using renewable energy and minimizing impact are crucial.

B. Economic Viability and Accessibility: Cost-effective technologies enhance accessibility for diverse users. Energy efficiency affects the economic viability of computing systems. High operational costs can impact resource accessibility.

C. Regulatory Frameworks and Ethical Dimensions: Ethical choices in design, procurement, and operation ensure sustainability. Adhering to standards promotes responsible energy consumption. Regulatory frameworks address ethical aspects of energy efficiency.

(VII) **Challenges and Future Directions**:

A. Technological Challenges: Overcoming energy efficiency challenges requires addressing technological bottlenecks. Exploring novel materials, device architectures, and cooling technologies enhances efficiency.

B. Interdisciplinary Collaborations: Interdisciplinary collaborations are vital for addressing energy efficiency complexities. Engineers, scientists, economists, and policymakers must collaborate for holistic solutions.

C. Educational Initiatives and Awareness: Promoting energy-conscious computing involves educational initiatives and awareness campaigns. Empowering users, developers, and decision-makers with knowledge cultivates a mindful approach.

2.2.2.1 Growing Importance of Low-Power Computing

Low-power solutions gain prominence in computing evolution (cf. Section 2.1.2 detailing power dissipation issues). Exploration examines ties with sustainability, economics, and technological sustainability. The evolution of low-power computing (ELPC) is as follows:

A. ELPC: Historical Context and Technological Progression: Low-power computing traces back to early electronic computing days. Evolution marked by energy-efficient architectures and power management integration. Transition from mainframes to mobile devices emphasizes low-power solution importance.

B. ELPC: Portable Device Proliferation: Ubiquity of devices like smartphones drives demand for low-power computing. Manufacturers innovate low-power architectures for prolonged device operation. Balancing performance and energy efficiency crucial for user experience and battery life.

C. ELPC: Embedded Systems and IoT Revolution: Specialized architectures needed for computational efficiency within on-demand low-power constraints. Rise of embedded systems and IoT expands computing beyond traditional devices. Demand surges for low-power embedded solutions in various applications.

2.2.2.2 Environmental and Economic Considerations

(I) **Environmental Considerations**:

A. Energy Consumption in Data Centers: Sustainable practices, renewable energy, and efficient server architectures gain attention. Data centers, driven by high computational demand, raise environmental concerns. Massive data centers supporting digital infrastructure draw attention due to energy consumption.

B. Emissions and Climate Impact: The need to address climate change emphasizes energy-efficient computing for sustainability. Carbon emissions from energy-intensive computing highlight climate impact. Global data center proliferation contributes to carbon footprints and climate change.

C. Green Computing Initiatives: Green computing initiatives aim to tackle environmental challenges. Optimization of server farms and use of energy-efficient hardware reduce ecological impact. Integrating renewable energy into data centers marks a significant step toward sustainability.

(II) **Economic Considerations**:

A. Operational Costs and Total Cost of Ownership (TCO): Economic aspects of low-power computing (cf. Section 2.2.2.1 briefly presenting growing importance of low-power computing) gain attention beyond the environment concerns. Energy-efficient solutions promise ecological benefits and economic advantages. Energy consumption forms a substantial part of data center TCO.

B. Energy-Efficiency in Business and Enterprise: Low-power hardware and cloud services become strategic choices for cost-effective businesses. Enterprises recognize economic benefits of energy-efficient computing. Optimization of energy consumption brings direct cost savings for organizations.

C. Energy-Efficiency as a Market Differentiator: Energy-efficient computing aligns with sustainability, influencing market perception and choices. Market dynamics highlight energy efficiency as a competitive advantage. Consumers and businesses prioritize environmentally conscious practices.

2.2.2.3 Necessity for Sustainable Technological Solutions

(I) **Technological Innovations and Sustainable Solutions**:

Innovations drove low-power computing forward. System-on-chip (SoC) designs further improved power efficiency The processors focus on energy optimization. From evolving reduced instruction set computing (RISC) to integrating heterogeneous computing models. The three key nodes are as follows:

A. Advancements in Processor Architectures

B. Dynamic Power Management (DPM) Techniques

C. Emergence of Low-Power Components

(II) **Challenges and Considerations in Low-Power Computing**:
   A. Performance Trade-Offs: Low-power computing involves trade-offs between performance and energy efficiency. Achieving balance requires considering application requirements and computational workloads. Innovations improving the performance-per-watt metric help overcome this challenge.
   B. Application-Specific Demands: Diverse applications pose varied demands on systems. Adapting low-power architectures required to meet specific demands.
   C. Technological Hurdles in Miniaturization: Miniaturization faces hurdles. Quantum effects, leakage currents, and heat challenges amplify with shrinking components. Overcoming these hurdles demands breakthroughs in materials, transistor design, and cooling technologies.

(III) **Future Trajectories and Perspectives**:
   A. Holistic Approaches to Sustainability: Low-power computing aligns with sustainability aspirations. Holistic approaches, covering hardware, software, and eco-friendly data centers. It helps shape an environmentally mindful computing ecosystem.
   B. Innovation as a Catalyst for Change: Sustainability drives innovation as a catalyst. Collaborations, research, and policy initiatives converge for an sustainable environment. Low-power computing translates to efficiency, economic, and ecological responsibility.
   C. A Call for Collective Responsibility: Embracing low-power computing requires collective responsibility. From consumers to organizations, stakeholders shape a sustainable technological

2.3 Technological and Economic Challenges

This section talks about the technological and economic challenges. This section briefs the technological barriers in Section 2.3.1, and describes the adaptability and innovation in Section 2.3.2.

The intersection of technology and economics unravels the symbiotic relationship that defines the evolution of computing. Lets understand the process ([components include [I–IV]):

(I) **Technological Challenges**:
   A. Proliferation of Emerging Technologies: Rapid spread of AI, quantum, and edge computing poses challenges. Integrating these technologies demands a paradigm shift for industries.

B. Security and Privacy Concerns: Interconnected systems raise security and privacy concerns. Balancing innovation with robust security measures is crucial.

C. Scalability and Performance Optimization: Constant demand for scalable and high-performance computing solutions persists. Data-intensive applications, cloud computing, and the IoT emerge day by day. Optimizing performance without compromising scalability is a challenge. Innovative architectures, parallel processing, and efficient algorithms are the focal points.

D. Legacy System Integration: Industries struggle with integrating new tech into legacy systems. Bridging the gap between outdated infrastructures and modern demands is challenging.

E. Environmental Sustainability: Energy consumption in data centers, electronic waste, and the carbon footprint of computing activities demand sustainable solutions. The environmental impact of these is a growing concern. Developing eco-friendly practices, from energy efficiency to e-waste management, is crucial.

(II) **Economic Challenges**:

A. Costs of Technological Adoption: Balancing investments with long-term gains is a delicate process. Economic challenges stem from adopting technology. Upgrading, investing, and training involve upfront expenses.

B. Digital Inequality and Accessibility: Initiatives for digital literacy, affordable hardware, and equal access are crucial. Bridging the gap involves addressing technology affordability. The digital divide worsens societal inequalities.

C. Monopolies and Market Dynamics: Regulatory frameworks encouraging competition are essential. Tech conglomerates' dominance poses economic challenges. Balancing innovation and preventing monopolies is crucial.

D. Workforce Transition and Reskilling: Economic challenges arise in facilitating a seamless transition. Evolving computing tech requires a workforce skills evolution. Reskilling initiatives and educational reforms address this challenge.

E. Intellectual Property and Innovation: Balancing intellectual property protection and innovation is challenging. Legal frameworks, patent regulations, and collaboration foster innovation and economic sustainability.

(III) **The Interplay Between Technology and Economics**:

A. Innovation and Economic Growth: Collaborative efforts support an innovation-friendly environment. Technological progress creates markets and jobs. Innovation drives economic growth in computing.

B. Tech Determinism and Economic Realities: Tech determinism interacts with economic realities. Economic constraints influence technological development.

C. Policy Interventions for Growth: Regulatory frameworks and R&D investments foster innovation. Governments shape policies for tech-economic symbiosis. Balancing ensures tech contribution to economic prosperity.

(IV) **Future Trajectories and Perspectives**:

A. Coordinated Strategies: Computing's evolution needs strategies aligning with economy. Crafting frameworks leverages technology for sustainable economic growth. Collaboration among leaders, policymakers, and economists is essential.

B. Ethics in Tech and Economy: Integrating ethics is essential to responsible and sustainable growth. Ethical considerations are crucial as it shapes society. Privacy-equity and accountability influence economic decisions.

C. Educational Paradigms for Holistic Competence: Holistic approaches blend expertise with economic literacy. Addressing challenges requires an educational paradigm shift. Lifelong learning and interdisciplinary education prepare a future-ready-workforce.

2.3.1 Technological Barriers

This section talks about the technological barriers throughout. This section briefs the materials and manufacturing constraints (MMC) in Section 2.3.1.1, describes the integration challenges with emerging technologies in Section 2.3.1.2, and presents in detail the economic viability in a changing landscape in Section 2.3.1.3. The following key components are essential to grasp the understanding (components include [I–V]).

Technological innovation faces various challenges and barriers. Barriers influence the pace and trajectory of innovation. Ranging from material limitations to complex system integration. This exploration analyzes the nature, impact, and overcoming strategies.

(I) **Materials and Semiconductor Challenges**:

A. Beyond Moore's Law: As we have been discussing, miniaturization, faces hurdles at atomic scales. Quantum effects like electron tunneling and leakage impact device reliability. Material innovations and novel architectures are vital for building next-generation computing platform.

B. Materials Limitations: Semiconductors face limits, especially silicon in shrinking devices. Exploring materials with better electrical properties is essential. Spintronics and 2D materials show promise but need extensive research.

C. Heat Dissipation: Traditional cooling struggles with escalating high-performance device heat. Densely packed components pose heat dissipation challenges. Innovations, advanced cooling, and thermally conductive materials are crucial.

(II) **System Integration Challenges**:

A. Heterogeneous Integration: Seamless integration demands advancements in heterogeneous integration technologies. Integrating diverse technologies requires attention in bonding and packaging. Challenges include thermal compatibility, interconnect density, and overall performance.

B. Interconnect Scaling: Shrinking feature sizes intensify resistance–capacitance (RC) delays. Scaling interconnects faces limits in resistance, capacitance, and signal integrity. Overcoming these challenges requires novel materials and designs including exploration of Graphene.

C. Power Integrity: Ensuring power integrity, especially in high-performance systems, is critical. Voltage drops, signal noise, and power delivery affect performance. Addressing power integrity involves innovations in power delivery, regulation, and DPM.

(III) **Emerging Technology Challenges**:

A. QC Limitations: QC faces challenges of maintaining coherence and error minimization. Progress depends on advancements in materials which displays quantum properties at operational temperatures. Scaling quantum bits (qubits) necessitates breakthroughs in error correction and fault-tolerant gates.

B. Neuromorphic Computing Complexity: Emulating synaptic connections and parallel processing demands exploration in hardware, including memristors. Neuromorphic computing, mirroring the brain's architecture, poses design and implementation hurdles. Unconventional materials replicating synaptic behavior are crucial for advancements.

C. Advanced Manufacturing and Nanotechnology: Reliable and scalable nanomanufacturing techniques are essential for unlocking nanotechnology's potential. Advancing miniaturization requires improvements in manufacturing technologies. Nanoscale fabrication struggles with challenges like defects, yield issues, and cost-effectiveness. Refer to IEEE 62659 IEC/IEEE International Standard for nanomanufacturing. It outlines procedures for integrating nanoelectronics into high-volume semiconductor production, ensuring yield and controlled introduction of nanomaterials.

(IV) **Cross-Cutting Challenges**:

A. Security and Trustworthiness: Ensuring security and trust in complex, interconnected systems is critical. Challenges in key management, side-channel attack protection, and secure hardware need interdisciplinary efforts.

B. Ethical and Societal Implications: Emerging technologies, like AI and biotech, bring ethical challenges. Balancing innovation with ethical principles and addressing algorithmic biases require broad regulatory-frameworks.

(V) **Strategies for Overcoming Technological Barriers**:

A. Interdisciplinary Research Collaborations: Breaking technological barriers needs expertise from multiple disciplines. Collaborations among physicists, materials scientists, electrical engineers, and computer scientists foster holistic problem-solving.

B. Investments in Research and Development: Governments, industries, and academia must invest significantly in R&D. Funding long-term, high-risk research projects supports exploring uncharted territories and overcoming challenges.

C. Education and Skill Development: Nurturing innovators involves comprehensive education and skill programs. Curricula evolving traditional boundaries and hands-on research experiences prepare a challenging workforce.

D. Agile Regulatory Frameworks: Regulatory frameworks must swiftly adapt to evolving technology. Agile regulations balancing innovation and addressing ethical and security concerns provide the necessary support.

Overcoming these technological challenges requires collective, persistent effort. Interdisciplinary collaboration, strategic investments, and ethical considerations pave the way forward. Researchers and innovators need to address complexities in materials, system integration, and emerging technologies. These technological breakthroughs will shape up computing innovations.

2.3.1.1 Materials and Manufacturing Constraints (MMCs)

In the evolving tech scenario, the interaction of materials, manufacturing, and emerging technologies brings challenges. This exploration looks at complexities in materials, manufacturing, and integrating emerging technologies. The economic viability is crucial in a constantly changing landscape.

MMC: A. Silicon Limitations and Beyond: Silicon, a standard in semiconductor manufacturing, faces limitations at nanoscale. Challenges include electron mobility, leakage currents, and thermal management. Emerging materials like gallium nitride (GaN) and silicon carbide (SiC) offer advantages. Yet, they bring their own set of manufacturing challenges.

MMC: B. Nanomanufacturing Challenges: Fabricating nanoscale devices introduces manufacturing challenges. Defects, variability in nanomaterial properties, and precision bring challenges. Innovations like directed self-assembly and atomic layer deposition address challenges.

MMC: C. Sustainability in Manufacturing: Environmental impact of manufacturing adds complexity. Hazardous materials, energy-intensive processes, and electronic wastes raise sustainability concerns. Eco-friendly process and recyclable materials are crucial for long-term viability.

2.3.1.2 Integration Challenges with Emerging Technologies
This includes as follows:

A. Heterogeneous Integration: Mismatched thermal properties and varying electrical characteristics demand innovative solutions. Integrating diverse technologies present challenges in system integration. Overcoming these challenges is critical.

B. QC Integration: QC components have unique integration requirements. Maintaining quantum coherence and minimizing interference demand specialized approaches. Hybrid integration, combining classical and quantum components, addresses these.

C. Biocompatible Electronics: Integrating electronics with biological systems poses questions related to biocompatibility. Traditional electronics may induce tissue damage or trigger immune responses. Developing biocompatible materials, like flexible electronics, is pivotal for bioelectronics applications.

2.3.1.3 Economic Viability in a Changing Landscape
This includes as follows:

A. Cost-Effective Scaling: Traditional scaling has reduced returns; transitioning has high capital costs. Cost-effective scaling for advanced technologies is a challenge. Balancing performance gains with economic viability is a key requirement.

B. Accessibility and Affordability: Reducing production costs, optimizing processes, and fostering competition enhance accessibility. Societal impact relies on the accessibility and affordability of emerging technologies. Government incentives and industry collaboration contribute to broader accessibility.

C. Economic Impacts of Sustainability: Integrating sustainable practices affects economic considerations. While eco-friendly technologies aid environmental friendly solutions, initial investments may challenge the economy.

Strategies for overcoming challenges (SfoC) are as follows:

SfoC: A. Materials Innovation Hubs: The key requirements are establishing research hubs focusing on material innovation. Scientists, engineers, and industry experts collaborate on new materials, accelerating development to address performance and manufacturing issues.

SfoC: B. Digital Twins in Manufacturing: Using digital twin tech in manufacturing for precision and efficiency. Simulating and optimizing workflows in a virtual environment. Reducing defects, accelerating prototyping, and ensuring cost-effective manufacturing are the way ahead.

SfoC: C. Public–Private Partnerships: Form strategic partnerships between public institutions and private industries. Foster innovation while addressing economic challenges. Joint initiatives share the financial burden for ambitious and high-risk projects.

SfoC: D. Education for Sustainable Practices: Equips future innovators with knowledge and ethical considerations. Integrates sustainable practices education into engineering curricula. Promotes a shift toward eco-conscious advancement.

Toward a Sustainable Computing: Overcoming materials, manufacturing, and integration challenges shows the path. Manufacturing precision and economic viability are crucial. Overcoming these constraints with interdisciplinary collaboration can lead to a future where the node aligns with environmental and economic sustainability.

2.3.2 Adaptability and Innovation

This section talks about the adaptability and innovation throughout and presents in detail the industry response to addressing challenges in Section 2.3.2.1. The following key components are essential to grasp the understanding.

Industries adapt to multifaceted challenges in the evolving technological landscape. Technology advancs consistently, seeking adaptability and innovation. This exploration focuses on adaptability, innovation, alternative computing, and industry responses.

A. Dynamic Nature of Technological Challenges:

Industries must be proactive in tackling complex technological issues. The tech landscape is ever-changing, posing a range of challenges. From scaling issues to quantum effects, adaptability is crucial.

B. Organizational Adaptability:

Empowering employees for innovative solutions is crucial for staying ahead. Meeting tech challenges requires organizational flexibility. Fostering a culture embracing change and continuous learning is vital.

C. Agile Development Methodologies:

Agile methods, with their iterative and collaborative approach, aid adaptability. Rapid prototyping and flexible project management respond leading to challenges. Ensuring the development process remains dynamic is facilitated by agility.

Need for Innovative Solutions such as:

A. Driving Progress with Innovation

B. Research and Development Initiatives

C. Innovative Business Models

2.3.2.1 Industry Response to Addressing Challenges
This includes as follows:

A. Collaborative Ecosystems: Combined efforts accelerate progress by pooling resources and expertise. Challenges aid industry cooperation. Partnerships among tech companies, startups, research institutions, and governments foster synergy.

B. Corporate Social Responsibility (CSR) Initiatives: CSR initiatives go beyond immediate challenges. Tech leaders recognize societal impacts. The focus is on promoting ethical, sustainable, and inclusive practices for responsible global citizenship.

C. Regulatory Engagement and Advocacy: Industries actively work with regulators. Advocacy for policies balancing innovation and ethical practices.

Adaptability, innovation, and exploring alternatives shape this resilient future. Collaboration in industry responses is crucial for sustainable progress. Challenges become catalysts for transformative change.

2.4 Bridging to Nanoscale Computing

This section talks about the bridging to nanoscale computing. This section briefs the introduction to nanomagnetic logic in Section 2.4.1.

In technology's vast evolution, nanoscale computing (NC) is transformative. It bridges conventional and nanoscale paradigms requiring technical understanding. It explores NC, unraveling intricacies. Let us explore (using these components [I–V]) how to address scaling challenges, and envision potential applications.

1) **Understanding the Nanoscale Landscape**:
   A. Foundations of NC: NC involves manipulating matter on an unprecedented scale. Essential is: understanding quantum mechanics, quantum dots, and nanomaterials. Quantum tunneling, superposition, and entanglement redefine computing rules.
   B. Quantum Computing Overview: Quantum bits (qubits) can exponentially boost computational power, solving complex problems. Delving into quantum gates, and algorithm reveals the promise and challenges.
2) **Scaling Challenges and Innovations**:
   A. Moore's Law in the Nanoscale Context: Nanoscale challenges involve quantum effects, coherence, and information management. Quantum error correction, fault-tolerant designs, and qubit coherence innovations are crucial.
   B. Beyond Quantum: NML: NML is an alternative using nanoscale magnetic properties. Keeping aside quantum challenges, it offers unique opportunities.

Exploring domain wall logic, magnetic tunnel junction logic, and spin wave-based logic. Focusing on material design, spintronics, and magnetic anisotropy engineering.

3) **Applications and Use Cases**:
A. NC Applications: NC finds uses across various fields. Case studies showcase its transformative potential in AI technologies. It impacts edge AI devices and enhances energy efficiency.
B. Hybrid Systems: Merging NML with CMOS: Merging NML with CMOS creates a hybrid paradigm. Hybrid memory and logic architectures illustrate the interplay between NML and traditional frameworks. Exploring synergies promises improved performance and energy efficiency.

4) **Overcoming Challenges: Reliability, Scalability, and Integration**:
A. Navigating the Nanoscale Challenges: NC presents reliability and scalability challenges. QC intersects with nanoscale, requiring integrated solutions. Analyzing error rates, scalability, and unique manufacturing issues reveals obstacles.
B. QC and Nanoscale Integration: Integrating QC with NC poses possibilities and challenges. Assessing implications for reliability, scalability, and synergy is crucial.

5) **Emerging Frontiers: Neuromorphic Computing, IoT, Bio-nano-electronics**:
A. Beyond Traditional Computing Horizons: NC expands conventional limits. Applications include neuromorphic computing, bio-nanoelectronics, and the IoT. Convergence promises advancements in healthcare, sensing technologies, and IoT.
In summary, bridging to NC is a multifaceted exploration. Journey covers quantum realms, magnetic domains, scaling challenges, and technology integration. NC offers enhanced power and redefines possibilities.

2.4.1 Introduction to Nanomagnetic Logic

This section talks about the introduction to nanomagnetic logic throughout. This section briefs the overview of nanoscale computing paradigms in Section 2.4.1.1, describes the comparative advantages Over CMOS in Section 2.4.1.2, and presents in detail the potential solutions to addressing limitations in Section 2.4.1.3. The following key components are fundamental to grasp the understanding.
A. Defining NC Paradigms:

- NC shifts our approach to computational processes.
- The focal point is NML, a key concept.
- Understanding requires exploring principles, historical development, and driving forces.

B. Historical Development of NC:

- Rooted in quantum mechanics and nanomaterials, NC evolves.
- Historical development reveals milestones leading to NML.
- The journey spans theoretical propositions to experimental validations.

C. Significance and Applications:

- "Why NC?" aids exploration of its significance and applications.
- Uncovering unique capabilities offers insights into contemporary problem-solving.
- From transforming edge AI devices to enhancing energy-efficient computing, applications vary.

2.4.1.1 Overview of Nanoscale Computing Paradigms

NML transforms modern computing devices. It disrupts traditional architectures and redefines computational boundaries.

1) **Overcoming Scaling Limitations**: NML addresses scaling issues in traditional CMOS technology. Challenges like miniaturization and power dissipation find solutions in NML. Magnetic properties at the nanoscale enable downsizing components, mitigating performance problems.
2) **Low-Power Dissipation and Energy Efficiency**: (cf. Section 2.1.2 briefing power dissipation issues) Unlike CMOS, NML utilizes energy-efficient magnetic phenomena, reducing heat generation. Integrating NML results in low-power dissipation, distinct from traditional CMOS. This approach presents a promising solution to heat dissipation challenges in high-performance computing (cf. Section 2.1.2.2 detailing heat dissipation challenges).
3) **NVMs**: Non volatile nature enhances energy efficiency and responsiveness. NML introduces NVMs. Data retention does not need a continuous power supply. Enables instant-on, reducing boot time.
4) **Parallelism and Scalable Architectures**: Characteristics different from sequential processing enhances computational throughput. NML magnetic interactions allow architecture parallelism. Magnetic coupling enables concurrent operations, enabling scalable architectures.
5) **Beyond von Neumann Architecture**: Magnetic domain-based computing shows promise in handling vast datasets and complex algorithms. Traditional von Neumann architectures struggle with data-intensive tasks. NML proposes non-von Neumann approaches to overcome limitations.
6) **Integration with Emerging Technologies**: NML complements diverse architectures in addressing modern computational challenges. Integrating NML has implications beyond immediate performance improvements. Synergies with emerging technologies like quantum, neuromorphic, and edge computing are notable.

7) **Reliability and Resilience**: NML's magnetic nature provides resilience in tough conditions. From IoT edge devices to aerospace and healthcare applications, reliability is vital. This resilience ensures reliability for devices in various environments.

8) **Exploration of Novel Device Structures**: NML aids for exploration of unconventional device structures. MTJs, DW logic, and SW architectures attract attention. Results toward diversifying design space and creating customized devices for specific tasks (ASIC).

In the evolving computing landscape, NML's potential to redefine possibilities in next gen computing plays a key role. Integrating NML marks a significant innovation. It propels the industry toward surpassing current performance limits.

2.4.1.2 Comparative Advantages Over CMOS

This includes as follows:

A. Limitations of Traditional CMOS Technology: Understanding NML requires recognizing CMOS limitations. Challenges in scaling, power dissipation, and performance bottlenecks set the stage.

B. A Game Changer: NML relies on nanoscale material magnetic properties for novel information processing. Exploring foundations like magnetic properties and spintronics reveals how it surpasses CMOS limitations.

C. Beyond CMOS Architectures: NML paves path to computing beyond traditional CMOS. Physical structures like DW logic and MTJ logic offer integration. Comparative analyses highlight the unique strengths of NML.

2.4.1.3 Potential Solutions to Addressing Limitations

This includes as follows:

A. Material Design for NC: Magnetic materials, spin-transfer torque, spin-orbit torque, and magnetic anisotropy engineering are key. In NC, material design is crucial. They form the foundation for designing substrates in nanoscale computations.

B. Edge-AI Devices: Practical Applications: NML finds practical use in edge-AI devices. Chapter 6 elaborately details Nanoscale Computing at the Edge: AI Devices and Applications, exploring into edge computing basics, investigating opportunities, challenges, and real-world cases. Moving beyond theory, it places NML in the tangible areas of AI-driven technologies.

C. Hybrid Computing Systems: Integrating the Best: Integrating NML with traditional CMOS creates hybrid computing systems (cf. Chapter 7, which details Hybrid Computing Systems and Emerging Applications). Synergies between these technologies and potential performance enhancements are investigated.

This integration hints at the future of computing architectures, combining the best of both technologies.

D. Challenges and Future Directions:

- **Challenges in Reliability, Scalability, and Integration**: NC encounters challenges in reliability, scalability, and integration. Examining error rates, scalability limits, and manufacturing issues is crucial.
- **Applications in Emerging Technologies: A Glimpse:** Concluding our exploration, we glimpse into future applications of NML. From neuromorphic computing to bio-nanoelectronics, IoT devices, and environmental sensing. Potential applications showcase the transformative power of NC.

At the intersection of theory and application, NC signifies both evolution and untapped potential. In summary, delving into NML reveals a diverse technological landscape. Historical roots, advantages over traditional CMOS, and potential solutions shape a comprehensive understanding. The ongoing journey promises a closer examination of complexities, challenges, and emerging frontiers.

2.5 Educational Emphasis

This section talks about the educational emphasis. This section briefs the accessible technical content in Section 2.5.1 and presents the Chapter End Quiz in Section 2.5.2.

Setting the Stage: In this book, an innovative educational model designed to simplify the complexities of NC specifically for undergraduate students is introduced. This model aims to mold readers into future tech experts within the educational landscape of NC. It focuses on understanding challenges and opportunities in nanotechnology and computing.

2.5.1 Accessible Technical Content

This section talks about the accessible technical content throughout. This section describes the addressing undergraduate audiences in Section 2.5.1.1. The following key component is crucial to grasp the understanding.

Balancing Technicality and Accessibility: The core of this model lies in crafting content that balances technical depth and accessibility. It ensures clarity in presenting NC concepts, simplifying complexities while preserving the subject's essence. Pedagogical innovations are included for an effective connection with the readers.

2.5.1.1 Addressing Undergraduate Audiences

This includes the following components:

A. Connecting Foundational and Advanced Concepts: For undergraduate students, this model becomes a bridge between basic and advanced NC concepts. Its integration into curricula aligns with educational standards, offering a comprehensive learning experience.

B. Closing Education-Industry Gaps: Crucial to link education and industry for real-world readiness. Collaborative initiatives, internships, and industry partnerships are pivotal.

C. Preparing for the Future: Recognizing the crucial link between education and industry, this model prepares readers for real-world challenges. Collaborative initiatives, internships, and industry partnerships play a pivotal role in one's educational journey.

D. Evolution of NC Education: Aims to: extend beyond traditional methods with accessible content; shape a new generation of tech experts through transformative education; and impact the future tech landscape.

2.5.1.2 Closing and Next Chapter

Now as we have examined CMOS limitations practically. Understanding these limitations pedagogically involves breaking down challenges for learners. From Moore's Law implications to scaling challenges, each limitation teaches computing evolution lessons. With these, let us understand NML's principles and basics detailed in Chapter 3 presenting fundamentals of nanomagnetic logic.

2.5.2 Chapter End Quiz

An online Book Companion Site is available with this fundamental text book in the following link: www.wiley.com/go/sivasubramani/nanoscalecomputing1

For Further Reading:

Kumar and Agrawal (2017), Razavieh et al. (2019), Abbas and Olivieri (2014), Mukhopadhyay et al. (2003), Wong et al. (1999), Mishra (2013), Himpsel et al. (1998), Huff (2008), Chen et al. (2007).

References

Zia Abbas and Mauro Olivieri. Impact of technology scaling on leakage power in nano-scale bulk cmos digital standard cells. *Microelectronics Journal*, 45 (2):179–195, 2014.

Julie Chen, Haris Doumanidis, Kevin Lyons, et al. Manufacturing at the nanoscale, National Nanotechnology Initiative Workshop Report, 2007. https://www.nano.gov/sites/default/files/pub_resource/manufacturing_at_the_nanoscale.pdf.

Franz J Himpsel, Jose Enrique Ortega, Gary J Mankey, and RF Willis. Magnetic nanostructures. *Advances in Physics*, 47(4):511–597, 1998. https://doi.org/10.1080/000187398243519.

Howard Huff. *Into the Nano Era: Moore's Law Beyond Planar Silicon CMOS*, volume 106. Springer Science & Business Media, 2008.

Gagnesh Kumar and Sunil Agrawal. CMOS limitations and futuristic carbon allotropes. In *2017 8th IEEE Annual Information Technology, Electronics and Mobile Communication Conference (IEMCON)*, pages 68–71. IEEE, 2017.

Priyank Mishra. Emerging nanotechnology and it's impact on electronic circuit designing. *IJECT*, 4(5):64–65, 2013.

Saibal Mukhopadhyay, Hamid Mahmoodi-Meimand, Cassandra Neau, and Kaushik Roy. Leakage in nanometer scale cmos circuits. In *2003 International Symposium on VLSI Technology, Systems and Applications. Proceedings of Technical Papers.(IEEE Cat. No. 03TH8672)*, pages 213–218. IEEE, 2003.

Ali Razavieh, Peter Zeitzoff, and Edward J Nowak. Challenges and limitations of CMOS scaling for FinFET and beyond architectures. *IEEE Transactions on Nanotechnology*, 18:999–1004, 2019.

Hon-Sum Philip Wong, David J Frank, Paul M Solomon et al. Nanoscale CMOS. *Proceedings of the IEEE*, 87(4):537–570, 1999.

3

Fundamentals of Nanomagnetic Logic

Specific, Measurable, Achievable, Relevant, and Time-bound (SMART) learning objectives and goals for chapter 3: after reading this chapter you should be able to:

- Grasp the basic principles of nanomagnetic logic (NML)
  Summarize the key concepts, including the basic principles and evolution of NML. Essential for building a solid understanding of the fundamental principles governing NML
- Analyze the design and operation of NML gates
  Discuss the specifics of each logic gate and comprehend their functionalities. Critical for understanding the practical aspects of implementing NML.
- Evaluate signal processing in NML
  Identify the key components of signal processing and their roles within NML. Crucial for understanding how information is processed within NML systems.
- Understand energy considerations and low-power design
  Summarize the key concepts related to energy efficiency and low-power design strategies. Integral for understanding the practical applications and sustainability of NML.

This chapter talks about the introduction to NML in Section 3.1, briefs the fundamentals of coupling mechanisms in NML in Section 3.2, describes the design and operation of NML Gates in Section 3.3, presents in detail the signal processing in NML in Section 3.4 and, elaborates the energy considerations and efficiency in Section 3.5, and details about the educational emphasis in Section 3.6.

NML is a modern nanotechnology paradigm. It shows promise for compact, energy-efficient computing devices. Traditional semiconductor technologies have physical and scalability limitations as we discussed in Chapter 2 Limitations of CMOS Technology. Researchers explore alternatives, and NML is a leading player.

Nanoscale Computing: The Journey Beyond CMOS with Nanomagnetic Logic,
First Edition. Santhosh Sivasubramani.
© 2025 The Institute of Electrical and Electronics Engineers, Inc. Published 2025 by John Wiley & Sons, Inc.
Companion website: www.wiley.com/go/sivasubramani/nanoscalecomputing1

It aims to address challenges and lead the next-generation computing architectures.

Background: Conventional transistor technologies, following Moore's Law, face physical constraints. NML originates from the search for new computing approaches. It addresses limitations set by miniaturization in conventional electronics. The need for increased computational power drives interest in alternatives. NML stands out as a promising candidate for efficient computing. NML originated from spintronics, which uses electron spin for processing. Conventional electronic logic relies on the flow of charge for information processing. Unlike traditional electronic logic, spintronics manipulates electron spin states. This shift underlies NML, where magnetic properties play a key role in binary information processing.

3.1   Introduction to Nanomagnetic Logic

This section talks about the introduction to NML, briefs the basic principles in Section 3.1.1, and describes the evolution and development in Section 3.1.2.

NML is an emerging field in nanotechnology and computing. It offers an alternative to traditional electronic logic circuits. Understanding its intricacies requires exploring basic underlying principles (cf. Figure 3.1). This section aims to establish a foundation by investigating into these fundamental concepts. The goal is to enable a comprehensive exploration of NML's applications and advancements.

3.1.1   Basic Principles

This section talks about the basic principles throughout, and briefs the theory and fundamental in Section 3.1.1.1.

NML centers on manipulating magnetic nanoparticles as the core elements for information processing. Sized at just a few nanometers, these nanoparticles showcase distinctive magnetic properties (cf. Figure 3.2) crucial for logical operations. External magnetic fields influence the magnetic moments of these nanoparticles, forming the foundation for information encoding and processing. The below set of equations and laws help define the basic principles. Additionally, understanding key concepts such as electron spin, magnetic bits and nanomagnets, spin transfer torque (STT), spin–orbit interaction (SOI), spin-dependent transport, logic operations, spin waves, and thermal stability and energy efficiency is paramount.

Electron spin, denoted by its quantum property, plays a crucial role in spin-based devices. Nanomagnets and magnetic bits are key elements in NML, where their spin orientations encode information. Spin transfer torque (STT) is a phenomenon used to manipulate the magnetization of nanomagnets in spintronics devices. SOI is another important aspect, influencing spin dynamics and behavior

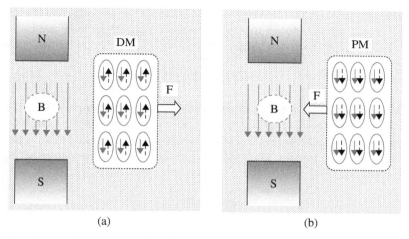

(a)　　　　　　　　　　　　(b)

Figure 3.1 Molecular power 39. Effect of magnetic field on diamagnets and paramagnetics. File:Molecular power 39. Effect of magnetic field on diamagnets and paramagnetics.jpg https://commons.wikimedia.org/wiki/File:Molecular_power_39._Effect_of_magnetic_field_on_diamagnets_and_paramagnetics.jpg. Source: Василь Іванович Сидоров/Wikimedia Commons/CC BY-SA 4.0.

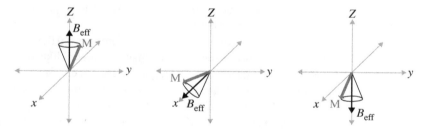

Figure 3.2 Spins precess around Beffective, and if Beffective meets the adiabatic condition, these spins will track the position of Beffective as it changes. https://commons.wikimedia.org/wiki/File:Adiabatic_Passage_of_Magnetization.png. Source: BearTurtle/Wikimedia Commons/CC BY-SA 4.0.

in materials. Spin-dependent transport refers to the spin polarization of charge carriers in materials, essential for spintronics applications. Logic operations in spin-based devices utilize spin states for data processing, paving the way for novel computing paradigms. Spin waves are collective excitations of magnetic moments, exploited for information transmission in spintronics. Thermal stability and energy efficiency are crucial considerations for practical implementation of spin-based devices. Understanding these concepts and their interconnections as detailed in the items 1–8 is vital for advancing spintronics and NML technologies, offering faster, more energy-efficient computing and information processing solutions [cf. Eqs (3.1)–(3.9)].

1) **Electron Spin**: Electron spin, denoted by s, is a quantum property cf. Eq. (3.1). It is intrinsic angular momentum, a fundamental characteristic of electrons. Spin quantum number, s, has values of $\frac{1}{2}$ and $-\frac{1}{2}$.

$$s = \frac{1}{2} \tag{3.1}$$

Two quantum states represent electron spin: ↑ (up) and ↓ (down). These states indicate the orientation of the magnetic moment. Equation describes electron spin quantization. It is crucial in various fields like quantum mechanics and condensed matter physics. Understanding electron spin is fundamental in atomic physics.

2) **Magnetic Bits and Nanomagnets:**

$$\mathbf{S} = \pm 1 \tag{3.2}$$

Nanomagnets exhibit spin orientation, denoted by the unit Vector $\mathbf{S}$ cf. Eq. (3.2). These nanoscopic systems display spontaneous magnetic order (magnetization) at zero applied magnetic field. The small size prevents magnetic domain formation, limiting nanomagnets for permanent information storage. Magnetization dynamics in sufficiently small nanomagnets, for example, single-domain magnets at low temperatures, show quantum phenomena like macroscopic spin tunneling.

At higher temperatures, magnetization experiences random thermal fluctuations, known as superparamagnetism. Grains of ferromagnetic metals like iron, cobalt, and nickel, along with single-molecule magnets, serve as examples of nanomagnets. The majority of nanomagnets consist of transition metal or rare earth magnetic atoms. In NML, binary information is represented by the orientation of magnetic moments in nanomagnets. The alignment of these magnetic moments encodes the binary states 0 and 1 (cf. Figure 3.3).

3) **Spin Transfer Torque (STT)**: STT is defined as a mechanism, where a spin-polarized current can modify the orientation of a magnetic layer in a magnetic tunnel junction (MTJ) or spin valve (cf. Eq. (3.3)). This effect finds application in spin-transfer torque (STT) magnetic random-access memory (STT-MRAM), representing non volatile memory with negligible leakage power consumption. STT allows the transfer of spin angular momentum. In NML, this is exploited to manipulate the orientation of spins in nanomagnets. The basic equation governing the spin transfer torque (τ_{STT}) is given by:

$$\tau_{STT} = \frac{\hbar}{2e}\mathbf{M} \times (\mathbf{S} \times \mathbf{S}') \tag{3.3}$$

where $\hbar$ is the reduced Planck constant, e is the elementary charge, $\mathbf{M}$ is the magnetization vector, and $\mathbf{S}$ and $\mathbf{S}'$ are the spin vectors of the electrons involved.

Figure 3.3 A chain of magnetic dots 110 nm in diameter. Information is passed from one dot to the next by magnetostatic interactions. File: Figure 1 https://ieeexplore.ieee .org/document/1226896/. Source: Cowburn, 2003/With permission of IEEE.

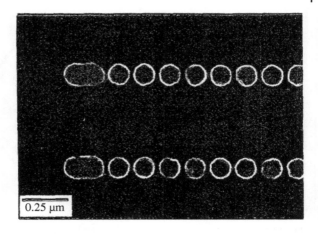

0.25 µm

4) **Spin–Orbit Interaction**: Spin can be manipulated using electric fields through the SOI (cf. Eq. (3.4)). This interaction allows the conversion between the spin (**S**) and orbital angular momentum (**L**) of electrons. The SOI Hamiltonian (H_{SOI}) can be expressed as:

$$H_{SOI} = \lambda \mathbf{L} \cdot \mathbf{S} \tag{3.4}$$

where λ is the strength of the spin–orbit coupling (constant), **L** is the orbital angular momentum vector, and **S** is the electron spin vector.

5) **Spin–Dependent Transport**: Current spin polarization cf. Eq. (3.5). Spin-dependent transport is harnessed to transmit and process information (cf. Figure 3.4). The spin polarization (*P*) of the current is defined as:

$$P = \frac{N_\uparrow - N_\downarrow}{N_\uparrow + N_\downarrow} \tag{3.5}$$

where $N_\uparrow$ and $N_\downarrow$ are the number of electrons with spin-up and spin-down orientations, respectively.

6) **Logic Operations:** Logic gates in nanomagnetic systems are constructed using nanomagnets. The manipulation of spin orientations within these gates allows for the execution of logical operations, cf. Eq. (3.6). The magnetic interaction energy E_{int} between two nanomagnets can be used to design logic gates. For example, the interaction energy for a pair of nanomagnets in an AND gate configuration can be expressed as:

$$E_{int} = -K \cdot (\mathbf{S}_1 \cdot \mathbf{S}_2) \tag{3.6}$$

where *K* is a coupling constant, and $\mathbf{S}_1$ and $\mathbf{S}_2$ are the spin vectors of the two nanomagnets.

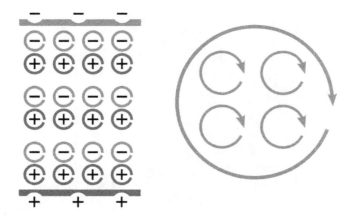

Figure 3.4 Schematic visualization of how microscopic dipoles assemble to form a macroscopic polarization and how microcosmic magnetic loop currents assemble to form a macroscopic magnetization. File:Polarization and magnetization.JPG https://commons.wikimedia.org/wiki/File:Polarization_and_magnetization.svg. Source: Marmelad/Wikimedia Commons/CC BY-SA 3.0.

7) **Spin-Wave**: Spin waves are collective excitations of magnetic moments. These moments align in a magnetic material. Quantum mechanics governs their behavior. They propagate through the material's lattice. Spin waves carry angular momentum and energy. Their dynamics are described by spin wave theory. Excitations occur due to perturbations in magnetization. These perturbations propagate as waves through the material. Spin waves can have various wavelengths and frequencies. Their behavior depends on the material's properties. Spin wave propagation involves spin precession. This precession is about the equilibrium direction. It is analogous to sound waves in materials. Spin waves can be manipulated and controlled. Researchers explore their potential for information processing. This includes applications in beyond-CMOS computing. Spin waves in NML propagate magnetic signals. These waves encode and transmit data. They replace traditional electronic signals for computation. Spin waves exploit the spin of electrons. They can travel long distances with minimal energy. This technology enables efficient beyond-CMOS computing. Spin waves promise faster and more energy-efficient devices.

Spin waves, collective excitations of spins, can be utilized for information propagation (cf. Eq. (3.7)). The spin wave equation is given by:

$$\omega = \gamma H_0 + \frac{2Dk^2}{\hbar} \tag{3.7}$$

where ω is the spin wave frequency, γ is the gyromagnetic ratio, H_0 is the effective magnetic field, D is the Dzyaloshinskii-Moriya interaction constant, k is the wave vector, and $\hbar$ is the reduced Planck constant (cf. Figure 3.5).

Figure 3.5 Visualization of the Dzyloshinskii Moriya antisymmetric exchange. https://commons .wikimedia.org/wiki/File:Dzyloshinskii_Moriya_ antisymmetric_exchange.jpg. Source: Macfenchel/Wikimedia Commons/CC BY-SA 3.0.

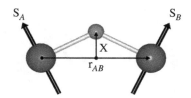

8) **Thermal Stability and Energy Efficiency:**
The Stoner–Wohlfarth Model (cf. Eq. (3.28)) provides a criterion for thermal stability in nanomagnets. It involves comparing the thermal energy $k_B T$ (where k_B is the Boltzmann constant and T is the temperature) with the magnetic anisotropy energy barrier (E_a) (cf. Figure 3.6). Ensuring thermal stability of spin configurations in nanomagnets is critical for reliable information storage and processing, cf. Eq. (3.8) (thermal stability criterion).

$$k_B T > E_a \tag{3.8}$$

The energy consumption in NML is lower compared to traditional electronic logic gates. It is crucial for energy-efficient computing, cf. Eq. (3.9) (energy consumption). The energy consumption ($E_{consumed}$) during a logic operation can be estimated by:

$$E_{consumed} = \alpha \cdot \beta \cdot k_B T \tag{3.9}$$

where α is the Gilbert damping parameter, β is the switching efficiency, and T is the temperature.

Exploring Nanomagnetic Logic: Foundations and Prospects:
As we have understood the basic principles as afore-stated, in the realm of NML, understanding spintronics and nanomagnetism is paramount. These fields explore the behavior of magnetic materials at the nanoscale. Magnetic nanoparticles play a crucial role in NML, serving as the building blocks for information processing. Logic gates and Boolean operations are fundamental components of NML architectures, enabling logical computations using magnetic states. Understanding the energy-efficiency and non volatility of NML systems is essential for developing sustainable computing technologies. Magnetic inter-actions in NML aids the behavior of nanomagnetic elements and their collective dynamics, influencing system performance. Exploring NML architectures is crucial for designing efficient and scalable computing systems. Equations explain mathematical foundations, offering quantitative insights into magnetic behavior. By combining theoretical foundations with practical advancements, learners can unlock the full potential of NML for information processing. This comprehensive understanding of NML and its components provides a solid foundation for advancing computing technologies. Through interdisciplinary research, scholars can pioneer a new era of computing using nanomagnetic materials, offering

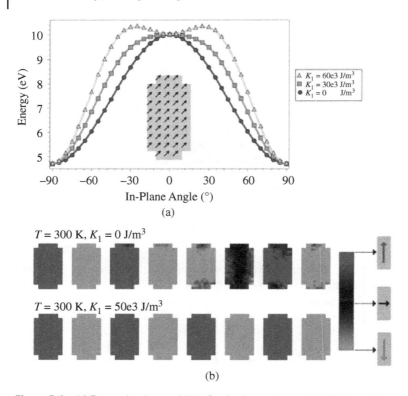

Figure 3.6 (a) Energy landscape $U(\theta)$ of a single nanomagnet with increasing amounts of biaxial anisotropy. (b) A line of rectangular magnetic bits align mutually antiparallel due to the dipole field coupling between them. When there is no biaxial anisotropy, aberrant nucleations in the cascade cause incorrect propagation of the signal.
File: Figure 1 https://ieeexplore.ieee.org/document/4634504/. Source: David Carlton et al., 2008/With permission of IEEE.

faster, more energy-efficient solutions. This academic book writing style provides a solid foundation for NML comprehensive understanding and its prospects in information processing in subsequent pages.

A) **Spintronics and Nanomagnetism:**

Spintronics, is a branch of physics and electronics. Its focus is on manipulating electron spin for information processing. Nanomagnetism is a subfield within nanoscience, focusing on the magnetic properties of nanoscale materials. NML relies on the precise alignment and manipulation of magnetic moments at the nanoscale to execute logical operations. In the context of NML, spintronics plays a crucial role in encoding and transmitting information through the spin states of electrons. Leveraging the inherent spin of electrons, NML devices attain low-power, non volatile operation. This characteristic positions them potential candidates for future computing technologies.

B) **Magnetic Nanoparticles:**
Magnetic nanoparticles are tiny particles with magnetic properties. They follow classical and quantum mechanics principles. Interactions between particles affect their collective behavior. Dipole–dipole interactions are significant among neighboring particles. Understanding these interactions is crucial for device design. Nanoparticles play a vital role in NML. They enable efficient data storage and processing. Their magnetic properties are essential for device functionality. Researchers study nanoparticles for various applications. These include biomedical imaging, environmental remediation, and information technology.

C) **Logic Gates and Boolean Operations:**
The foundation of any computing paradigm involves executing logical operations, and NML adheres to this principle. Essentially, NML utilizes the manipulation of magnetic states for implementing logic gates and conducting Boolean operations. Grasping the broader context of nanomagnetic information processing necessitates understanding the functionality of these logical elements.

In NML gates, interactions between magnetic nanoparticles are harnessed to carry out logical operations like AND, OR, NOT, and Universal Logic Gates (ULGs). Binary states are represented by the magnetic configurations of these nanoparticles, with distinct orientations corresponding to logical values, for instance, 0 and 1. Engineering the interactions and magnetic properties of nanoparticles enables the creation of NML gates capable of executing reliable and energy-efficient computations.

D) **Energy-Efficiency and Non volatility:** NML excels in energy efficiency, non volatility. Traditional circuits rely on charge carriers' movement; NML manipulates magnetic moments (cf. Figure 3.7). Minimal energy is needed for

Figure 3.7 Spins precess around Beffective, and if Beffective meets the adiabatic condition, these spins will track the position of Beffective as it changes. https://commons.wikimedia.org/wiki/File:Damped_Magnetization_Precession.jpg. Source: Chakageorge/Wikimedia Commons/Public Domain.

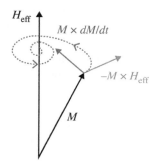

information processing. Magnetic states retain information when power is off. It tackles challenges linked to volatile electronic memory. Unique advantages distinguish NML from traditional circuits.

One fundamental equation (cf. Eq. (3.10)) that characterizes the magnetic moment (μ) of a nanoparticle is given by the expression:

$$\mu = \gamma \cdot \chi \cdot \vec{M} \tag{3.10}$$

where γ is the gyromagnetic ratio, (χ) is the particle's magnetic susceptibility, and $\vec{M}$ is the magnetization vector.

An equation defining the non volatility of nanomagnets (cf. Eq. (3.11)):

$$\frac{dM}{dt} = 0 \tag{3.11}$$

where $\frac{dM}{dt}$ represents the rate of change of magnetization over time. Since non-volatile nanomagnets retain their magnetization state without external power, the rate of change of magnetization $\{\frac{dM}{dt}\}$ is equal to zero.

Nonvolatility in nanomagnets refers to their ability to retain magnetization without external power. This means their magnetic state remains unchanged over time. Nanomagnets exhibit this property due to their inherent stability. They maintain their magnetization even in the absence of a magnetic field. This characteristic is crucial for data storage applications. Nonvolatility ensures that information encoded in nanomagnets remains intact without continuous power supply. It allows for persistent storage of data in memory devices. Nanomagnets retain their magnetization state due to energy barriers. These barriers prevent spontaneous changes in magnetization, ensuring data reliability. Nonvolatility is advantageous for low-power electronics. It enables energy-efficient operation of devices without constant refreshing of data. Nanomagnetic memory devices utilize nonvolatility for long-term data retention. This property contributes to the development of non-volatile memory technologies. Understanding nonvolatility is essential for designing reliable nanomagnetic devices. It ensures stable performance and longevity of nanomagnetic systems. In summary, nonvolatility is a fundamental property of nanomagnets, enabling stable and persistent data storage in low-power electronic devices. Nonvolatility in computing ensures data retention without power. It enables energy-efficient operation. This property reduces power consumption in electronic devices. Overall, nonvolatility enhances the performance and efficiency of computing systems.

E) **Magnetic Interactions in Nanomagnetic Logic**: To understand the interaction of nanoparticles in NML, we will explore the magnetic interactions that govern its behavior (cf. Eq. (3.12)). The Dipole–Dipole interaction energy (E_{dip}) between two magnetic nanoparticles can be expressed using the following equation:

$$E_{\text{dip}} = \frac{\mu_0}{4\pi} \frac{\mu_i \cdot \mu_j - 3(\mu_i \cdot \vec{r}_{ij})(\mu_j \cdot \vec{r}_{ij})}{|\vec{r}_{ij}|^3} \tag{3.12}$$

Here, μ_0 is the permeability of free space, μ_i and μ_j are the magnetic moments of the nanoparticles, and $\vec{r}_{ij}$ is the vector separating the two nanoparticles. Fundamental understanding is crucial for designing NML devices. Equation model, predicts magnetic ensemble behavior, enabling efficient logic gates.

F) **Nanomagnetic Logic Architectures**: NML principles evolve, shaping intricate computational architectures. Cascaded and parallel designs arise for distinct computational goals. Exploring into NML architecture uncovers trade-offs and advantages. Equations governing the magnetic interactions in NML architectures become increasingly complex as the number of nanoparticles and their arrangement grow (cf. Eq. (3.13)). For example, the total magnetic energy (E_{total}) in a cascaded architecture can be expressed as:

$$E_{\text{total}} = \sum_{i=1}^{N} E_{\text{gate},i} + \sum_{j=1}^{N-1} E_{\text{int},j,j+1} \tag{3.13}$$

Here, N represents the number of logic gates, $E_{\text{gate},i}$ is the energy associated with the i-th logic gate, and $E_{\text{int},j,j+1}$ is the interaction energy between adjacent gates. Understanding these are essential to optimize NML architecture' performance. It helps tailoring to application specific computational tasks.

NML faces challenges; researchers strive for solutions. Fabrication issues persist; scalability improvements needed for devices. Active efforts address hurdles, unlocking NML's potential. Challenges in NML, such as the influence of thermal fluctuations on magnetic stability ($\Delta E_{\text{thermal}}$), are essential for quantifying the impact of these challenges (cf. Eq. (3.14)). For instance:

$$\Delta E_{\text{thermal}} = k_B T \ln\left(\frac{\tau_0}{\tau_{\text{flip}}}\right) \tag{3.14}$$

Here, k_B is the Boltzmann constant, T is the temperature, τ_0 is the attempt time, and τ_{flip} is the characteristic flipping time.

G) **Radiation hardening of nanomagnets**
Another key aspect for consideration is radiation hardening of nanomagnets. It ensures stability under radiation. Nanomagnetic computing withstands harsh radiation environments reliably. Radiation-hardened nanomagnets resist radiation-induced errors and disturbances. They maintain data integrity in space and nuclear applications. Shielding and material selection enhance radiation tolerance in nanomagnets. Robust nanomagnetic devices offer reliable computing solutions in extreme conditions. Radiation-hardened

nanomagnets contribute to space exploration and nuclear technology. Their resilience ensures accurate data processing in radiation-prone environments. Understanding radiation hardening is crucial for reliable computing systems. It enables the development of resilient NML devices. Overall, radiation-hardened nanomagnets play a vital role in advancing computing technology for challenging environments.

1. Equation for radiation damage:

$$D = \Phi \cdot \sigma \tag{3.15}$$

where: D is the radiation damage. Φ is the fluence of incident radiation. σ is the cross section for radiation damage.

2. Equation for material degradation under radiation:

$$\frac{dC}{dt} = -kC \tag{3.16}$$

where: $\frac{dC}{dt}$ is the rate of change of material concentration. k is the degradation constant. C is the material concentration.

3. Equation for the effectiveness of shielding materials:

$$I = I_0 \cdot e^{-\mu x} \tag{3.17}$$

where: I is the intensity of radiation after passing through the shielding material. I_0 is the initial intensity of radiation. μ is the linear attenuation coefficient of the shielding material. x is the thickness of the shielding material.

4. Equation for magnetic properties and stability of nanomagnets:

$$E_{\text{total}} = \sum_{i=1}^{N} E_{\text{gate},i} + \sum_{j=1}^{N-1} E_{\text{int},j,j+1} \tag{3.18}$$

where: E_{total} is the total magnetic energy. $E_{\text{gate},i}$ is the energy associated with the i-th logic gate. $E_{\text{int},j,j+1}$ is the interaction energy between adjacent gates. These equations (cf. Eqs.(3.15)–(3.18)) provide a glimpse into the mathematical aspects related to radiation hardening of nanomagnets and its implications in NML design.

Understanding these foundational equations (cf. Eqs.(3.10)–(3.14)) allows readers to develop strategies for mitigating challenges and take forward NML based design and development.

3.1.1.1 Theory and Fundamentals

Magnetic domain theory explains nanomagnet behavior. Spin continuity equation governs spin polarization evolution. MTJ and TMR play crucial roles in NML gates. Understanding these concepts is fundamental in NML design. Let us walk through:

Magnetic Domain Theory: is the basic for understanding the behavior in NML. Tiny magnets, uniform magnetization define nanoscale material domains (cf. Eq. (3.19)). Magnetic domain theory explains magnetic material behavior. Materials are divided into small regions called domains. Each domain has uniform magnetization direction. Domains align to create overall material magnetization. Understanding domain dynamics is crucial in magnetism. The dynamics of these domains are governed by the Landau-Lifshitz-Gilbert (LLG) equation:

$$\frac{d\mathbf{M}}{dt} = -\gamma \mathbf{M} \times \mathbf{H}_{\text{eff}} + \alpha \mathbf{M} \times \frac{d\mathbf{M}}{dt} \qquad (3.19)$$

where $\mathbf{M}$ is the magnetization vector, γ is the gyromagnetic ratio, $\mathbf{H}_{\text{eff}}$ is the effective magnetic field, and α is the damping parameter. Gyromagnetic ratio: Quantifies ratio of magnetic moment to angular momentum. Damping parameter: Measures dissipation of energy in system.

Spin Continuity – Governing: A key concept is spin transport and manipulation. It is the process of controlling spin polarization in materials. The spin continuity equation, governs the evolution of spin polarization $\mathbf{P}$ in a material (cf. Eq. (3.20)), essential for understanding spin behavior and designing spintronic devices:

$$\frac{\partial \mathbf{P}}{\partial t} + \nabla \cdot \mathbf{J}_s = \frac{\mathbf{P}_0 - \mathbf{P}}{\tau_s} \qquad (3.20)$$

where $\mathbf{J}_s$ is the spin current density, $\mathbf{P}_0$ is the equilibrium spin polarization, and τ_s is the spin relaxation time. This equation captures the spin continuity in materials. It is crucial for designing and analyzing spintronic elements in NML (cf. Figure 3.8).

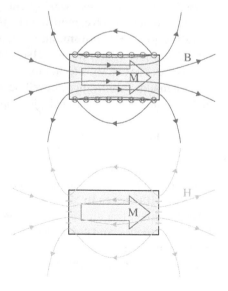

Figure 3.8 Magnetic fields B and H created by a bar magnet. Top: the magnetic currents $\nabla \times M$ (in magenta) create a B field similar to the one produced by a coil. Bottom: the magnetic charges $-\nabla \cdot M$ (i.e. the magnet poles) create an H field similar to the electric field in a parallel-plate capacitor. B and H are the same outside the magnet but differ inside. https://commons.wikimedia .org/wiki/File:Fields_bar_magnet_correct .png. Source: Geek3/Wikimedia Commons/CC BY-SA 4.0.

MTJ and TMR – NML Gates: NML gates rely on manipulating magnetic elements. The MTJ is a key component, crucial in magnetic logic circuits, cf. Eq. (3.21). MTJ: consists of two ferromagnetic layers separated by insulator. Tunneling magnetoresistance (TMR) effect: variation in resistance based on magnetization orientation. It is utilized in magnetic logic circuits. TMR calculated by comparing resistances in different alignments. MTJ resistance varies with ferromagnetic layers' magnetization orientation, defined by the TMR effect.

$$\text{TMR} = \frac{R_{\text{AP}} - R_{\text{P}}}{R_{\text{P}}} \times 100\% \tag{3.21}$$

where R_{P} is the resistance with parallel alignment and R_{AP} is the resistance with antiparallel alignment. Magnetic logic gates rely on the relative alignment of magnetic moments to encode information.

Exploring NML involves understanding of magnetic domain dynamics. An overview of fundamental laws and its governance concludes this discussion. Equations provided iterate essential mathematical frameworks for comprehension. Designing NML devices requires understanding these frameworks.

3.1.2 Evolution and Development

This section talks about the evolution and development throughout, briefs the historical context of NML in Section 3.1.2.1, describes the pioneering research and innovations in Section 3.1.2.2, presents in detail the transition from theory to practical implementations in Section 3.1.2.3. NML transforms computing with magnetic material properties. It diverges from traditional silicon, driven by Moore's Law challenges. Seeking energy-efficient alternatives, NML emerges. Compact systems evolve, motivated by the need for innovation.

1. Emergence of Nanomagnetic Logic:

NML's origins lie in the early 21st century. Researchers sought new computing approaches due to CMOS technology constraints. Faster, efficient computing solutions demanded innovation, unveiling NML. This technique manipulates nanomagnets' magnetic states for logical operations. It offers a viable alternative to charge-based computation.

2. Developmental Phases:

NML evolves through distinct phases. Theoretical foundations explore nanomagnetic materials for logical operations. Pioneering research develops prototypes, paving practical implementations.

3. Key Milestones in Nanomagnetic Logic:

NML achieved significant milestones in development. Material science, fabrication techniques, and device design advanced crucially. Success in manipulating nanomagnetic states for logic gates. Integration of nanomagnetic elements into scalable architectures achieved. Viability of NML demonstrated for real-world applications.

3.1.2.1 Historical Context of Nanomagnetic Logic

Grasping NML's evolution requires understanding the historical context. Challenges, breakthroughs, and paradigm shifts shaped its journey.

A. Early Visions and Concepts:

Let us get into early visions foreseeing semiconductor limitations. Researchers sought alternatives, exploring magnetic materials for processing. Groundbreaking concepts like quantum cellular automata shaped NML.

B. Role of Spintronics and Magnetic Technologies:

Spintronics focusing on manipulating electron spin shaped NML's history. Magnetic technologies, such as giant magnetoresistance (GMR) and tunnel magnetoresistance (TMR), contribute for reading and writing information using magnetic states. With these, NML emerged as a viable computing paradigm.

C. Influential Research and Researchers:

NML's history thrives on influential researchers. IBM developed MTJs, advancing the field. One of the key person- Stuart Parkin explored nanomagnetic materials. Academia and industry collaborated to enrich NML technology.

D. Evolutionary Journey: Conceptualization to Practical Implementations, Shaping NML's Development

NML emerges in the beginning of 21st century, challenging silicon computing (cf. Table 3.1). Theoretical foundations laid, exploring physics of nanomagnetic materials. Logic gates demonstrated using nanomagnetic elements, and proving feasibility. Advances in nanomagnetic materials: nanoparticles, thin films, and nanowires. Integration of nanomagnetic elements into scalable architectures achieved. Fabrication techniques innovate, achieving necessary resolution for devices. Researchers tackle scalability, developing strategies for larger circuits. Energy-efficient design and thermal dissipation address power consumption (cf. Table 3.1). Hybrid systems integrate NML with conventional computing. Ongoing advancements will lead to commercial applications, mainstream computing.

3.1.2.2 Pioneering Research and Innovations

1) **Theoretical Foundations**: NML's evolution saw crucial theoretical translation. Exploring key innovations advanced practical feasibility. Theoretical Foundations strengthens NML's advancement. Physicists studied Superparamagnetism and domain wall motion phenomena. These models predict nanomagnets behavior and laid groundwork for experiments.

2) **Experimental Demonstrations of Nanomagnetic Logic Gates:**

    A significant milestone in NML's evolution was the successful demonstration of logic gates using nanomagnetic elements. Researchers illustrated the execution of Boolean operations by manipulating the magnetic states

Table 3.1 Key Evolution in the History of NML

S.N.O.	Year	Milestone
		Category: Nanomagnetic Logic Evolution
1	2008	Nanorobot architecture for medical target identification
2	2010	Logic operations and data storage using vortex magnetization states in mesoscopic permalloy rings, and optical readout
3	2011	Non-majority magnetic logic gates: experiments and future prospects for shape-based logic
4	2012	Enhanced functionality in magnonics by domain walls and inhomogeneous spin configurations
5	2013	Voltage-induced switching in MTJs with perpendicular magnetic anisotropy
6	2014	Majority logic gate for 3D magnetic computing
7	2015	Dual-tip magnetic force microscopy with suppressed influence on magnetically soft samples
		Category: Materials and Fabrication Techniques
8	2016	Fast switching of magnetic vortex state under an alternating magnetic field
9	2017	Voltage pulse induced repeated magnetization reversal in strain-mediated multiferroic nanomagnets: size and material dependent
10	2018	Flexible magnetic thin films and devices
		Category: Applications and Devices
11	2019	Coupling of the skyrmion velocity to its breathing mode in periodically notched nanotracks
12	2020	Dipole-coupled magnetic quantum-dot cellular automata-based efficient approximate nanomagnetic subtractor and adder design approach
13	2021	Introduction of Interlayer exchange coupled nanomagnetic universal logic gate design
14	2022	Unidirectional spin-wave propagation and devices
15	2023	Controlled evolution of three-dimensional magnetic states in strongly coupled cylindrical nanowire pairs
		Category: Experimental Techniques and Studies
16	2013	Optimal ferromagnetically coated carbon nanotube tips for ultra-high resolution magnetic force microscopy
17	2014	Focus on artificial frustrated systems
18	2015	Perpendicular-anisotropy MTJ switched by spin-Hall-assisted spin-transfer torque

Table 3.1 (Continued)

S.N.O.	Year	Milestone
		Category: Theoretical and Computational Approaches
19	2014	Ultra–low-energy computing paradigm using giant spin Hall devices
20	2015	Limitations in artificial spin ice path selectivity: the challenges beyond topological control
		Category: Logic Gates and Computing Architectures
21	2017	A magnetic shift register with out-of-plane magnetized layers
22	2018	Computational logic with square rings of nanomagnets
23	2019	Ultrafast low-power magnetization switching of anisotropic nanoparticle
24	2020	Emerging neuromorphic devices
25	2021	High Sensitivity mapping of magnetic induction fields
26	2022	Applications of nanomagnets as dynamical systems
27	2023	Skyrmion based 3D low complex runtime reconfigurable architecture design methodology of universal logic gate
		Category: Magnonics and Spintronics
28	2014	Spin dependent transport between magnetic nanopillars through a nano-granular metal matrix
29	2015	Tailoring the nucleation of domain walls along multi-segmented cylindrical nanoelements
30	2017	Nanomagnets with geometry-tuned domain wall nucleation
31	2018	A multi-state synthetic ferrimagnet with controllable switching near room temperature
32	2019	Voltage control of magnetization in nanomagnetic devices for low-power computing
33	2020	Inhomogeneous domain walls in spintronic nanowires
34	2021	Chiral switching and dynamic barrier reductions in artificial square ice
35	2022	Magnonics with engineered spin textures
36	2023	Magnetic domain walls: processes, and applications
		Category: Non-Boolean Computing and Novel Systems
37	2014	Straintronic spin-neuron
38	2015	Cascade-able spin torque logic gates with input–output isolation
39	2016	The magnetic synapse: a nanomagnetic device based on the electronic and magnetic properties of neuron-like nanocolumns
40	2017	Non-Boolean computing with nanomagnetic devices
41	2018	A majority gate with chiral magnetic solitons

Table 3.1 (Continued)

S.N.O.	Year	Milestone
42	2019	Unusual perpendicular anisotropy in Co_2TiSi films
43	2020	Nanomagnetic logic based runtime reconfigurable area efficient and high-speed adder design methodology
44	2021	Improvement on the manipulation of a single nitrogen-vacancy spin and microwave photon at single-quantum level
45	2022	Applications of nanomagnets as dynamical systems – technology
46	2023	Skyrmion-based majority logic gate by voltage-controlled magnetic anisotropy
Category: General Topics		
47	2015	Subnanosecond magnetization reversal of nanomagnets
48	2016	Machine learning to understand and control complex systems
49	2017	Magnonic interconnects for logic circuits
50	2018	Spin transfer torque switching in synthetic antiferromagnetic racetracks with domain wall pinning defects
51	2019	Chirality in magnetic nanowires with zinc-blende crystal structure
52	2020	Fabrication of hybrid artificial spin ice arrays with periodic site-specific local magnetic fields
53	2021	Interlayer exchange couple-based reliable and robust 3-input adder design methodology
54	2022	QCA-based design techniques reliability and security
55	2023	The magneto-optics design
Category: Nanomagnetic Reservoir Computing		
56	2019	Simulated annealing with surface acoustic wave in a dipole-coupled array of magnetostrictive nanomagnets for collective ground state computing
57	2020	Implementation of an efficient MTJ based stochastic neural network with application to iris data classification
58	2022	Computational capability of a nanomagnetic reservoir computing platform
59	2023	Machine learning using magnetic stochastic synapses
Category: Other		
60	2022	The new SI and the fundamental constants of nature
61	2024	Magnetic skyrmions in chiral ferromagnets

of nanomagnets. This achievement marked a crucial advancement in validating the practicality of employing nanomagnetic materials for computing applications.

3) **Materials Advancements for Nanomagnetic Logic:**
   The prosperity of NML relies heavily on progress in nanomagnetic materials. Researchers explored a range of materials, such as magnetic nanoparticles, thin films, and nanowires, aiming to identify those with optimal properties for logic operations. The primary emphasis is on customizing materials at the nanoscale. It facilitates precise control of magnetic states and improves the overall performance of NML devices.

4) **Integration Challenges and Solutions:**
   Scaling up NML from individual gates to functional circuits presents several integration challenges. Overcoming these hurdles involves addressing issues related to reliable interconnects, thermal stability, and compatibility with existing technologies. Fabrication techniques like self-assembly and top-down lithography play a crucial role, facilitating progress toward implementations.

3.1.2.3 Transition from Theory to Practical Implementations

Moving from theoretical concepts to practical applications is a crucial stage in the development of emerging technologies. In NML, this progression involves overcoming challenges related to (a) fabrication, (b) scalability, (c) power consumption, and (d) integration with existing computing architectures.

1) **Fabrication Techniques for Nanomagnetic Logic:**
   Precision and reproducibility in fabricating NML devices exhibit significant challenges. The shift from small-scale experimental prototypes to large-scale integrated circuits necessitates the advancement of fabrication techniques. Key methods such as electron-beam lithography, ion-beam etching, and self-assembly played essential roles in achieving the required resolution and precision for NML.

2) **Scalability and Complexity:**
   Ensuring scalability is pivotal for implementing any computing technology practically. NML encounters challenge in device and circuit scalability. Potentially could be addressed through innovations in (a) device design, (b) interconnect technologies, and (c) parallelization strategies. The development of scalable architectures is more important for creating extensive and intricate NML circuits.

3) **Power Consumption and Thermal Management**: Although NML offers potential for low-power operation. Addressing power consumption and thermal issues became imperative with the scaling up. Innovations in (a) energy-efficient-design, (b) dynamic power management, and (c) thermal

dissipation strategies contributes to the practicality of NML across various applications.

4) **Integration with Conventional Computing Architectures**: To position NML as a mainstream computing tech, integration with existing architecture is essential. Developing hybrid systems capable of leveraging the strengths of both conventional semiconductor technology and NML is crucial. This integration facilitates creation of hybrid computing systems. It harness NML's advantages for application specific tasks. It also maintain compatibility with known computing paradigms.

This evolution of NML from theoretical concepts to practical applications involves overcoming challenges. Historical context, pioneering research, and innovations in materials science and fabrication techniques collectively advance NML into reality. As the technology progress, its integration with conventional computing architectures opens up new possibilities for energy-efficient and compact computing systems.

3.2 Fundamentals of Coupling Mechanisms in Nanomagnetic Logic

This section talks about the fundamentals of coupling mechanisms in NML, it briefs the magnetic materials for logic operations in Section 3.2.1, and describes the spin dynamics and magnetization reversal in Section 3.2.2.

Designing NML circuits pose challenges on understanding coupling between nanomagnets. This coupling has a substantial impact on the overall performance of logic devices. This exploration will walk through various types of coupling in NML. It includes (a) dipole-dipole, (b) exchange, (c) chiral, and (d) magneto-electrostatic coupling, providing relevant equations and explaining the underlying physics. Item I in this section details dipole coupling, item II details interlayer exchange coupling, item III details chiral coupling, and item IV in section details magneto-electric coupling.

1) **Dipole Coupling:**

Dipole coupling is a one of the interaction between magnetic dipoles. It plays a significant role in the behavior of magnetic nanoparticles (cf. Eq. (3.22)). In NML, dipole coupling is often between neighboring nanomagnets. The interaction energy (E_{dd}) between two magnetic dipoles can be expressed as:

$$E_{dd} = -\frac{\mu_0}{4\pi} \frac{\mathbf{m}_1 \cdot \mathbf{m}_2 - 3(\mathbf{m}_1 \cdot \hat{\mathbf{r}})(\mathbf{m}_2 \cdot \hat{\mathbf{r}})}{r^3} \qquad (3.22)$$

where
- μ_0 is the permeability of free space,
- $\mathbf{m}_1$ and $\mathbf{m}_2$ are the magnetic moments of the two nanomagnets,

- $\hat{r}$ is the unit vector pointing from one nanomagnet to the other, and
- r is the separation between the nanomagnets.

Dipole coupling influences the alignment of magnetic moments in neighboring nanomagnets. The Dipole–Dipole Coupling has a long-range nature (cf. Eq. (3.22)), and it strongly depends on the distance between the nanomagnets. The energy landscape of Dipole–Dipole Coupling can influence the stability of magnetic states (E_{dd}), affecting the reliability of NML devices.

2) **Interlayer Exchange Coupling**: Interlayer coupling is an interaction between magnetic layers in a multilayered structure. In nanomagnetic systems, it is associated with synthetic antiferromagnets, (cf. Eq. (3.23)), where two ferromagnetic layers are separated by a non-magnetic spacer. The energy associated with interlayer coupling can be expressed as:

$$E_{\text{interlayer}} = -J_{\text{interlayer}} \sum_i \mathbf{S}_{1i} \cdot \mathbf{S}_{2i} \tag{3.23}$$

Here:

- $J_{\text{interlayer}}$ is the interlayer coupling constant,
- $\mathbf{S}_{1i}$ and $\mathbf{S}_{2i}$ are the spin vectors in the two ferromagnetic layers at position i.

IEC tends to align or anti-align the magnetic moments in adjacent layers, depending on the sign of the coupling constant. This phenomenon is crucial in designing devices like MTJs.

3) **Chiral Coupling:**

Chiral coupling involves the interaction between magnetic moments in a helical or chiral magnetic structure. This can be witnessed in broken inversion symmetry (cf. Eq. (3.24)) including multilayered structures or magnetic nanowires. The energy associated with chiral coupling can be expressed as:

$$E_{\text{chiral}} = -J_{\text{chiral}} \sum_{\langle i,j \rangle} \mathbf{S}_i \cdot (\mathbf{S}_j \times \hat{r}_{ij}) \tag{3.24}$$

where

- J_{chiral} is the chiral coupling constant,
- $\mathbf{S}_i$ and $\mathbf{S}_j$ are the spin vectors at positions i and j,
- $\hat{r}_{ij}$ is the unit vector pointing from i to j.

Chiral coupling influences the rotation or helicity in the magnetic structure. It introduces properties that can be exploited for specialized applications such as magnetic skyrmions.

4) **Magneto-electrostatic Coupling**: Magneto-electrostatic coupling refers to the interaction between the magnetic state of a nanomagnet and an external electric field (cf. Eq. (3.25)). This coupling can be described by the following energy term:

$$E_{me} = -\vec{M} \cdot \vec{E} \tag{3.25}$$

Here,

$\vec{M}$ is the magnetization vector of the nanomagnet,

$\vec{E}$ is the external electric field.

The energy stored in a magnetically coupled circuit (cf. Eq. (3.26)) can be expressed as:

$$w = \frac{1}{2}L_1 i_1^2 + \frac{1}{2}L_2 i_2^2 \pm M i_1 i_2 \qquad (3.26)$$

In this equation:
- L_1 and L_2 are inductances,
- i_1 and i_2 are currents,
- M is a constant representing mutual inductance,
- The positive or negative sign in the last term depends on the relative direction of the currents.

In nanomagnetic systems, the magnetoelectric coupling energy (cf. Eq. (3.27)) can be expressed as:

$$E_{ME} = \frac{1}{2}\alpha_{ij} E_i H_j \qquad (3.27)$$

In this equation:
- E_i and H_j are components of the electric and magnetic fields, respectively,
- α_{ij} is the magnetoelectric coupling tensor.

It also provides a nonmagnetic means of controlling logic operations. This coupling help achieving energy efficient and reconfigurable NML devices.

The overall behavior of NML circuits is a combination and permutation result of (a) dipole-dipole coupling, (b) exchange coupling, (c) chiral coupling, and (d) magneto-electric coupling. Designing efficient NML device requires balanced optimization of these coupling mechanisms. The goal is to achieve reliable logic operations, also minimizing energy consumption and interference between neighboring nanomagnets.

3.2.1 Magnetic Materials for Logic Operations

This section talks about the magnetic materials for logic operations throughout, briefs the selection criteria for magnetic materials in Section 3.2.1.1, describes the magnetic anisotropy and stability in Section 3.2.1.2, and presents in detail the role of nanomagnetic materials in logic devices in Section 3.2.1.3.

Understanding primary concerns and requirements guides effective NML system design. Thinking upon the key pointers below shape the foundation for NML systems. Subsequent sections will describe into details for clarity.

1) **Magnetic Materials for Logic Operations**:
   a) **Overview of Magnetic Materials in Logic Operations**:
      - Significance of magnetic materials in logic operations.
      - Brief historical context and development.
   b) **Understanding Magnetic Concepts**:
      - Explanation of magnetic domains and their relevance.
      - Impact of magnetic hysteresis on logic operations.
   c) **Magnetic Material Categories**:
      - Ferromagnetic, antiferromagnetic, and ferrimagnetic materials overview.
      - Role of magnetic susceptibility in logic operations.
   d) **Magnetic Logic Gates**:
      - Application of magnetic materials in logic gates.
      - Comparative analysis with semiconductor-based logic gates.
   e) **Addressing Challenges in Logic Operations**:
      - Challenges using magnetic materials for logic operations.
      - Ongoing research and advancements to overcome challenges.
2) **Selection Criteria for Magnetic Materials**:
   a) **Importance of Magnetic Material Selection**:
      - Choosing materials tailored for specific applications.
      - Material properties' impact on overall performance.
   b) **Relevant Magnetic Properties for Selection**:
      - Saturation magnetization, coercivity, and remanence.
      - Role of anisotropy in the material selection process.
   c) **Material Parameters Influencing Applications**:
      - How material parameters affect various applications.
      - Real-world examples of criteria-based Material Selection.
   d) **Balancing Performance Trade-offs**:
      - Managing material properties effectively.
      - Considering temperature and external factors in selection.
   e) **Advancements in Material Selection Methods**:
      - Exploration of emerging materials for future applications.
      - Utilizing computational methods to optimize material choices.
3) **Magnetic Anisotropy and Stability**:
   a) **Magnetic Anisotropy and Stability Overview**:
      - Definition and significance of magnetic anisotropy.
      - Stability's crucial role in magnetic device performance.
   b) **Diverse Types of Magnetic Anisotropy**:
      - Shape, magnetocrystalline, and magnetoelastic anisotropy.
      - Influence of each type on stability in materials.
   c) **Anisotropy's Impact on Device Stability**:
      - Anisotropy's role in preventing undesired switching.
      - Techniques to enhance stability in practical applications.

    d) **Measuring and Characterizing Anisotropy**:
- Methods for quantifying magnetic anisotropy.
- Importance of accurate characterization for stability predictions.

    e) **Applications and Future Trends**:
- Improved magnetic device design through anisotropy understanding.
- Current research trends for enhancing stability through anisotropy control.

4) **Role of Nanomagnetic Materials in Logic Devices**:

    a) **Definition and Characteristics of Nanomagnetic Materials**:
- Introduction to nanomagnetic materials and their behaviors.
- Motivation behind incorporating them into logic devices.

    b) **Spintronics and Nanomagnetic Logic Devices Overview**:
- Spintronic devices and their role in NML.
- Comparative advantages over traditional semiconductor devices.

    c) **Nanomagnetic Material Fabrication Techniques**:
- Techniques for manufacturing nanomagnetic materials.
- Challenges and advancements in the fabrication process.

    d) **Integration with Current Technologies**:
- Compatibility with existing technology infrastructure.
- Potential for hybrid devices combining nanomagnetic and semiconductor elements.

    e) **Performance Metrics and Energy Efficiency Analysis**:
- Evaluation of NML devices' performance metrics.
- Energy efficiency comparisons with conventional semiconductor.

    f) **Challenges and Future Prospects**:
- Identification of challenges in NML device development.
- Exploration of future prospects and possible advancements.

Magnetic materials have advantages in logic operations. They generate and respond to magnetic fields, storing information in spin orientation. Unlike conventional electronic devices using electrical charges, magnetic devices provide a non-volatile energy-efficient approach. This involves controlling magnetic domains, regions where magnetic moments align. Switching between magnetic states forms the basis for logic operations. It also creates a binary system encoding information as "0 or 1".

Magnetic materials suitable for logic operations include ferromagnetic, antiferromagnetic, and ferrimagnetic types. Ferromagnetic materials feature parallel-aligned magnetic moments. Antiferromagnetic ones have anti-parallel alignment. Ferrimagnetic materials combine both alignments. The choice of material depends on application-specific requirements – (a) speed, (b) energy efficiency, and (c) scalability. This helps improve magnetic anisotropy, determining the preferred magnetization direction. Minimizing energy dissipation during state transitions is primary to low-power NML devices. A domain wall's movement signifies a logical operation, forming a magnetic shift registers' basis. MRAM stores data as magnetic moments, providing non volatile memory with rapid read/write.

3.2.1.1 Selection Criteria for Magnetic Materials

Choosing magnetic materials for logic involves considering criterias as specified above. The process ties to application characteristics. Following criteria play a crucial role in determining the suitability of a magnetic material:

A. Magnetic Anisotropy:

Magnetic anisotropy is a fundamental property influencing the stability of magnetic states. High magnetic anisotropy ensures that the material maintains a specific magnetic orientation. It reduces the risk of unintentional state transitions.

B. Magnetic Coercivity:

Magnetic coercivity determines the strength of an external magnetic field. It is a key to induce a change in the material's magnetic state. Low coercivity is desirable for energy-efficient logic operation. Allows easy switching between magnetic states with minimal energy.

C. Magnetic Saturation:

The maximum magnetization with an applied magnetic field determines reliable information storage and retrieval. Materials with high saturation magnetization yields strong magnetic signals, improving the performance.

D. Spin Dynamics:

Fast spin dynamics enhance efficient magnetic moment manipulation. Quick switching in materials speeds logic operations significantly. Understanding material's relaxation times and coherence lengths for spin dynamics comprehension.

E. Temperature Stability:

The performance of magnetic materials is temperature dependent. Choosing stable magnetic materials is vital for diverse temperatures in practical applications (cf. item 8 in Section 3.1.1 detailing basic principles of thermal stability). Ensures reliable operation with varied environmental conditions.

F. Stoner-Wohlfarth Model: (cf. item 8 in Section 3.1.1 detailing basic principles of thermal stability) The Stoner–Wohlfarth model describes the behavior of magnetic hysteresis in a material. It incorporates anisotropy energy and temperature to study magnetic material's response for external magnetic field. The behavior of magnetic hysteresis is described by the Stoner-Wohlfarth model as (cf. Eq. (3.28)):

$$M = M_s \frac{\cos(\theta - \phi)}{\cos(\theta - \phi) + e^{-K_u V/(k_B T)}} \tag{3.28}$$

Here, M is the magnetization, M_s is the saturation magnetization, θ is the angle between the magnetization and the applied field, ϕ is the angle between the anisotropy axis and the applied field, K_u is the magnetic anisotropy constant, V is the volume, k_B is the Boltzmann constant, and T is the temperature.

G. Material Selection Examples:

Ferrites, known for their high magnetic anisotropy, find application in microwave devices due to their unique properties. These materials offer excellent performance in high-frequency applications. Their ability to maintain a stable magnetic field makes them ideal for microwave circuitry. Permalloy, a soft magnetic material, possesses low coercivity, making it suitable for NML devices. Permalloy's low coercivity allows for easy magnetization reversal. This property is crucial for efficient operation in NML circuits. Soft magnetic materials like permalloy are preferred for their magnetic properties. Their ability to respond quickly to magnetic fields makes them ideal for NML. Material selection plays a vital role in optimizing device performance. Choosing the right material ensures the desired functionality of NML devices. Factors such as magnetic anisotropy and coercivity influence material selection. Understanding the properties of different materials is essential for effective design. Material selection examples highlight the importance of tailored choices. In NML, materials must meet specific criteria for optimal performance. Each material offers unique advantages and limitations in NML applications. By carefully selecting materials, engineers can enhance the efficiency of NML devices. (cf. Figure 3.9).

3.2.1.2 Magnetic Anisotropy and Stability

Magnetic anisotropy shapes magnetic materials, impacting applications. It influences stability and reliability, crucial for functionality. Anisotropy defines

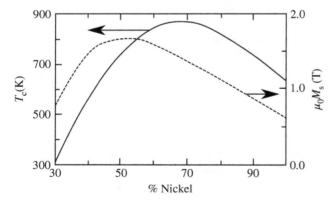

Figure 3.9 Curves for Curie temperature (solid) and saturation magnetization (dashed) dependence from Nickel concentration in permalloy File:Permalloy Curie and Saturation.svg https://commons.wikimedia.org/wiki/File:Permalloy_Curie_and_Saturation .svg. Source: Alex-engraver/Wikimedia Commons/Public Domain.

property dependence on measurement direction. In magnetic materials, it determines moment orientation. This orientation preference is vital for functionality. It affects various applications, ensuring stability and reliability.

i) **Types of Magnetic Anisotropy**:
   Understanding types of magnetic anisotropy and its origin is essential to tailor application specific material properties. Shape anisotropy is understood from geometry of magnetic domains. Magnetocrystalline anisotropy is understood from crystal structure. Strain-induced anisotropy is understood from mechanical deformation. These provides an additional means of controlling magnetic properties.

ii) **Manipulating Anisotropy for Stability**:
   Manage magnetic anisotropy with engineering techniques. Adjust material composition for magnetic anisotropy control. Introduce crystal structures or apply external stress. Shape and magnetocrystalline anisotropy combine in recording media. Stable magnetization orientations are engineered for data storage.

iii) **Stability and the Role of Domain Walls**:
   Magnetic stability is vital for reliable logic devices. Unwanted fluctuations can corrupt crucial data in devices. Domain walls, interfaces with distinct magnetic orientations, are pivotal. Their dynamics, influenced by anisotropy, play a critical role. Maintaining magnetic state integrity depends on these dynamics. Stability in logic devices relies on well-maintained integrity.

iv) **Anisotropy Energy**:
   The anisotropy energy ($E_{anisotropy}$) in a magnetic material (cf. Eq. (3.29)) can be described using the following equation:

$$E_{anisotropy} = -K_u \sin^2(\theta) \tag{3.29}$$

   Here, K_u is the magnetic anisotropy constant, and θ is the angle between the magnetization direction and the easy axis of anisotropy. This equation highlights how the anisotropy energy depends on the orientation of magnetic moments, providing insights into stability.

v) **Temperature Dependence of Anisotropy**:
   Magnetic anisotropy's temperature dependence is crucial for stability. At higher temperatures, thermal energy reduces material stability. Designing materials with stable magnetic properties is essential.

vi) **Examples of Stable Magnetic Materials**:
   Neodymium–iron–boron (NdFeB) magnets exemplify hard magnetic materials. High anisotropy ensures long-term magnetization stability. These magnets used in storage devices. Magnetic sensors benefit from exchange-coupled structures' enhanced stability. Interaction among layers provides additional stability in these structures.

3.2.1.3 Role of Nanomagnetic Materials in Logic Devices

Nanomagnetic materials redefine logic devices in information processing. Harnessing nanoscale properties revolutionizes speed, energy efficiency, scalability. Distinct magnetic behaviors emerge at the nanoscale level.

i) **Nanoscale Effects on Magnetic Properties**:
Magnetic nanoparticles display decreased coercivity with size reduction. Nanoscale materials reveal distinct magnetic behaviors versus macroscopic counterparts. Pronounced quantum effects and size-dependent phenomena like superparamagnetism emerge. The shift from classical to quantum behavior is observed in reducing magnetic coercivity with nanoparticle size.

ii) **Superparamagnetism and Energy Efficiency**:
Superparamagnetism in nanosized magnetic particles essential for energy-efficient computing. Reduced magnetic element size lowers energy barrier, enabling controlled switching. This advantage proves crucial for low-power logic operations.

iii) **Fabrication Techniques for Nanomagnetic Materials**:
Engineered nanomagnetic materials demand precise control of size, shape, and composition. Utilizing techniques like lithography, chemical vapor deposition, self-assembly is crucial. These methods create nanoscale structures with customized magnetic properties. Self-assembly forms regular arrays of magnetic nanoparticles.

iv) **Quantum Properties of Nanomagnetic Materials**:
Nanomagnetic materials show potential in classical and quantum computing. Quantum states of nanomagnetic elements encode qubits for information processing. Inherent quantum properties like spin coherence enable new possibilities. Entanglement in nanomagnetic elements further enhances quantum logic.

v) **Challenges and Solutions in Nanomagnetics**:
Challenges in nanomagnetics involve stability, reliability, and manufacturability. Addressing these requires robust control mechanisms, error-correction techniques. Spintronic device advancements, like STT devices, tackle stability challenges.

vi) **Examples of NML Devices**:
Nanoscale MTJs are fundamental to spintronic devices. In MRAM, these MTJs manipulate nanomagnetic elements for storage and Read/Write circuitry. Single-domain magnetic nanowires promise compact, efficient NML circuits. Nanomagnetic materials in logic devices revolutionize information processing. Their unique properties, along with fabrication advancements, promise faster, energy-efficient, highly scalable logic devices. Ongoing research may crucially shape the future of computing.

3.2.2 Spin Dynamics and Magnetization Reversal

This section talks about the spin dynamics and magnetization reversal throughout, briefs the spin transport and manipulation in Section 3.2.2.1, describes the switching dynamics in magnetic devices in Section 3.2.2.2, and presents in detail the magnetization reversal mechanisms in Section 3.2.2.3.

i) **Spin Dynamics and Magnetization Reversal**:
Understanding spin dynamics is crucial in NML. It directly influences magnetization reversal, a fundamental operation. Spin dynamics involves angular momentum behavior in electron spins. Crucial for manipulating magnetization in nanomagnetic systems.

ii) **Introduction to Spin Dynamics**:
Nanoscale spin manipulation is crucial for efficient computing. Spin dynamics includes precession, relaxation, coherent electron motion. The Landau-Lifshitz-Gilbert (LLG) (cf. Eq. (3.19)) equation defines these dynamics. It captures interactions between magnetic moments and external fields.

iii) **Magnetization Reversal and Spin-Transfer Torque**:
Magnetization reversal is pivotal for data storage and processing. STT, a result of angular momentum transfer, plays a key role. Equations like the Slonczewski equation depict torque effects.

iv) **Ultrafast Spin Dynamics and All-Optical Switching**:
Ultrafast spin dynamics advances magnetization reversal techniques. All-optical switching emerges through ultrafast laser pulses. Incident lasers induce rapid spin dynamics in materials. Phenomena like inverse Faraday effect govern this process. Result: ultrafast, non volatile control of magnetization achieved.

v) **Applications and Examples**:
Understanding spin dynamics and magnetization reversal is essential. Magnetic memory, like MRAM, enables controlled magnetization switching. STT devices serve magnetic sensors and logic gates. They promise energy-efficient computing solutions in various applications.

Spin dynamics influence magnetization reversal of the magnetic materials'. NML relies on comprehending spin dynamics. Efficient, reliable logic devices emerge from this understanding (cf. Section 3.1.1.1 detailing theory and fundamentals). Spin dynamics is governed by the precession of electron spins under the influence of an external magnetic field (cf. Eq. (3.20)). The Larmor precession (cf. Eq. (3.30)) frequency ω_L describes the rotation of spins and is given by the equation:

$$\omega_L = \gamma B \tag{3.30}$$

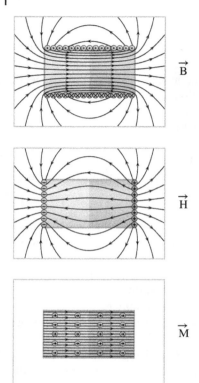

$\vec{B}$

$\vec{H}$

$\vec{M}$

Figure 3.10 Comparison of the magnetic B-field and H-field inside and outside of a cylindrical bar magnet. The fields are accurately computed with Ampère's model and a magnetic-charge model, respectively, which both yield precise solutions that coincide perfectly outside of the magnet. For the B-field in Ampère's model, effective current around the magnet is considered, which is displayed here with symbols for ring currents (dot symbols for current out of the image plane, cross symbols for current into the image plane). For the H-field in the magnetic-charge model, magnetic charges are considered at the magnetic poles (plus and minus symbols). Discrete field lines are drawn. The difference between B and H is the magnetization field M, which is homogeneous inside the cylindrical bar magnet. The magnetization is considered as the average field of many equally distributed atomic magnetic dipoles. File:VFPt magnets BHM symbols.svg. https://commons.wikimedia.org/wiki/File:VFPt_magnets_BHM_symbols.svg. Source: Geek3/Wikimedia Commons/CC BY-SA 4.0.

Here, γ is the gyromagnetic ratio, and B is the applied magnetic field (cf. Figure 3.10). The equation emphasizes the direct proportionality between precession frequency and magnetic field strength. Magnetization reversal is vital in NML. Spintronic devices, such as MRAM, use this process. Achieving reversal occurs through magnetic fields or currents. This method is essential for writing and erasing data.

3.2.2.1 Spin Transport and Manipulation

Spin transport, key in NML, controls electron flow. Manipulating spins is crucial for efficient information processing. Physics behind spin transport explored in this section.

A. Spin Transport in Nanomagnetic Systems:

Efficient spin information transport vital in nanomagnetic systems. Spin currents, with polarized electrons, manipulated using diverse mechanisms. Spin Hall effect and Rashba–Edelstein effect play key roles. Effects induce transverse spin currents in response to fields.

B. Spin Manipulation in Spintronic Devices:

Spin manipulation alters spin states in NML. Spin valves use giant magnetoresistance to determine resistance. Control spin configuration to switch resistance states efficiently.

C. Spin–Orbit Torque and Efficient Spin Manipulation:

Spin–orbit torque aids spin manipulation in NML devices. (cf. item 4 in Section 3.1.1 detailing basic principles of SOI; Eq. (3.4)). Torque results from the interaction of electron spins and orbit. STT influences magnetic moment orientation. Spin transport transfers spin angular momentum across materials.

In NML, spin transport ensures coherent information transfer. Efficient spin transport maintains coherent spin information in nanostructures. Control of spin orientation and coherence in spin manipulation is crucial.

3.2.2.2 Switching Dynamics in Magnetic Devices

Switching dynamics involve time-dependent processes in state change of magnetic devices. In NML, understanding and optimizing dynamics are crucial. Helps achieving high-speed and energy-efficient logic operations. The Landau–Lifshitz–Bloch (LLB) (cf. Eq. (3.31)) equation describes the dynamics of magnetization in the presence of an external magnetic field and spin current:

$$\frac{d\mathbf{M}}{dt} = -\gamma \mathbf{M} \times \mathbf{H}_{\text{eff}} + \alpha \mathbf{M} \times \frac{d\mathbf{M}}{dt} + \beta \mathbf{M} \times (\mathbf{M} \times \mathbf{H}_{\text{ext}}) - \frac{\beta}{M_s} \mathbf{M} \times \mathbf{I}_{\text{spin}} \quad (3.31)$$

Here, β is the non adiabatic parameter, $\mathbf{H}_{\text{ext}}$ is the external magnetic field, and $\mathbf{I}_{\text{spin}}$ is the spin current density. The LLB equation accounts for additional torques involved in switching dynamics.

Devices like MTJs demonstrate switching dynamics used in MRAM. External magnetic fields induce switching between magnetic states, enabling information storage and retrieval.

3.2.2.3 Magnetization Reversal Mechanisms

Diverse nanostructure magnetization depends on material properties and geometry. One common mechanism is coherent rotation, where the entire magnetization vector rotates collectively (cf. Eq. (3.32)). This process is described by the macrospin model:

$$\frac{d\theta}{dt} = -\gamma \mu_0 H_{\text{eff}} \sin(\theta) \quad (3.32)$$

Here, θ is the angle between the magnetization and the external magnetic field. Coherent rotation is observed in nanomagnetic elements like nanodisks and nanowires.

Domain wall motion is defined as: propagation of interface between magnetic domains through a material (cf. Eq. (3.33)). The Thiele equation governs the dynamics of domain wall motion:

$$\alpha \mathbf{v}_{\mathrm{DW}} = \gamma \mathbf{G} \times \mathbf{v}_{\mathrm{DW}} + \beta \mathbf{v}_{\mathrm{DW}} \qquad (3.33)$$

Here, $\mathbf{v}_{\mathrm{DW}}$ is the domain wall velocity, $\mathbf{G}$ is the gyromagnetic tensor, and β is the non adiabatic parameter. Understanding magnetization reversal mechanisms is crucial. To tailor memory storage, sensors, or logic operations.

In summary, spin dynamics, spin transport, switching dynamics, and magnetization reversal mechanisms are fundamental aspects of NML. Equations (cf. Eqs. (3.30)–(3.33)) reveal physics and engineering principles in NML. Deeper into nanoscale, principles guide faster, efficient, scalable devices. Refer to Algorithm 1: Nanomagnetic Logic Integrated Chip Design Flow detailing Algo:NML1.

3.3 Design and Operation of Nanomagnetic Logic Gates

This section talks about the design and operation of NML gates, briefs the majority gate in Section 3.3.1, and describes the inverter and majority-inverter logic in Section 3.3.2.

The majority gate is a fundamental building block in NML. The output state is determined by the majority input magnetic state. This gate leverages the superparamagnetic behavior of nanomagnetic particles. It is a key component in the design of magnetic logic circuits.

Eq. (3.34) governing the magnetic field in the context of majority gate is given by:

$$\nabla \times \mathbf{H} = \mu \mathbf{M} \qquad (3.34)$$

where μ is the magnetic permeability, and $\mathbf{M}$ is the magnetization. Assessing feasibility in memory applications requires understanding information retention and stability in majority gate with magneto-electrostatic coupling (cf. Figure 3.11).

3.3.1 Majority Gate

This section talks about the majority gate throughout, and briefs the logic operations in Section 3.3.1.1. NML gates promise advancements in information processing technology. Here we explore the fundamental physics and engineering principles of design and operation. Refer to Listing 3.1 Majority Gate detailing design flow for a NML majority gate – design.

Algorithm 3.1 Nanomagnetic Logic-Integrated Chip Design Flow

1: Define Specifications:
   Determine functional and performance requirements.
2: Conduct Research:
   Investigate existing nanomagnetic logic designs and technologies.
3: Develop Conceptual Design:
   Create a high-level conceptual design based on specifications.
4: Establish System Architecture:
   Define overall chip structure and interconnections.
5: Implement Logic Design:
   Design logical components using nanomagnetic principles.
6: Design Circuitry:
   Develop circuitry for logic implementation.
7: Conduct Simulation:
   Verify functionality and performance through simulation.
8: Create Layout Design:
   Produce physical chip layout.
9: Design Magnetic Elements:
   Design nanomagnetic elements.
10: Select Materials:
   Choose appropriate materials for nanomagnetic elements.
11: Choose Fabrication Process:
   Select compatible fabrication process.
12: Design Masks:
   Create masks for etching and deposition.
13: Assess Manufacturability:
   Evaluate manufacturability constraints.
14: Develop Prototypes:
   Build prototype chips for testing.
15: Conduct Characterization:
   Assess performance and reliability.
16: Iterative Design:
   Iterate based on testing results.
17: Optimize Design:
   Optimize design for performance and manufacturability.
18: Document Design:
   Document design process and details.
19: Conduct Compliance Testing:
   Ensure compliance with standards and regulations.
20: Perform Integration Testing:
   Test compatibility with other systems.
21: Plan Production:
   Plan production process.
22: Initiate Mass Production:
   Manufacture chips in large quantities.
23: Implement Quality Control:
   Implement quality control measures.
24: Design Packaging:
   Design protective packaging.
25: Manage Distribution:
   Distribute chips to customers or partners.
26: Provide Customer Support:
   Offer assistance with chip integration.
27: Collect Feedback:
   Gather feedback for improvement.
28: Continuous Improvement:
   Use feedback to improve design and process.
29: Lifecycle Management:
   Manage chip lifecycle.
30: Intellectual Property Protection:
   Protect intellectual property rights.

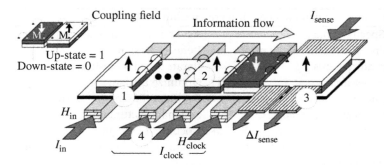

Figure 3.11 The four building blocks for electrically embedded ferromagnetic logic gates: Electrical input (1), field-coupled magnetic wire (2), electrical output (3), and current driven clocking structures (4) File: Figure 1 https://ieeexplore.ieee.org/document/5331551/. Source: M. Becherer et al., 2009/With permission of IEEE.

```
1   # Design flow for a nanomagnetic logic majority gate
2   # 1. Design Concept
3   # Utilize the magnetic orientations of nanomagnets to
      implement the majority logic function.
4   # 2. Magnetic Material Selection
5   # Choose a suitable magnetic material with properties
      conducive to nanomagnetic logic operations.
6   # Consider factors such as magnetic anisotropy,
      coercivity, and stability.
7   # 3. Nanomagnet Geometry
8   # Define the geometry of the nanomagnets to achieve the
      desired functionality.
9   # Consider parameters such as size, shape, and aspect
      ratio.
10  # 4. Magnetic Interaction
11  # Determine the magnetic interactions between nanomagnets
       to implement the majority logic function.
12  # Explore concepts such as dipole-dipole interaction and
      exchange coupling.
13  # 5. Input/Output Configuration
14  # Define the input and output configurations of the
      majority gate.
15  # Determine the number of input nanomagnets required to
      implement the majority function.
16  # Design the output detection mechanism to accurately
      determine the majority output.
17  # 6. Control Mechanism
```

```
18  # Implement a control mechanism to manipulate the
       magnetic states of the nanomagnets.
19  # Utilize external magnetic fields, spin-polarized
       currents, or spin waves for control.
20  # 7. Signal Amplification
21  # Incorporate signal amplification techniques to enhance
       the reliability and robustness of the majority gate.
22  # Explore methods such as spin-transfer torque
       amplification or magnetic coupling amplification.
23  # 8. Power Consumption Optimization
24  # Optimize the design to minimize power consumption.
25  # Explore techniques such as adiabatic switching or
       energy-efficient control schemes.
26  # 9. Temperature Stability
27  # Ensure the majority gate operates reliably across a
       range of temperatures.
28  # Consider thermal stability and thermal noise effects on
       the magnetic states of nanomagnets.
29  # 10. Noise Immunity
30  # Enhance the noise immunity of the majority gate to
       mitigate errors due to external disturbances.
31  # Explore techniques such as error correction coding or
       noise-tolerant logic designs.
32  # 11. Fabrication Process
33  # Define a fabrication process compatible with
       nanomagnetic technology.
34  # Consider techniques such as electron beam lithography,
       focused ion beam milling, or self-assembly methods.
35  # 12. Characterization and Testing
36  # Develop characterization techniques to evaluate the
       performance of the majority gate.
37  # Conduct comprehensive testing under various operating
       conditions to validate functionality and reliability.
38  # Simulation and CAD Structure Creation
39  # Same applies for simulation:
40  # - Utilize simulation tools such as OOMMF, MuMax3,
       MagCAD, Nmag, MATLAB/Simulink, or SPICE to simulate the
       majority gate.
41  # - Set up simulation environments to model magnetic
       behavior, logical operations, and interconnect
       performance.
42  # - Create the CAD structure using MIF coding or CAD
       model software like CleWin for hierarchical layout
       design.
```

Listing 3.1 Design flow for a nanomagnetic logic majority gate – Design.

A. Fundamental Physics of Nanomagnetic Logic Gates: The building blocks of NML are magnetic nanoparticles. These nanoparticles exhibit unique magnetic properties due to their small size. This size makes them suitable for nanoscale information storage and processing (cf. Section B) (cf. Eq. (3.35))

$$\vec{M} = \chi \vec{H} \tag{3.35}$$

The magnetization ($\vec{M}$) of the nanoparticle is expressed as a function of the magnetic susceptibility (χ) and the applied magnetic field ($\vec{H}$). It is crucial to understand behavior of magnetic nano particles in logic-gate operations. **Superparamagnetism**: At the nanoscale, thermal fluctuations can significantly affect the magnetic behavior. Superparamagnetism is a phenomenon where the nano particle's magnetic-moment fluctuates spontaneously among two direction (cf. Eq. (3.36))

$$\tau = \frac{1}{f_0} \tag{3.36}$$

The characteristic time (τ) for superparamagnetic relaxation is expressed where f_0 is the attempt frequency. Time constant crucial for magnetic states stability in NML gates.

B. Engineering Aspects of Nanomagnetic Logic Gates

1) **Gate Design**:
   NML gates design requires arrangement of magnetic nanoparticles performing logical operations. Majority gate, output is determined by the majority state of the input:

$$\text{Output} = \text{Majority}(\text{Input}_1, \text{Input}_2, \text{Input}_3) \tag{3.37}$$

   The majority gate operation is expressed above (cf. Eq. (3.37)). Output state is dependent on the majority magnetic state among the input.

2) **Controlled Switching**
   Switching magnetic states is vital for nanomagnetic gates. Use fields and torque to control magnetic orientation. Nanomagnetic gates show promise but face thermal challenges. Scalability hurdles persist, leading to the exploration of novel materials. Nanomagnetic gates blend physics and engineering, offering possibilities. Integrating nanoparticles and switching mechanisms advances computing tech.

3.3.1.1 Logic Operations

Logic operations, such as AND, OR, NOT, are implemented by arrangement and control of the magnetic orientations.

1) **AND Gate Operation**:
   Consider a simple nanomagnetic AND gate with two inputs (Input_1 and Input_2) and an output (Output):

   $$\text{Output} = \text{Input}_1 \cdot \text{Input}_2 \tag{3.38}$$

   Equation (3.38) represents the logic operation of the AND gate. The output is true only when both inputs are in the magnetic state corresponding to logical "1".

2) **OR Gate Operation**
   For a nanomagnetic OR gate with two inputs (Input_1 and Input_2) and an output (Output):

   $$\text{Output} = \text{Input}_1 + \text{Input}_2 \tag{3.39}$$

   Equation (3.39) describes the logic operation of the OR gate. The output is true when at least one of the inputs is in the magnetic state corresponding to logical "1".

3) **NOT Gate Operation**:
   A nanomagnetic NOT gate with a single input (Input) and an output (Output) can be expressed as:

   $$\text{Output} = \overline{\text{Input}} \tag{3.40}$$

   Equation (3.40) denotes the logic operation of the NOT gate. The output is the opposite of the input magnetic state.

4) **Dipole Coupling Influence on Decision Threshold**:
   Expanding on the earlier dipole coupling discussion, here we see how dipole alignment affects the decision threshold of the majority gate (cf. Eq. (3.41)). The mathematical representation of the gate output under dipole coupling is given by:

   $$V_{\text{out}} = \begin{cases} 1, & \text{if } V_{\text{in}} > V_{\text{th}} \\ 0, & \text{otherwise} \end{cases} \tag{3.41}$$

   where V_{out} is the output voltage, V_{in} is the input voltage, and V_{th} is the threshold voltage.
   Examining threshold modulation reveals gate's adaptability to environments. Insights gained provide potential advantages in robustness, reliability.

3.3.2 Inverter and Majority-Inverter Logic

This section talks about the inverter and majority-inverter logic throughout, briefs the functionality and design considerations in Section 3.3.2.1, describes the comparisons with traditional CMOS inverters in Section 3.3.2.2, and presents in detail the implementation challenges in Section 3.3.2.3. Key player: Nanomagnetic Logic Inverter (NLI). Let us explore functionality, design of groundbreaking NML technologies.

3.3.2.1 Functionality and Design Considerations

NLI uses magnetic domains for logical operations. CMOS inverter employs complementary transistors to invert signals (cf. Eq. (3.42)). NLI relies on nanomagnet orientation for logic operations.

The functionality of NLI can be mathematically described using the following equation:

$$V_{out} = \overline{V}_{in} = V_{in} \tag{3.42}$$

Here, V_{in} and V_{out} represent the input and output voltages, respectively. The bar over V_{in} denotes the logical NOT operation.

NLI design considers magnetization dynamics, thermal stability, and energy dissipation (cf. Eq. (3.43)). Balancing factors optimizes performance in NLI devices. One of the key design parameters is the shape anisotropy, which can be expressed as:

$$E_{anis} = K \cdot V \cdot (\sin^2\theta - \sin^2\theta_0) \tag{3.43}$$

Here, E_{anis} represents the Anisotropy Energy, K is the anisotropy constant, V is the volume of the magnetic element, θ is the angle of magnetization, and θ_0 is the equilibrium angle.

3.3.2.2 Comparisons with Traditional CMOS Inverters

The comparison between NML inverters and traditional CMOS inverters is crucial to understand advantages and limitations (cf. Eq. (3.44)). While CMOS technology has dominated the semiconductor industry, the emergence of NML brings several potential benefits.

Energy Efficiency Comparison: NLI offers the prospect of ultra-low power consumption due to the absence of resistive losses associated with traditional CMOS technology (cf. Eq. (3.44)). The energy consumption (E_{NLI}) of NLI can be expressed as:

$$E_{NLI} = \frac{1}{2}MVH_{eff}\cos\theta \tag{3.44}$$

where M is the magnetization, H_{eff} is the effective magnetic field, and θ is the magnetization angle.

3.3.2.3 Implementation Challenges

1) **Speed and Size Considerations**:
   NLI devices exhibit faster switching speeds (cf. Figure 3.12) compared to CMOS inverters. This is due to the absence of electron transit times. The size reduction is also notable. The scaling potential of nanomagnetic devices allows higher integration densities.

2) **Thermal Stability Challenges**:
   Thermal stability is a critical concern. It directly affects the reliability of NLI devices (cf. Eq. (3.45)). The thermal stability factor (Δ) can be expressed as:

$$\Delta = \frac{E_{\text{anis}}}{k_B T} \tag{3.45}$$

where k_B is the Boltzmann constant and T is the temperature. Incorporating materials with higher anisotropy constants is a key to enhance thermal stability.

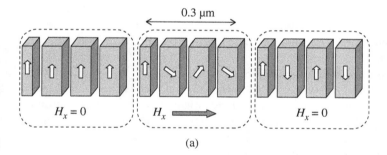

(a)

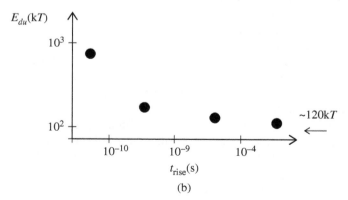

(b)

Figure 3.12 Adiabatic switching of the nanomagnet line. Part (a) illustrates the switching process and part (b) is the switching speed – dissipation curve for the entire structure. File: Figure 4 https://ieeexplore.ieee.org/document/1392346/. Source: G. Csaba et al., 2004/With permission of IEEE.

3.4   Signal Processing in Nanomagnetic Logic

This section talks about the signal processing in NML, it briefs the signal generation and detection in Section 3.4.1, and describes the signal propagation and interconnects in Section 3.4.2.

A. Single-Domain Ferromagnetic Elements:
NML relies on the use of single-domain ferromagnetic elements. These are magnetic domains with uniform magnetization. These elements exhibit distinct magnetic properties of (a) saturation magnetization, (b) coercivity, and (c) anisotropy.

i) **Saturation magnetization** M_s represents maximum magnetic moment per unit volume attainable in a material.

ii) **Coercivity** H_c is the measure of magnetic field required to demagnetize the material.

iii) **Anisotropy** determines the preferred direction of magnetization.

B. Magnetic Nanoparticles and Their Properties:

i) **Magnetic tunnel junction (MTJs)**: is pivotal in NML. An MTJ consists of two ferromagnetic layers separated by an insulating layer. The relative alignment of magnetizations in these layers affects electrical resistance of the junction.

ii) **Representation of Information**:
In NML, information is encoded in the magnetic state of the nanomagnets. Binary information is represented by orientation of the magnetization. Example, "0" is represented by the magnetization pointing in one direction, and "1" is represented by opposite direction.

iii) **Signal Propagation and Amplification**:
In NML circuits, signals are propagated through the manipulation of magnetic states. Magnetic signals are transmitted through interaction between nanomagnet. Orientation of one magnet influences the state of its neighbors. Amplification of signals in NML is achieved by exploiting inherent properties of materials. Magnetic elements with high saturation magnetization are used to amplify weak signals. It enables robust signal processing.

iv) **Thermal Stability and Reliability**:
Key challenges in NML is ensuring thermal stability of magnetic states. As the size of nanomagnets decreases, thermal fluctuations is significant. It leads to the unintended state transitions. Materials engineering and advanced fabrication techniques help enhance the thermal stability.

v) **Propagation Delays and Signal Integrity**:
Signal propagation delays and maintaining signal integrity are concerns. Magnetic interactions between nanomagnets lead to signal delay. It impacts overall performance. Methods to optimize signal paths, reduce delays, and enhance signal reliability are a key.

vi) **Micromagnetic Simulations:** To model behavior of nanomagnet, micromagnetic simulations are used (cf. Eq. (3.19)). These simulations solve the LLG equation in discretized magnetic volumes (cf. Section 3.1.1.1 briefing theory and fundamentals).

C. Chirality Effects on Signal Propagation:

i) **Signal Modulation:**
Chirality-induced effects can be harnessed to modulate signals within the MG (cf. Eq. (3.46)). The mathematical representation of chiral-modulated signals is expressed as:

$$V_{mod} = A \cdot \sin(\omega t + \phi) \tag{3.46}$$

where V_{mod} is the modulated voltage, A is the amplitude, ω is the angular frequency, t is time, and ϕ is the phase.

3.4.1 Signal Generation and Detection

This section talks about the signal generation and detection throughout, briefs the input signal encoding (ISE) in Section 3.4.1.1, describes the output signal detection in Section 3.4.1.2, and presents in detail the Role of Magnetic Sensors in Section 3.4.1.3.

A) **Signal Generation:**
The fundamental of signal generation in NML involves manipulating magnetization state of FM layers within MTJ. Application of an external magnetic field or spin-polarized currents, it toggles the orientation of magnetic moments. This magnetization change induces variation in the tunneling resistance. The signal generated leads to encoding and processing. The precise control ability of magnetization state contributes to reliability and speed of NML operations.

B) **Signal Detection:**
Detecting generated signal is crucial for functionality of NML. The change in tunneling resistance within the MTJ measured by passing a small current through junction. This current experiences a modulation in resistance, detecting magnetic state. Advanced signal detection techniques utilize GMR or TMR effects. These amplify the change in resistance. The detection sensitivity and efficiency are measurable factors to analyze the overall performance of NML circuits.

C) **STT Effect-Based Signal generation:**
Signal Generation and manipulation in NML is also achieved by the STT effect (cf. Eq. (3.3) detailed in item 3 in Section 3.1.1 presenting basic principles of STT). Transfer of angular momentum between electrons and magnetization of a magnetic layer leads to this.

In NML, STT is used for switching magnetic orientation of a ferromagnetic layer. Rotation is achieved, with spin-polarized current applications through MTJ. The angular momentum is transferred to the magnetization. This rotation changes the resistance, basis for signal generation.

STT offers advantages of lower-power consumption and faster switching speeds. It enhances the efficiency of NML devices. For signal detection, the STT effect is leveraged in reverse. The change in resistance is detected by measuring current passing through the MTJ. STT-based efficient signal detection contributes to overall reliability and performance of NML circuits.

3.4.1.1 Input Signal Encoding (ISE)

In NML, encoding information into the generated signals is important. This involves representing binary data with magnetization states of ferromagnetic layers in MTJs.

Binary Encoding:

The common approach for ISE in NML is binary encoding. The two magnetization states, typically represented as "0"and "1," correspond to binary logic.

Magnetic State Representation:

In a binary encoding, "1 of 2" ferromagnetic layer's state represent "0," and opposite state corresponds to "1." The control and manipulation of these "2 of 2" states enable binary-encoded information processing.

3.4.1.2 Output Signal Detection

Output signal detection in NML is by interpreting magnetization states of MTJs. It extract results of logical operations. The detection process is important to obtaining computational output.

(I) **Readout Circuits**:

Readout circuit essential for accurate detection of encoded information. These are designed to interface with the MTJ. Also, translate the magnetization state into electrical signals.

a) **Sensing Amplifiers**: Sensing Amplifiers enhance signal-to-noise ratio (SNR) during readout process. It amplifies the weak signals from the MTJs. Also ensures reliable and accurate detection of magnetization states.

b) **Signal Processing**: Error correction and noise filtering-based techniques are employed to enhance accuracy. Integration of advanced signal processing algorithms increases robustness of NML devices.

(II) **Parallel Readout**:

High-throughput computation, is achieved by parallel readout mechanisms. Unlike traditional computing paradigms where bits are processed sequentially. In NML simultaneous readout of multiple MTJs, enablie parallel computation.

a) **Array Architectures**: NML devices comprises arrays of MTJs, in a grid-like structure. Each MTJ represents a bit of information. Parallel readout architecture performs concurrent analysis of multiple bits. This parallelism significantly enhance the processing speed.

b) **Crossbar arrays**: Crossbar arrays is one other configuration in NML circuits. The intersection points of the rows and columns in an array has individual MTJs. By selectively addressing these intersections, parallel readout is achieved. Paves way for efficient and high-performing NML devices.

3.4.1.3   Role of Magnetic Sensors

Magnetic sensors are integral components in NML circuits. Functions include from signal detection to external-field sensing. The precise and reliable operation is important. It facilitates transfer of information within the system.

(I) **Signal Detection**: As discussed, magnetic sensors plays a vital role in detecting signals alongside MTJs. The sensors must be sensitive to small variations in magnetic fields. It also possess ability to convert these variations into measurable electrical signals. Giant Magnetoresistance (GMR) Sensors, TMR Sensors, and STT Sensors are used.

a) **Giant Magnetoresistance (GMR) Sensors**: GMR sensors detect magnetic field changes. The GMR effect relies on magnetic moment alignment. Layers of ferromagnetic and non magnetic materials are involved. Applied magnetic field alters sensor resistance. This change yields a measurable output. GMR sensors excel at small magnetic field detection. Ideal for nanoscale magnetic logic circuits.

b) **Tunnel Magnetoresistance (TMR) Sensors**: TMR sensors detect magnetic changes using quantum tunneling. Two layers, insulating barrier, magnetic field applied. Electron tunneling probability depends on orientation of magnetic moments. TMR sensors are highly sensitive, suitable for NML.

c) **Spin-Transfer Torque (STT) Sensors**: STT sensors alter magnetic orientation with STT (cf. Eq. (3.3), detailed in item 3 in Section 3.1.1 presenting basic principles of STT). Electrical current transfers electron spin, changing free layer orientation. STT sensors aid NML, offering electric control for efficient information processing.

Integration with Readout Circuits: Magnetic sensors are seamlessly integrated into readout circuits of NML devices. This ensures MTJs signals detected and processed, leading to reliable output.

(II) **External Field Sensing**: Beyond signal detection, magnetic sensors in NML circuits sense external magnetic fields. This targets applications such as magnetic memory storage and navigation systems.

    a) **Fluxgate Sensors**: Fluxgate sensors are used for measuring external magnetic fields. These sensors operate on the principle of modulating the magnetic flux within a core material. The resulting changes are detected as variations in electrical signals. The integration enhances adaptability of NML devices in diverse environments.

    b) **Hall effect sensors**: Hall Effect Sensors detect magnetic fields. They measure voltage generated in a current-carrying conductor exposed to a magnetic field. These sensors are simple and versatile, having applications in NML.

This interplay of signal generation, detection, input signal encoding, output signal detection, and magnetic sensors are foundations of NML computing. Addressing challenges of signal reliability, multi valued logic control, and external field interference is crucial. The integration of advanced materials and fabrication techniques, with understanding of the underlying physics, is important, paving way to realize practical and robust NML devices.

3.4.2 Signal Propagation and Interconnects

This section talks about the signal propagation and interconnects throughout, briefs the nanomagnetic waveguides in Section 3.4.2.1, describes the challenges in signal transmission in Section 3.4.2.2, and presents in detail the strategies for efficient interconnects in Section 3.4.2.3.

3.4.2.1 Nanomagnetic Waveguides

Concept and Design: Nanomagnetic waveguides is a specialized form of interconnects in NML. Facilitates efficient transmission of signals between different components. Guides, magnetic signals in predetermined paths. It aids minimizing energy loss: mitigating challenges of signal transmission.

1) **Types of Nanomagnetic Waveguides**:

    a) **Magnetic Nanowires**: It is the simplest form of nanomagnetic waveguides. Magnetic domain walls are manipulated to carry signals along the wire.

    b) **Spin Wave Waveguides**: Utilizes spin waves for signal transmission. Achieves long-distance communication in NML circuits.

    c) **Resonant Waveguides**: Resonant structures, such as magnonic crystals, as waveguides. Enables the control of frequencies and enhances signal integrity.

2) **Challenges and Opportunities in Nanomagnetic Waveguides**: While nanomagnetic waveguides present efficient signal transmission. Challenges exist in design and implementation. These challenges include:
   a) **Waveguide Losses**: Minimizing losses in nanomagnetic waveguides to maintain signal integrity over extended distances.
   b) **Integration with Other Components**: Integration with other NML components is crucial for overall functionality.
   c) **Nonlinear Effects**: Soliton formation and interactions impact reliability of nanomagnetic waveguides. It needs careful attention.

3.4.2.2 Challenges in Signal Transmission

Importance of Signal Propagation and Interconnects: Efficient signal transmission is important for NML-devices operation. As the components in NML get smaller, challenge to transmit signals. Design and optimizing interconnects are key to performance and reliability of NML circuits.

(I) **Signal Propagation in Nanomagnetic Logic**:
   a) **Magnetic Domain Wall Motion**:
   Signal in NML relies on magnetic domain walls. Magnetic domain boundaries move via external magnetic fields. Nanowires use domain walls as carriers for binary information. Essential to understand and control domain wall motion. Designing efficient, reliable signal paths in NML devices.
   b) **Spin Waves in Nanomagnetic Logic**:
   Magnons, or spin waves, aid signal transfer in NML. Spins collectively oscillate, transmitting data over short distances. Understanding spin wave challenges optimizes NML circuits.

(II) **Challenges in Signal Transmission**: Though having advantages, several challenges exist for efficient signal transmission at the nanoscale. Some of these challenges include:
   a) **Damping and Energy Loss**: The small sized nanomagnetic components lead to increased damping and energy loss. It affects the distance, up-to signals can be reliably transmitted.
   b) **Temperature Effects**: Temperature fluctuations influence the magnetic properties. Introducing variability in signal transmission.
   c) **Crosstalk**: The proximity of elements increases with size reduction, leading to crosstalk between neighboring signals.
   d) **Fabrication Variability**: Nanomagnetic devices are susceptible to fabrication variations. Impact on signal propagation and device performance is noted.

3.4.2.3 Strategies for Efficient Interconnects

A) **Material Selection**:
Optimal materials for nanomagnetic interconnects ensure efficient circuit performance. Low damping, high magnetic anisotropy, and stability under temperature variations are crucial.

B) **Geometric Optimization**:
Efficient interconnect design depends on geometric considerations. Nanowire width, thickness, and component spacing must minimize crosstalk and signal loss.

C) **Signal Coding and Error Correction**:
Essential to implement advanced coding and error correction techniques. Redundancy and error detection enhance signal transmission reliability in nanomagnetic circuits.

D) **Hybrid Approaches**:
Combining nanomagnetic interconnects with emerging technologies boosts overall circuit performance. Synergies with photonic / superconducting interconnects overcome individual limitations.

NML has potential to lead rebooting computing. Success lies on addressing nanoscale signal propagation and interconnect challenges. Mastering magnetic domain wall motion and spin waves is crucial. Nanomagnetic waveguides promise efficient signal transmission, overcoming challenges. Damping, energy loss, temperature effects, crosstalk, and variability demand innovative solutions. Advancements in material science, fabrication, and signal processing essential. Efficient interconnect strategies include material selection, optimization, coding, error correction, and hybrids. A roadmap guides overcoming challenges in signal transmission for NML.

3.5 Energy Considerations and Efficiency

This section talks about the energy considerations and efficiency, briefs the low-power design in NML in Section 3.5.1, describes the energy consumption for NML in Section 3.5.2, and describes the input – output interface – signals perspective overview in Section 3.5.3

A) **Energy Consumption in Nanomagnetic Logic**: (cf. Figure 3.13)
   i. **Switching Energy**:
   Nanomagnets demand efficient energy for state transitions. Switching energy depends on size, anisotropy, and mechanism. Nanomagnet size impacts speed and thermal stability challenges arise. LLG equation controls magnetization dynamics (cf. Eq. (3.47)), revealing switching energy.

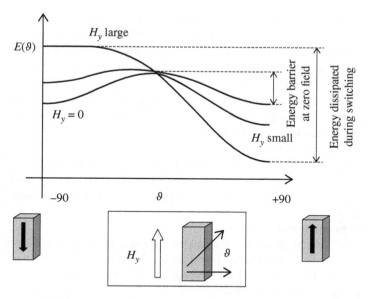

Figure 3.13 The energy of a nanomagnet as a function of the angle between magnetization direction and the easy axis. The three curves correspond to different easy-axis fields. File: Figure 1 https://ieeexplore.ieee.org/document/1392346/. Source: G. Csaba et al., 2004/With permission of IEEE.

The energy dissipation during switching can be expressed as:

$$E_{switch} = \alpha \cdot \frac{KV}{\gamma M_s} \cdot H_c \cdot \sin(\theta) \tag{3.47}$$

where E_{switch} is the switching energy, α is the Gilbert damping constant, K is the magnetic anisotropy constant, V is the volume of the nanomagnet, γ is the gyromagnetic ratio, M_s is the saturation magnetization, H_c is the coercive field, and θ is the angle between the magnetization and the applied field.

ii. **Energy Dissipation Mechanisms**:

In addition to switching energy, energy dissipation occurs in NML. The primary sources of energy dissipation include:

a. **Eddy Currents**: Rapid magnetic field changes induce eddy currents. Joule heating and energy loss are concerns. Crucial to minimize resistive losses in conductive layers.

b. **Damping Losses**: Gilbert damping causes energy dissipation in nanomagnetic materials. Lower damping constants (α) essential for minimizing losses.

   c. **Spin-transfer Torque**: Applying spin-polarized currents manipulates magnetization states. Transfer of angular momentum leads to energy dissipation (cf. Eq. (3.3) Detailed in item 3 in Section 3.1.1 presenting Basic Principles of STT). Efficient control critical for reducing energy consumption in NML devices.

B) **Efficiency Enhancement Strategies**: NML's potential requires addressing energy issues and improving efficiency. Strategies to mitigate energy dissipation and improve the overall efficiency of NML systems include:

  i) **Material Optimization**:
Choose materials with favorable magnetic properties to reduce energy consumption. Focus on low damping constants, high thermal stability, and tunable magnetic anisotropy. Enable efficient switching while minimizing energy losses.

  ii) **Design Optimization**:
NML circuit design important for determining energy efficiency. Careful layout design, minimized interconnect lengths, and optimized nanomagnet arrangement. Three-dimensional integration for compact, energy-efficient NML architectures a key.

  iii) **Advanced Switching Schemes**:
Investigate innovative switching schemes:adiabatic and resonant switching. Adiabatic switching minimizes energy dissipation: gradually changing magnetic states. Resonant switching exploits frequencies: efficient and selective magnetization reversal.

C) **Challenges and Future Directions**: NML faces challenges for mainstream computing adoption. Some key challenges include:

  i) **Thermal Stability**:
Shrinking nanomagnets require enhanced thermal stability without sacrificing speed. Research focuses on materials and techniques to address undesired state transitions.

  ii) **Interference and Crosstalk**:
Densely packed circuits lead to interference and crosstalk between nanomagnets. Mitigation involves shielding techniques and advanced signal processing methods.

  iii) **Scalability**:
Achieving scalability with energy efficiency is challenging in NML. Scalable fabrication and novel materials- meet miniaturization demands.

  iv) **Power Supply and Integration**:
Efficient power supply and integration with existing technologies are the key. Attention needed for compatible interfaces and power-efficient control-circuitry development. Addressing integration challenges: semiconductor technologies for adoption.

3.5.1 Low-Power Design in Nanomagnetic Logic

This section talks about the low-power design in NML throughout, it briefs the power consumption analysis in Section 3.5.1.1, and describes the strategies for minimizing energy dissipation in Section 3.5.1.2.

NML explores low-power design. Here we discuss: analyzing power consumption and minimizing energy dissipation strategies. Comparative assessment of energy efficiency with traditional CMOS technology. A distinct approach for computation without power-hungry electric currents. Contributes to the potential for ultra-low power consumption.

3.5.1.1 Power Consumption Analysis

Let us understand the sources of power dissipation and its characteristics. It is important for assessing the energy efficiency: NML circuits.

A) **Sources of Power Dissipation in NML**:
- **Switching Energy**: The energy required to change the magnetic state of a nanomagnet during a logic operation.
- **Signal Propagation**: Energy dissipated during propagation of signals through nanowires.
- **Clocking Overhead**: Power consumed during clocking of logic gates.

B) **Mathematical Models for Power Consumption**: Mathematical models to quantify the power consumption in NML is presented.

i) **Switching Energy Model**: The energy required for changing the magnetic state (cf. Eq. (3.48)) of a nanomagnet during a logic operation. Switching energy (E_{switch}), is modeled as:

$$E_{\text{switch}} = \alpha \cdot V \cdot M_s \cdot H_k \cdot \cos(\theta) \tag{3.48}$$

where: α is the Gilbert damping constant, V is the volume of the nanomagnet, M_s is the saturation magnetization, H_k is the anisotropy field, and θ is the angle between the initial and final magnetization states.

ii) **Signal Propagation Energy Model**: The energy dissipated (cf. Eq. (3.49)) during signal propagation through nanowires. Propagation energy ($E_{\text{propagation}}$) is expressed as:

$$E_{\text{propagation}} = C \cdot V_{\text{dd}}^2 \tag{3.49}$$

where: C is the total capacitance of the nanowire, and V_{dd} is the supply voltage.

iii) **Clocking Overhead Model**: (cf. Figure 3.14) The power consumed during the clocking of logic gates (cf. Eq. (3.50)). Clocking overhead energy (E_{clock}) is modeled as:

$$E_{\text{clock}} = P_{\text{clock}} \cdot T \tag{3.50}$$

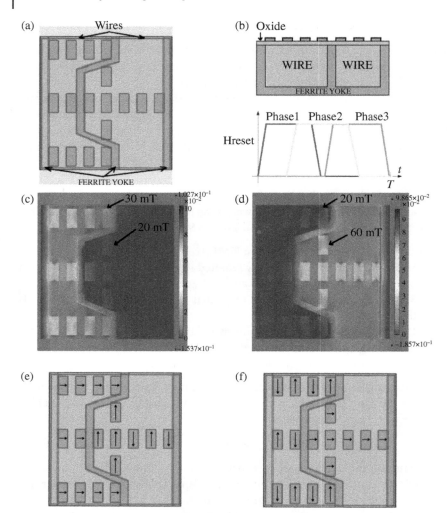

Figure 3.14 Comsol simulation of clock wires. (a) Clock wires model, upper view. (b) Clock wires model, section view. (c) Clock signals waveform. (d) Simulation results with current flowing in the first clock wire. Color gradations represent the horizontal component of the magnetic flux density (b) expressed in Tesla. (e) Simulation results with current flowing in the second clock wire. (f) and (g) Magnetization schematic representation of Comsol simulations. File: Figure 9 https://ieeexplore.ieee.org/document/6237531/. Source: Marco Vacca, 2012/With permission of IEEE.

where P_{clock} is power consumed by the clocking mechanism, and T is the clock period.

3.5.1.2 Strategies for Minimizing Energy Dissipation

Here we discuss key approaches and optimizations tailored for energy efficiency in NML circuits. Strategies to minimize power dissipation follows.

(I) **Material and Device Optimization**:
  a) **Magnetization Orientation**: Materials with optimal magnetization orientations reduce the switching energy (cf. Eq. (3.48)).
  b) **Anisotropy Field Reduction**: Lowering anisotropy field (H_k) by material engineering decrease switching energy (cf. in item i of Section 3.5.1.1 detailing power consumption analysis).
  c) **Voltage Scaling**: Operating at lower-supply voltages reduces signal propagation energy.
(II) **Circuit-Level Optimizations**:
  a) **Majority Gate Optimization**: Fine-tuning the design of majority gates improve efficiency, reducing power consumption.
  b) **Clock Gating**: Introducing clock gating techniques minimizes clocking overhead – idle periods.
  c) **Parallelism**: Leveraging parallelism in NML circuits enhances overall performance and reduces energy dissipation per operation.

3.5.2 Energy Consumption for Nanomagnetic Logic

This section talks about the energy consumption for NML throughout.

The total energy consumption (cf. Eq. (3.51)) E_{total} in NML is the sum of switching energy E_{switch} and thermal dissipation energy E_{thermal}:

$$E_{\text{total}} = E_{\text{switch}} + E_{\text{thermal}} \tag{3.51}$$

Focuses on energy usage in NML. "Switching energy = energy to change magnetic state." "Thermal dissipation energy = energy lost as heat." High energy efficiency desired for NML devices. Minimizing energy consumption critical for performance. In addition to this: achieving high-energy efficiency is crucial for the performance of NML devices, making it essential to minimize energy consumption. Various techniques and strategies are employed to optimize energy usage in NML, including the development of low-power design methodologies and the exploration of novel materials with favorable magnetic properties. Additionally, system-level optimization approaches are utilized to further enhance energy efficiency by optimizing the overall architecture and operation of NML-based systems. Understanding and mitigating energy consumption challenges in NML play a significant role in advancing the practical implementation and widespread adoption of NML technology across various applications, ranging from computing to energy harvesting and beyond.

3.5.3 Input – Output Interface – Signals Perspective Overview

This section talks about the input – output interface – signals perspective overview throughout, briefs the electric signals to magnetic signals in Section 3.5.3.1, presents in detail the magnetic to electric signals in Section 3.5.3.2, and illustrates the notations, algorithms, python codes and modeling in Section 3.5.3.3.

NML relies on complex electrical–magnetic interfaces and vice-versa for signal conversion. Explore NML computing, revealing electrical-to-magnetic ($E \rightarrow M$) and magnetic-to-electric ($M \rightarrow E$) interfaces. Underlying principles, tech advancements, and possible applications are covered throughout.

3.5.3.1 Electric Signals to Magnetic Signals

Let us assume electric signals generate magnetic signals in NML. The relationship between electric signals (E) and magnetic signals (M) can be represented by a nonlinear function (cf. Eq. (3.52)).

Electric Signals to Magnetic Signals (ESMS): Let $E(t)$ represent the electric signal at time t, and $M(t)$ represents the magnetic signal at time t (cf. Eq. (3.52)). The relationship can be expressed as a function f:

$$M(t) = f[E(t)] \tag{3.52}$$

(I) **ESMS: Principles and Mechanisms**: Convert electrical signals to magnetic states in NML. Key for initializing, manipulating, and reading nanomagnets. Mechanism: spin-polarized current through a magnetic layer. Torque on nanomagnet's magnetization, controlled for desired logic states.

(II) **ESMS: Challenges and Solutions**: Challenge: Minimize energy for nanomagnet manipulation. Energy dissipation brings down NML's efficiency advantages. Advanced materials and optimized device architectures are the way. Variability due to temperature fluctuations and manufacturing differences. Robust control mechanisms and error-correction techniques to be investigated. Enhance reliability of the electrical to magnetic interface.

(III) **ESMS: Applications**: Electromagnetic interface (EMI) finds extensive use in diverse applications. Magnetic random-access memory (MRAM) to be noted as a viable solution. MRAM merges magnetic storage's non volatility with RAM's speed. A substitute for conventional memory technologies. Beyond MRAM, NML devices have varied applications. Serves in reconfigurable logic circuits, signal processing. Also, they play a role in neuromorphic computing advancements.

3.5.3.2 Magnetic to Electric Signals

The conversion of magnetic signals back to electric signals (cf. Eq. (3.53)) can be modeled using Faraday's Law of electromagnetic induction.

$$E = -\frac{d\Phi}{dt} \tag{3.53}$$

where:

E is the induced electromotive force (EMF),

Φ is the magnetic flux.

The magnetic flux $\Phi(t)$ is related to the magnetic signal $M(t)$ through the material properties. Magnetic signals to electric signals (MSES) interfaces includes the following components to understand:

(I) **MSES: Principles and Mechanisms**: Detect and convert magnetic states to electrical signals. Crucial for NML device output interpretation. Interface with conventional electronics efficiently. GMR and TMR techniques employed. Electrical resistance alters with magnetization orientation in adjacent layers. Integration of GMR/TMR into device architecture facilitates magnetic state detection. Magnetic states of nanomagnets translates into electrical resistance shifts. Facilitates effective reading of NML outputs.

(II) **MSES: Challenges and Solutions**: Reliable magnetic to electric conversion has difficulties. Sensitivity and scalability of detection mechanisms present challenges. Shrinkage of devices complicates accurate magnetic state detection. Ongoing exploration – advanced sensor technologies. Continuous research on minimizing disturbance to neighboring elements. Developing readout schemes to address ongoing challenges.

(III) **MSES: Applications**: Connects NML to conventional electronics seamlessly. Forms hybrid systems merging NML's advantages with traditional electronics. Potential applications include magnetic sensors, magnetic field imaging, and integrated circuits for signal processing.

3.5.3.3 Notations, Algorithms, Python Codes, and Modeling
Some Notations to remember:

$E(t)$: Electric signal at time t,

$M(t)$: Magnetic signal at time t,

f: Function describing the conversion from electric to magnetic signals,

γ: Gyromagnetic ratio,

H: External magnetic field,

α: Damping parameter,

M_s: Saturation magnetization,

$\Phi(t)$: Magnetic flux at time t.

A) **Algorithm and Modelling**:
- **Electric Signals to Magnetic Signals**: defines how electric signals are transformed into magnetic signals using a nonlinear function.
- **Computation Based on Magnetic Signals**: dynamics of magnetization under influence of external magnetic field and damping. It represents computation process within NML.
- **Magnetic to Electric Signals**: magnetic signals are converted back to electric signals through electromagnetic induction, completing the cycle.

B) **Basic Framework**: This algorithm provides a representation of NML system for understanding how electric signals can be converted to magnetic signals, undergo computation, and then be converted back to electric signals in a NML system. (cf. Listing 3.2, Listing 3.3 and Listing 3.4 for details.)

  i) **Sensor Input**: The electric signal is generated and noise is introduced to simulate real-world conditions.

  ii) **Electrical to Magnetic Interface**

  iii) **Physical Model**: Material properties and hardware characteristics, anisotropy and exchange terms, included in computation.

  iv) **Mathematical Model:** Includes equations of anisotropy term, exchange term, and noise (Eq. (3.54)).

  v) **Magnetic Signal Computation**: Use the Landau-Lifshitz-Gilbert equation to compute the magnetization over time. Update the magnetization at each time step based on the previous magnetization, external magnetic field, damping factor, and gyromagnetic ratio.

  vi) **Output**: Faraday's Law is applied to compute the induced EMF, accounting for noise in the magnetic signal.

  vii) **Plotting**: Visualize the sensor input, computed magnetic signals, and the output magnetic to electric signal over time.

$$\frac{dM}{dt} = -\gamma M \times H - \frac{\alpha\gamma}{M_s} M \times (M \times H) + \text{anisotropy\_term}$$
$$+ \text{exchange\_term} + \text{noise\_term} \tag{3.54}$$

C) **Verbatim**:

```
1  # Introduce noise to the electric signal
2  noise_amplitude = 0.1
3  electric_signal_with_noise = electric_signal +
     noise_amplitude
4  * np.random.normal(size=len(time))
```

Listing 3.2 Introducing Real World Noise to Sensor Input.

```
1  # Introduce material and hardware characteristics
2  anisotropy_constant = 0.02
3  exchange_constant = 0.005
4
5  def compute_magnetic_signal_with_properties(
     electric_signal):
6  magnetization_with_properties = np.zeros_like(
     electric_signal)
7  for i in range(1, len(time)):
8  delta_t = time[i] - time[i-1]
9  # Introduce material properties
10 anisotropy_term = -anisotropy_constant *
     magnetization_with_properties[i-1]
11 exchange_term = exchange_constant * np.gradient(
     magnetization_with_properties[i-1], time)
12 # Update magnetization with material properties
13 magnetization_change_with_properties  # Update the
     computation based on material properties
14 magnetization_with_properties[i] =
     magnetization_with_properties[i-1] +
     magnetization_change_with_properties * delta_t
15 return magnetization_with_properties
16
17 magnetic_signal_with_properties =
     compute_magnetic_signal_with_properties(
     electric_signal_with_noise)
```

Listing 3.3 Introducing material and hardware characteristics (Physical Model: Signals).

```
1  # Introduce noise in the magnetic signal
2  noise_amplitude_magnetic = 0.01
3  magnetic_signal_with_properties_and_noise =
     magnetic_signal_with_properties + \
4  noise_amplitude_magnetic * np.random.normal(size=
     magnetic_signal_with_properties.shape)
5
6  # Magnetic to Electric Signal with noise
7  induced_emf_with_noise = -np.gradient(
     magnetic_signal_with_properties_and_noise, time)
```

Listing 3.4 Magnetic to Electric Signal with Noise: Python.

3.6 Educational Emphasis

This section talks about the educational emphasis, briefs the accessible explanations for undergraduates in Section 3.6.1, and describes the practical examples and applications in Section 3.6.2.

3.6.1 Accessible Explanations for Undergraduates

This section talks about the accessible explanations for undergraduates throughout, and briefs the simplifying complex concepts in Section 3.6.1.1.

3.6.1.1 Simplifying Complex Concepts

Refer to : (a) item A detailing real-time analogy to understand Dipole-Dipole Coupling, (b) item B detailing real-time analogy to understand Inter-layer Exchange Coupling, and (c) item C detailing real-time analogy to understand Chiral Coupling presented in.

A) **Dipole–Dipole Coupling**: To understand dipole–dipole nanomagnetic coupling, let's use an analogy involving magnets. Imagine two small bar magnets placed near each other. It represents nanomagnetic elements in a Dipole–Dipole Coupling. The behavior of these illustrate: principles of magnetic interaction.

  i) **Magnetic Dipoles**:
  Each bar magnet represents a magnetic dipole. It is a fundamental concept in nanomagnetism. Magnetic Dipoles have both north (N) and south (S) poles.

  ii) **Orientation**:
  Initially, assume that two bar magnets have random orientations.

  iii) **Dipole-Dipole Interaction**:
  When magnets are close to each other: they experience dipole-dipole interactions. Similar to nanomagnetic elements, these interactions occur by magnetic forces.

  iv) **Alignment**:
  Magnetic Dipoles have a tendency to align themselves to minimize energy. If one magnet's north pole is close to the other magnet's south pole, it shows preference to align.

  v) **Energy Minimization**:
  The system seeks a state of minimum energy, achieved when magnets align. This is analogous to Dipole-Dipole Interaction, where Energy Minimization influences overall behavior of the system.

  vi) **Coupling Strength**:
  The strength of dipole-dipole coupling depends on proximity and orientation of the magnets. The closer and more aligned, the stronger the coupling.

  vii) **Dynamic Behavior**:
  If the system is dynamic, with changing external conditions the states, transition between orientations. This dynamism is similar to the nature of nanomagnetic elements in response to external stimuli.

viii) **Collective Behavior**:

Nanomagnetic elements collectively exhibit Dipole–Dipole Coupling effects. This interaction leads to emergent phenomena like magnetic patterns. Dipole coupling among nanomagnets induces magnetic domain formations.

Understanding dipole–dipole nanomagnetic coupling is essential to design and manipulating NML systems. Applications include information storage, sensors, and logic devices. The analogy helps visualize (Refer to item I in Section 3.2 briefing fundamentals of coupling mechanisms in NML - walking through Dipole-Coupling) fundamental principles of nanomagnetic interactions (cf. Eq. (3.22)) and their impact on collective behavior of magnetic elements.

B) **Interlayer Exchange Coupling** Let us create a playful and metaphorical : analogy using Hyderabadi Biryani : to explain interlayer-exchange coupling.

i) **Layers of Biryani**:

Consider a plate of Hyderabadi Biryani with distinct layers – rice, meat, and spices. These layers represent different magnetic materials in a nanomagnetic system.

ii) **Interlayer Exchange**:

In Hyderabadi Biryani, the layers interact through exchange of flavors. This exchange is similar to interlayer exchange in nanomagnetic systems. Adjacent layers influence each other's magnetic properties.

iii) **Spices as Magnetic Moments**:

Consider spices in Biryani as magnetic moments within each layer. Each spice represents a magnetic element in nanomagnetic materials.

iv) **Layer Alignment**:

Just as different layers in Biryani align to create a harmonious taste, nanomagnetic layers align to create a coordinated magnetic state through interlayer exchange.

v) **Flavor Transfer**:

The flavor and aroma from the meat layer in Biryani can represent transfer of magnetic information (alignment preferences) between adjacent layers.

vi) **Heat and Temperature**:

Introduce heat as an external factor that can influence layers of Biryani. Similarly, temperature changes in nanomagnetic systems affect exchange and magnetic alignment.

vii) **Collective Behavior**:

Just as layers collectively contribute to the overall taste of Biryani. NM layers collectively influence magnetic behavior of entire system via interlayer exchange.

Remember, while this analogy provides a creative way to understand concept (refer to item II in Section 3.2 briefing fundamentals of coupling mechanisms in NML – walking through IEC) the actual physics and mathematics of inter-layer exchange in nanomagnetic systems (cf. Eq. (3.23)) involve complex quantum mechanical interactions as discussed. Nevertheless, this metaphor serve as a fun and memorable way to introduce fundamental principles of IEC.

C) **Chiral Coupling**: Let us use a fun and metaphorical real-time analogy: involving dance to explain chiral-coupling.

  i) **Dance Floor Setup**:
Picture a dance floor where pairs of dancers represent magnetic moments in a chiral nanomagnetic system. Each dancer has a unique spin, alike magnetic orientation in nanomagnetic elements.

  ii) **Choreography as Chirality**:
The dance routine symbolizes chiral nature of nanomagnetic coupling. Chirality, in this analogy, is the specific direction of spin or movement that each dancer adopts.

  iii) **Right and Left-Handed Dancers**:
Label dancers as either right-handed or left-handed to represent chiral preference in nanomagnetic materials. Each pair consists of a right-handed and a left-handed dancer.

  iv) **Partner Interactions**:
The dance partners interact based on their chirality. For example, right-handed dancers might prefer to spin in one direction. Left-handed dancers might have a preference for alternate direction.

  v) **Chiral Dance Patterns**:
Just as chiral nanomagnetic elements exhibit specific spin patterns. Dance partners follow circular dance patterns that influence their movement.

  vi) **Mirror Dance**:
In chiral nanomagnetic coupling, the mirror image of a structure behave differently. Similarly, dancers may mirror each other's movements, but dance outcome might be unique for right-handed and left-handed pairs.

  vii) **External Influence (DJ)**:
Introduce an external influence, such as the DJ's music choices. It represents an external magnetic field affecting the chiral/dance. The dance partners adjust their movements based on this external factor. Similar to the external application of field in chiral systems.

  viii) **Harmony and Coherence**:
When the dance partners with matching chirality come together, they exhibit a harmonious dance routine. This mirrors the coherence observed in chiral-nanomagnetic systems.

While this analogy provides a playful way to grasp (refer to item III in Section 3.2 briefing fundamentals of coupling mechanisms in NML – walking through Chiral-Coupling) concept, it's essential to note that the actual physics and mathematics of chiral nanomagnetic coupling involve intricate quantum mechanical interactions (cf. Eq. (3.24)).

3.6.2 Practical Examples and Applications

This section talks about the Practical Examples and Applications throughout, it briefs the Case Studies Illustrating NML in Section 3.6.2.1, describes the Projects for Hands-on Understanding in Section 3.6.2.2, presents in detail the Encouraging Student Engagement and Exploration in Section 3.6.2.3.

Low-Power Computing:
NML has potential to enable low-power computing exhibiting non-volatile nature. Magnetic elements retain information without continuous power, making them suitable for energy-efficient applications like IoT devices.

Memory Devices:
NML for non volatile magnetic memory devices. This leads to faster and more energy-efficient storage solutions compared to traditional memory technologies.

Magnetic Sensors:
Utilizing sensors using nanomagnetic elements to detect magnetic fields. High sensitivity; applications in medical diagnostics and environmental monitoring.

Reconfigurable Logic Gates:
NML gates can be reconfigured dynamically, allowing flexible, adaptable operations. This is useful where logic requirements change frequently.

3.6.2.1 Case Studies Illustrating Nanomagnetic Logic

Spintronic Devices: Investigate case studies on spintronic devices that use NML. Explore how these devices – integrated into existing electronic circuits – enhancing functionality and reducing power consumption.

Magnetic Tunnel Junctions: Study case examples of MTJs as building blocks for NML circuits. Understand how MTJs manipulate to perform logical operations.

Magnetic Domain Wall Logic: Explore case studies focusing on magnetic domain wall logic, where the movement of magnetic domain walls used for information processing. Understand how these devices can be applied to create energy-efficient logic circuits.

3.6.2.2 Projects for Hands-On Understanding

1. Nanomagnetic Logic Circuit Design:
Measurable Tasks:
Task 1: Design a simple NML circuit on paper or using simulation tools.

Task 2: Specify the magnetic elements and their configurations within the circuit.

Task 3: Simulate the circuit to observe its behavior and logical operations.

Task 4: Measure and record the power consumption and response time of the circuit.

Constraints:

Constraint 1: Use a predefined set of nanomagnetic elements with specified properties. Specify the coercivity, anisotropy, and switching speed of nanomagnetic elements.

Constraint 2: The circuit should perform basic logical operations (AND, OR, NOT). Define the required truth tables for AND, OR, and NOT operations.

Constraint 3: Power consumption should be below a defined threshold. Limit the power consumption to, for example, ⌐1 microwatt during normal operation.

Constraint 4: Response time should meet a specified speed requirement. Require a response time of, for example, less than ⌐100 nanoseconds for logical operations.

2. Magnetic Memory Devices: Measurable Tasks:

Task 1: Build a prototype of a magnetic memory device using nanomagnetic elements.

Task 2: Implement write, read, and erase operations on the memory device.

Task 3: Measure the stability and retention of stored information over time.

Task 4: Record the energy consumption during each operation.

Constraints:

Constraint 1: Use specific nanomagnetic materials suitable for memory applications. Specify the saturation magnetization, anisotropy, and thermal stability of the chosen materials.

Constraint 2: Write, read, and erase operations should be reliable and repeatable. Demand a write/read/erase reliability of at least ⌐99.9%

Constraint 3: Information retention should meet a predefined standard. Set a retention standard of, for example, maintaining data integrity for 10 years.

Constraint 4: Power consumption during operations should be minimized. Limit the average power consumption to, for example, ⌐100 microwatts during operations.

3. Magnetic Sensor Construction:

Measurable Tasks:

Task 1: Construct a magnetic sensor using nanomagnetic elements.

Task 2: Calibrate the sensor to detect and measure magnetic fields of varying strengths.

Task 3: Record sensitivity levels and response times for different magnetic field strengths.

Task 4: Demonstrate the sensor's application in a specific scenario (e.g., detecting magnetic anomalies).

Constraints:

Constraint 1: Use nanomagnetic materials suitable for high sensitivity. Define the magnetic sensitivity range and coercivity of the materials.

Constraint 2: Calibration should result in accurate and repeatable measurements. Specify calibration accuracy, for example, within ∽ 1% of the true magnetic field strength

Constraint 3: Response times should be within an acceptable range. Require a response time of, for example, less than ∽50 milliseconds for detecting magnetic anomalies.

Constraint 4: The sensor should operate within defined magnetic field strength limits. Set limits, for example, between ∽0.1 and ∽10 mT for the operating range.

4. Reconfigurable Logic Demonstrations:

Measurable Tasks:

Task 1: Design a NML circuit with reconfigurable elements.

Task 2: Implement multiple logical configurations and demonstrate their functionality.

Task 3: Measure the time and energy required for reconfiguration.

Task 4: Showcase the adaptability of the circuit to different logical operations.

Constraints:

Constraint 1: The circuit should support easy and quick reconfiguration. Limit the reconfiguration time to, for example, ∽1 microsecond.

Constraint 2: Logical operations should be reliable in each configuration. Specify a reliability standard of ∽99.5% for each logical operation.

Constraint 3: Reconfiguration time and energy consumption should be minimized. Set a maximum allowable energy consumption of, for example, ∽100 nanojoules during reconfiguration.

Constraint 4: The circuit should adapt to at least three different logical operations. Demonstrate adaptability to at least three distinct logical operations (AND, OR, NOT).

3.6.2.3 Encouraging Student Engagement and Exploration

This includes:

A. Nano-storytelling Contest:

Objective: Write creative stories integrating NML concepts.

Details: Craft imaginative stories where NML plays a significant role. This task allows to explore practical applications and implications of nanomagnetic technology in a narrative format.

Implementation: Entries can be in written or multimedia format. Tag with the **#9781394263554**

B. Magnetic Poetry Challenge:
Objective: Create poetry using terms related to NML.
Details: with a set of magnetic words (e.g., spin, magnetism, logic). Arrange them to form poetry or short phrases.
Implementation: Digital platform. Tag with the **#9781394263554**
C. Nano-logic Comic Strip Creation:
Objective: Illustrate NML fundamentals through a comic strip.
Details: Create comic strips featuring characters as nanomagnetic elements, logical operations, and outcomes. This visual storytelling approach enhances comprehension and retention.
Implementation: Graphic design tools or even pen and paper to create comic strips, which can be shared digitally. Tag with the **#9781394263554**

3.6.3 Chapter End Quiz

An online Book Companion Site is available with this fundamental text book in the following link: www.wiley.com/go/sivasubramani/nanoscalecomputing1
For Further Reading:
Alawein et al. (2019), Dadjouyan et al. (2020), Sayedsalehi and Azadi Motlagh (2020), Zhou et al. (2021), Favaro et al. (2023), Vacca et al. (2014a), Mishra et al. (2023), Gonelli et al. (2018), Shuty et al. (2018), Amarú et al. (2014), Machado et al., Gu et al. (2015), Vacca et al. (2014b), Eichwald (2016), Vacca et al. (2012), Turvani et al. (2018), Paul et al. (2023), and Coughlin (2018).

References

Meshal Alawein, Selma Amara, and Hossein Fariborzi. Multistate nanomagnetic logic using equilateral permalloy triangles. *IEEE Magnetics Letters*, 10:1–5, 2019.
Luca Amarú, Pierre-Emmanuel Gaillardon, and Giovanni De Micheli. Majority-inverter graph: a novel data-structure and algorithms for efficient logic optimization. In *Proceedings of the 51st Annual Design Automation Conference*, pages 1–6, 2014.
Thomas M. Coughlin. *Digital Storage in Consumer Electronics*. Engineering, Engineering (R0). Springer Cham, 2 edition, 2018. ISBN 978-3-319-69906-6. doi: 10.1007/978-3-319-69907-3.
Ali Akbar Dadjouyan, Samira Sayedsalehi, Reza Faghih Mirzaee, and Somayyeh Jafarali Jassbi. Design and evaluation of clocked nanomagnetic logic conservative fredkin gate. *Journal of Computational Electronics*, 19:396–406, 2020.
Irina Eichwald. *Perpendicular nanomagnetic logic: three-dimensional devices for non-volatile field-coupled computing*. PhD thesis, Technische Universität München, 2016.

Diego Favaro, Luca Gnoli, Valentin Ahrens et al. Enabling logic computation between ta/cofeb/mgo nanomagnets. *IEEE Transactions on Magnetics*, 2023.

Marco Gonelli, Samuele Fin, Giovanni Carlotti et al. Robustness of majority gates based on nanomagnet logic. *Journal of Magnetism and Magnetic Materials*, 460:432–437, 2018.

Zheng Gu, Mark E.Nowakowski, David B. Carlton et al. Sub-nanosecond signal propagation in anisotropy-engineered nanomagnetic logic chains. *Nature communications*, 6(1):6466, 2015.

Felipe L. Machado, Vinicius N. Possani, Augusto S. Neutzling et al. From and-inverter graphs to majority-inverter graphs.

Pinkesh Kumar Mishra, Meenakshi Sravani, M.V.V. Narayana, and Swapnil Bhuktare. Acoustically assisted energy efficient field free spin orbit torque switching of out of plane nanomagnet. *Journal of Applied Physics*, 133(13), 2023.

Ambarish Paul, Mitradip Bhattacharjee, and Ravinder Dahiya. *Solid-State Sensors*. Wiley, 2023. ISBN 9781119473077. Published on 6 October 2023.

Samira Sayedsalehi and Zeinab Azadi Motlagh. Characterisation of a perpendicular nanomagnetic cell and design of reversible xor gates based on perpendicular nanomagnetic cells. *IET Circuits, Devices & Systems*, 14(1): 17–24, 2020.

Anatolij M Shuty, Svetlana V Eliseeva, and Dmitrij I Sementsov. Dynamics of the magnetic nanoparticles lattice in an external magnetic field. *Journal of Magnetism and Magnetic Materials*, 464:76–90, 2018.

Giovanna Turvani, Laura D'Alessandro, and Marco Vacca. Physical simulations of high speed and low power nanomagnet logic circuits. *Journal of Low Power Electronics and Applications*, 8(4):37, 2018.

Marco Vacca, Mariagrazia Graziano, and Maurizio Zamboni. Nanomagnetic logic microprocessor: hierarchical power model. *IEEE Transactions on very large scale integration (VLSI) systems*, 21(8):1410–1420, 2012.

Marco Vacca, Mariagrazia Graziano, Luca Di Crescenzo,et al. Magnetoelastic clock system for nanomagnet logic. *IEEE Transactions on nanotechnology* 13(5):963–973, 2014a.

Marco Vacca, Mariagrazia Graziano, Juanchi Wang et al. Nanomagnet logic: an architectural level overview. *Field-Coupled Nanocomputing: Paradigms, Progress, and Perspectives*, pages 223–256, 2014b.

Peng Zhou, Luca Gnoli, Mustafa M. Sadriwala et al. Multilayer nanomagnet threshold logic. *IEEE Transactions on Electron Devices*, 68(4):1944–1949, 2021.

4

Nanomagnetic Logic Architectures

Specific, Measurable, Achievable, Relevant, and Time-bound (SMART) learning objectives and goals. After reading this chapter you should be able to:

- Grasp the architectural designs in nanomagnetic logic (NML):
  Summarize various architectural paradigms in nanomagnetic logic: parallel processing, pipelined architectures, reconfigurable designs, and hybrid architectures. This is essential for gaining a comprehensive understanding of NML's architectural landscape.
- Analyze the computational throughput of parallel processing:
  Examine the principles of parallel processing in nanomagnetic logic and its impact on computational speed. This is critical for understanding how parallelism enhances performance in NML systems.
- Explore strategies for efficient data flow in pipelined architectures:
  Investigate strategies in pipelined architectures for efficient data flow in nanomagnetic logic. This is crucial for comprehending optimization techniques used to enhance computational efficiency.
- Understand the aplications and limitations of reconfigurable designs:
  Identify real-world applications and limitations of reconfigurable designs in nanomagnetic logic. This is integral for recognizing the adaptability and constraints of NML in dynamic environments.
- Summarize key concepts and achievements in nanomagnetic logic architectures:
  Summarize key concepts in this chapter, emphasizing achievements and implications of different NML architectures. This is critical for consolidating knowledge and appreciating advancements in NML.

This chapter details about the overview of nanomagnetic logic architectures in Section 4.1, briefs the major nanomagnetic logic architectures in Section 4.2,

Nanoscale Computing: The Journey Beyond CMOS with Nanomagnetic Logic,
First Edition. Santhosh Sivasubramani.
© 2025 The Institute of Electrical and Electronics Engineers, Inc. Published 2025 by John Wiley & Sons, Inc.
Companion website: www.wiley.com/go/sivasubramani/nanoscalecomputing1

talks about the fundamentals of NML architecture in Section 4.3, describes the parallel and pipelined architectures in Section 4.4, presents in detail the reconfigurable nanomagnetic architectures in Section 4.5, and elaborates the conclusion in Section 4.6.

This chapter serves as an fundamental understanding of nanomagnetic logic (NML) developments from 2009 to 2023. NML uses nanoscale magnets for efficient computation. It includes achievements in architectures, circuit designs, and material/fabrication advancements. In 2023, digital designs were effectively simulated using NML, showcasing its potential for low-power, high-density computation. The exploration of Bennett Clocking aimed to enhance power consumption efficiency. L-BANCS introduces a multi-phase tile design by integrating CMOS technology. NML for edge inference utilizes Tsetlin Machine, demonstrating high learning accuracy within a reasonable timeframe. Advancements in voltage-controlled nanoscale magnetic devices, edge inference, and quantum computing are observed.

Earlier breakthroughs include nonvolatile matrix multipliers and straintronic NML, with a focus on in-plane wire crossing structures. The development of NML Standard Cell Library and exploring sensitivity of spin–orbit torque (SOT) perpendicular NML to process variation are significant aspects. Research extend to shape engineering for custom NML circuits, stochastic aspects in performance, and reliability-aware design. The chapter covers studies on power dissipation, energy-efficient design, and temperature-dependent reliability analysis of NML circuits.

In addition to a chronological presentation, discussion on principles and trends in quantum nano-electronics plays a key role. It explores the transmission of bits using non uniform states in nanomagnets and explore MagCAD for designing 3D magnetic circuits. In 2022, a nonvolatile all-spin non-binary matrix multiplier is proposed as an efficient hardware accelerator for machine learning. Straintronic nanomagnetic logic employed self-biased dipole-coupled elliptical nanomagnets, outlining thermally stable, self-biased structures. An NML in-plane wire crossing structure presents reinforcing NML's potential for low-power devices.

In 2021, the methodology for an interlayer exchange-coupled nanomagnetic multiplier architecture is explored. Neuromorphic architectures with low-energy barrier nanomagnetic devices were analyzed, and a tempered and realistic vision of Nanomagnetic Boolean Logic was introduced. Shape engineering for custom NML circuits emphasis, along with the introduction of NMLSim 2.0 CAD and simulation tools. FUNCODE was introduced for effective device-to-system analysis. Field-coupled nanocomputing explored devices, circuits, and architectures, offering a comprehensive study of power dissipation in NML (Table 4.1).

Experimental demonstrations included a four-terminal magnetic majority gate. A SPICE-compatible compact model for NML circuits. Energy-efficient majority gate design and temperature-dependent reliability analysis of NML

Table 4.1 Nanomagnetic Logic Review

Year	Observation/Milestone Achievement
2024	Nanomagnetic Logic Simulation of Digital Design. Nanomagnetic Logic (NML) is one of the most promising technologies at present (doi: 10.1109/TETC.2024.3434723)
2023	Exploring Nanomagnetic Logic with Bennett Clocking. Improving power consumption in integrated circuits can have significant benefits (doi: 10.1109/SBCCI60457.2023.10261955)
2023	L-BANCS: A Multi-Phase Tile Design for Nanomagnetic Logic. The CMOS (Complementary Metal Oxide Semiconductor) technology is the industry standard (doi: 10.1109/ISVLSI59464.2023.10238640)
2023	Nano-Magnetic Logic Based Architecture for Edge Inference using Tsetlin Machine. High learning accuracy with less time, low energy (doi: 10.1109/NEWCAS57931.2023.10198204)
2023	Voltage Controlled Nanoscale Magnetic Devices for Non-Volatile Memory, Scalable Quantum Computing and Giant magnetoresistance (GMR) (doi: 10.1109/ECTC51909.2023.00308)
2023	A Deep Dive into the Computational Fidelity of High-Variability Low Energy Barrier Magnet Technology for Accelerating Optimization and Bayesian Problems (doi: 10.1109/LMAG.2023.3274051)
2023	Enabling Logic Computation Between Ta/CoFeB/MgO Nanomagnets. Dipolar coupled magnets proved to have the potential to be capable of logic computation (doi: 10.1109/TMAG.2023.3255306)
2023	Generation of Microwaves with Tunable Frequencies in Ultracompact Magnon Microwave Antenna via Phonon-Magnon-Photon Coupling. (doi: 10.1109/TED.2022.3221026)
2023	A Novel Approach to Design Multiplexer Using Magnetic Quantum-Dot Cellular Automata. MQCA is the key technology (doi: 10.1109/LES.2022.3207193)
2022	A Nonvolatile All-Spin Nonbinary Matrix Multiplier: An Efficient Hardware Accelerator for Machine Learning (doi: 10.1109/TED.2022.3214167)
2022	Straintronic Nanomagnetic Logic Using Self-Biased Dipole Coupled Elliptical Nanomagnets, thermally stable and self-biased design (doi: 10.1109/TMAG.2022.3199589)
2022	An NML In-plane Wire Crossing Structure. NML helps to build low-power devices at room temperature (doi: 10.1109/LASCAS53948.2022.9789062)
2021	Interlayer Exchange Coupled Based Nanomagnetic Multiplier Architecture Design Methodology. IEC is thermally stable - high temperatures (doi: 10.1109/TNANO.2021.3115936)
2021	NMLib: A Nanomagnetic Logic Standard Cell Library (doi: 10.1109/ISCAS51556.2021.9401107)

Table 4.1 (Continued)

Year	Observation/Milestone Achievement
2021	Process Variation Sensitivity of Spin-Orbit Torque Perpendicular Nanomagnets in DBNs. Neuromorphic architectures with low energy barrier nanomagnetic devices (doi: 10.1109/TMAG.2021.3075391)
2021	Shape Engineering for Custom Nanomagnetic Logic Circuits in NMLSim 2.0. NMLSim 2.0 a CAD simulation tool for design (doi: 10.1109/MDAT.2020.3043381)
2021	FUNCODE: Effective Device-to-System Analysis of Field-Coupled Nanocomputing Circuit Designs. Based on different switching mechanisms (doi: 10.1109/TCAD.2020.3001389)
2020	Geometry-Based Optimization of In-Plane Nanomagnetic Majority Circuit. Nanomagnetic circuits are characterized - propagate information - dipole coupling (doi: 10.1109/IEEECONF35879.2020.9329638)
2020	Investigating the Impact of Device Variability and Noise on Nanomagnetic Logic Circuits. Effect of device variability and noise (doi: 10.1109/SBCCI50935.2020.9189930)
2020	Optimization of Nanomagnetic Logic Circuits for Area and Power: A Design Perspective (doi: 10.1109/MWSCAS48704.2020.9184556)
2020	Nanomagnetic implementation can scale to achieve petaflips per second with millions of neurons (doi: 10.1109/ACCESS.2020.3018682)
2020	Global clocking is used to control a nanomagnetic logic device (NMLD) (doi: 10.1109/ACCESS.2020.2989930)
2019	Magnetic dots were arranged to exhibit carry ripple based full adder circuits. One bit and two bit full adder circuit was designed, transient energy profile is used to propose footprint, delay and energy models for ML based higher order adder subsystems (doi: 10.1109/NANO46743.2019.8993892)
2019	Power-Performance Trade-offs in Nanomagnetic Logic Circuits. Delay approximation for nanomagnetic logic based combinatorial circuits (doi: 10.1109/NEMS.2019.8915607)
2019	Field-Coupled Nanocomputing: Devices, Circuits, and Architectures. Beyond CMOS, novel device concepts (doi: 10.1145/3312661)
2019	Developed Ropper, a technology-independent framework for placement and routing challenges (doi: 10.1109/SBCCI.2019.8862291)
2019	NMLSim 2.0 enhances simulation for Nanomagnetic Logic with improved efficiency (doi: 10.1109/SBCCI.2019.8862290)
2019	Introduced an automatic routing method for nanomagnetic logic circuit integration (doi: 10.1109/SBCCI.2019.8862310)
2019	Proposed 3-D coprocessor design utilizing perpendicular nanomagnet logic technology (doi: 10.1109/TVLSI.2019.2905686)

Table 4.1 (Continued)

Year	Observation/Milestone Achievement
2019	Investigated multistate logic using equilateral permalloy triangles in nanomagnetic circuits (doi: 10.1109/LMAG.2019.2899819)
2019	Developed voltage-induced stress scheme for efficient dipole-coupled nanomagnetic logic (doi: 10.1109/LED.2018.2889707)
2018	Introduced a crossing device for efficient in-plane signal routing in NML circuits (doi: 10.1109/NANO.2018.862628)
2018	Proposed stick diagram representation to aid in NML circuit design methodology (doi: 10.1109/NANO.2018.8626369)
2018	Presented BANCS, a three-phase clocking scheme to improve nanomagnetic logic scalability (doi: 10.1109/SBCCI.2018.8533251)
2018	Developed area-efficient magnetic quantum-dot cellular automata design methodology for adders (doi: 10.1109/TNANO.2018.2874206)
2018	Designed stochastic nanomagnets for probabilistic spin logic applications in computing (doi: 10.1109/LMAG.2018.2860547)
2017	Several approaches to nanomagnetic devices focus on storage-class memory and logic (doi: 10.1002/9783527698509.ch11)
2017	Discusses perpendicular nanomagnetic logic as a beyond-CMOS digital computation method (doi: 10.1109/S3S.2017.8309232)
2017	CAM/TCAM — NML: (Ternary) content addressable memory implemented with nanomagnetic logic (doi: 10.1145/3109984.3110004)
2017	Reviews trends in quantum nanoelectronics and nanomagnetics for beyond-CMOS computing (doi: 10.1109/ESSDERC.2017.8066577)
2017	Investigates digital information transmission using non-uniform states in nanomagnets (doi: 10.1109/RTSI.2017.8065985)
2017	Presents MagCAD, a design tool for 3-D magnetic circuits, enhancing circuit analysis (doi: 10.1109/JXCDC.2017.2756981)
2017	Models spintronic circuits using stochastic dynamics for ultra-low energy operations (doi: 10.1109/INTMAG.2017.8008050)
2017	Describes the synthesis of magnetic nanoparticles and their potential biomedical applications (doi: 10.1109/LMAG.2017.2726505)
2017	Explores wave pipelining for majority-based technologies, improving throughput in nanotech (doi: 10.23919/DATE.2017.7927195)
2017	Demonstrates low-barrier nanomagnets functioning as p-bits for efficient spin logic (doi: 10.1109/LMAG.2017.2685358)
2017	Shows tuning of nanomagnet switching behavior via geometry changes and 3-D simulations (doi: 10.1109/TMAG.2017.2654969)
2016	Presents a 3-D model of nanomagnetic devices to tailor switching fields in pNML (doi: 10.1109/CEFC.2016.7815977)

Table 4.1 (Continued)

Year	Observation/Milestone Achievement
2016	Discusses a monolithic 3D pNML circuit integrating memory and logic on different layers (doi: 10.1109/ICRC.2016.7738700)
2016	Investigates controlled domain wall pinning in ferromagnetic nanowires using nanoparticles (doi: 10.1109/TMAG.2016.2564947)
2016	Designs stochastic computing circuits with nanomagnetic logic for low-energy operations (doi: 10.1109/TNANO.2015.2511072)
2016	Analyzes error-free magnetization reversal in coupled nanomagnets through dipolar coupling (doi: 10.1109/TMAG.2015.2475426)
2015	What makes logic devices different from memory in spintronics? (doi: 10.1002/9781118958254.ch02)
2015	Spintronics offers alternatives to CMOS for logic devices (doi: 10.1002/9781118958254.ch17)
2015	Magnetic tunnel junctions enable spin-based logic and circuits (doi: 10.1002/9781118869239.ch5)
2015	New paradigm of multi-valued logic using wave interference functions (doi: 10.1002/9781118869239.ch10)
2015	Spin-functional MOSFETs combine spintronics with low-power logic systems (doi: 10.1002/9781118869239.ch3)
2015	Spin orbit torque controls magnetization for low power memory (doi: 10.1002/9781118869239.ch6)
2015	Magnonic devices show potential for efficient spin wave logic (doi: 10.1002/9781118869239.ch7)
2015	Straintronics utilize mechanical strain for nanomagnetic operations (doi: 10.1002/9781118869239.ch9)
2015	Electric control of spin polarization in quantum point contacts (doi: 10.1002/9781118869239.ch2)
2015	Overview of spintronic devices and their energy-efficient applications (doi: 10.1002/9781118869239.ch1)
2015	Straintronics enables extremely low energy logic/memory devices (doi: 10.1109/NMDC.2015.7439253)
2015	Programmable ALU in nanomagnetic logic for parallel processing (doi: 10.1109/NMDC.2015.7439269)
2015	Study of coherent vs incoherent magnetization switching mechanisms (doi: 10.1109/NMDC.2015.7439270)
2015	Testable reversible adder circuits for nanomagnetic computing (doi: 10.1109/iNIS.2015.27)
2015	Compact modeling of nanomagnetic logic for benchmarking efficiency (doi: 10.1109/NANO.2015.7388796)

Table 4.1 (Continued)

Year	Observation/Milestone Achievement
2015	Full adder design using single domain out-of-plane nanomagnetic logic (doi: 10.1109/NANO.2015.7388798)
2015	Anomalous properties of sub-10-nm MTJs show potential for energy-efficient applications (doi: 10.1109/E3S.2015.7336785)
2015	Compact model for single nanomagnet using Verilog-A is verified through simulation (doi: 10.1109/ESSDERC.2015.7324722)
2015	Nanomagnet geometry affects reliability and energy dissipation in DC-NML systems (doi: 10.1109/TED.2015.2453118)
2015	Spin Hall effect enables programmable logic gate designs for low power applications (doi: 10.1109/INTMAG.2015.7157254)
2015	Ferromagnetic resonance modes inform nanomagnetic logic element dynamics for information propagation (doi: 10.1109/INTMAG.2015.7156974)
2015	Domain wall profile manipulation allows programmable logic operations in magnetic devices (doi: 10.1109/LMAG.2015.2434801)
2015	Systematic design approach improves 3D nanomagnetic logic layout for functionality (doi: 10.7873/DATE.2015.0673)
2015	Low-power 3D integrated ferromagnetic computing enhances performance and energy efficiency (doi: 10.1109/ULIS.2015.7063788)
2015	Analysis of feedback effects in QCA circuits aids design of complex systems. (doi: 10.1109/TVLSI.2014.2358495)
2014	Leading semiconductor companies are exploring new devices and circuit architectures beyond traditional FETs, including tunneling FETs and nanomagnetic devices, which promise advancements in low-power design and memory-logic integration (doi: 10.1109/SNW.2014.7348525)
2014	Nanomagnetic logic circuits, designed with a VHDL model, offer low-power consumption and retain information without power, demonstrating improved efficiency over CMOS and molecular QCA in combinational circuits (doi: 10.1109/ICICES.2014.7034025)
2014	The reliability of perpendicular nanomagnetic logic is linked to domain wall nucleation time, with experimental results confirming theoretical models and highlighting conditions for reliable computation (doi: 10.1109/NANO.2014.6968096)
2014	Strain-clocked Boolean nanomagnetic logic demonstrates non-volatility and energy-efficiency, with experimental results showing strain-induced magnetization switching in nanomagnets for logic applications (doi: 10.1109/DRC.2014.6872355)
2014	Hybrid spintronics–straintronic nanomagnetic logic uses two-state and four-state magnetostrictive nanomagnets to enhance logic density and enable advanced applications like image recognition and neuromorphic computing (doi: 10.1109/DRC.2014.6872321)

Table 4.1 (Continued)

Year	Observation/Milestone Achievement
2014	Cellular automata designs for out of plane nanomagnetic logic utilize the low-power and compact nature of NML devices, optimizing fabrication-friendly layouts for improved circuit performance (doi: 10.1109/IWCE.2014.6865879)
2014	Nanomagnetic logic gates with multiple weighted inputs reduce device footprint and interconnection complexity, demonstrated through a full adder design using threshold logic gates (doi: 10.1109/TNANO.2014.2342659)
2014	Three-dimensional nanomagnetic logic circuits utilizing perpendicular magnetic anisotropy simplify signal routing and reduce area footprint, as demonstrated in a full adder case study (doi: 10.7873/DATE.2014.132)
2014	Nano magnetic STT-logic partitioning in MRAM cells combines logic and memory functions, achieving significant reductions in energy, delay, and cell count in logic circuits like XOR and majority gates (doi: 10.1109/TVLSI.2012.2236690)
2013	Perpendicular nanomagnetic logic devices, clocked in the MHz regime, use an on-chip inductor for high-frequency operation, with power density estimates provided for scaled computing systems (doi: 10.1109/ESSDERC.2013.6818872)
2013	Large-scale fabrication of nanomagnetic logic devices using nanoimprint lithography is demonstrated as a fast and cost-effective method, verified through various microscopy techniques (doi: 10.1109/NANO.2013.6720988)
2013	Magnetic tunnel junction-based logic circuits, performing multiple logic operations, demonstrate non-volatility and lower power, with experimental results highlighting electrical input/output for integrated logic modules (doi: 10.1109/DRC.2013.6633850)
2013	Thermal noise impacts the reliability of stress-induced switching in multiferroic nanomagnets, with factors like dipole coupling strength and stress levels affecting switching probability and error rates (doi: 10.1109/TNANO.2013.2284777)
2013	A Place&Route engine for nano-magnetic logic circuits enables design at logic and physical levels, demonstrating potential advancements over CMOS in complex combinational circuits (doi: 10.1109/NanoArch.2013.6623045)
2013	The first experimental demonstration of a 1-bit full adder in perpendicular nanomagnetic logic, using majority gates and inverters, validates non volatile, field-coupled logic principles (doi: 10.1109/TMAG.2013.2243704)
2013	Building blocks for two-layer signal crossing in nanomagnetic logic are presented, facilitating the development of more complex and efficient nanomagnetic logic circuits (doi: 10.1109/TMAG.2013.2238898)
2012	Nanomagnetic logic (NML) error reduction techniques based on anisotropy engineering, verified via PEEM (doi: 10.1109/IEDM.2012.6479025)
2012	Theoretical and numerical formalism for the analysis and design of spintronic integrated circuits (doi: 10.1109/TCSI.2012.2206465)

Table 4.1 (Continued)

Year	Observation/Milestone Achievement
2012	Direct measurement of coupling fields between lithographically fabricated nanomagnets for NML applications (doi: 10.1109/TMAG.2012.2202219)
2012	Study of enhanced permeability dielectric samples and their magnetic properties for NML circuits (doi: 10.1109/TMAG.2012.2204236)
2012	A new clocking method for error-free, directed signal transmission in nanomagnetic logic inverter chains (doi: 10.1109/TMAG.2012.2196030)
2012	Majority gate with perpendicular magnetic anisotropy using focused ion beam irradiation for NML (doi: 10.1109/TMAG.2012.2197184)
2012	ToPoliNano: a synthesis and simulation tool for NML circuits, enabling high-level analysis of complex circuits (doi: 10.1109/NANO.2012.6321982)
2012	Hybrid spintronic/straintronics for energy-efficient computing using multiferroic nanomagnets (doi: 10.1109/NANO.2012.6321958)
2012	Addressing the layout constraint problem in NML due to magnetic coupling for high-density low-power circuits (doi: 10.1109/NANO.2012.6322168)
2012	Novel design concept for high-density hybrid CMOS-nanomagnetic circuits using Shannon's expansion theorem (doi: 10.1109/NANO.2012.6322169)
2012	Proposal of CMOS-MTJ integrated architecture for low-power magnetic logic computation using spin transfer torque (doi: 10.1109/TCSI.2012.2185311)
2012	Hybrid straintronics and spintronics for ultra energy-efficient logic and memory (doi: 10.1109/DRC.2012.6257020)
2012	Power reduction in NML clocking through the use of high permeability dielectrics (doi: 10.1109/DRC.2012.6256998)
2012	Balancing stress and dipolar interactions for reliable, low-power switching in multiferroic logic (doi: 10.1109/DRC.2012.6256929)
2012	Design of a scalable systolic pattern matcher for out-of-plane nanomagnetic logic (doi: 10.1109/IWCE.2012.6242837)
2012	Thorough analysis of the majority voter for NML circuits through micromagnetic simulations (doi: 10.1109/TNANO.2012.2207965)
2012	Investigation of defects and errors in nanomagnetic logic circuits with suggested design improvements (doi: 10.1109/TNANO.2012.2196445)
2011	A novel CMOS integrated nanomagnetic logic architecture using magnetic tunnel junctions (MTJs) improves bit controllability and reduces power consumption (doi: 10.1109/NANO.2011.6144605)
2011	A compact model based on experiments with field-coupled nanomagnets for nanomagnetic logic is introduced, highlighting error sources and the role of coupling fields (doi: 10.1109/NANO.2011.6144542)
2011	Evaluation of pipelined signal propagation in nanomagnetic logic using multiferroic single-domain nanomagnets demonstrates ultra-low energy consumption and high clock frequency (doi: 10.1109/NANO.2011.6144546)

Table 4.1 (Continued)

Year	Observation/Milestone Achievement
2011	Spin torque majority gates in nanomagnetic circuits are explored, showing non volatile, reconfigurable logic with low switching energy and noise margin (doi: 10.1109/NANO.2011.6144490)
2011	Analysis of error sources in Co/Pt multilayer-based nanomagnetic logic reveals strategies to reduce computing errors in these NML devices (doi: 10.1109/NANO.2011.6144465)
2011	Development of a nanomagnet full adder circuit with majority gates and internal interconnections highlights strategies to eliminate computational error (doi: 10.1109/NANO.2011.6144445)
2011	Demonstration of directed signal flow in nanomagnetic logic using perpendicular magnetization and homogeneous clocking fields for reliable information propagation (doi: 10.1109/ESSDERC.2011.6044169)
2011	Proposal of all-spin logic (ASL) devices with inherent nonreciprocity for low-power logic implementation, demonstrated through simulations of cascaded ASL units (doi: 10.1109/TMAG.2011.2159106)
2011	Magnetic quantum cellular automata (MQCA) using multilayer structures and spin-transfer torque current-induced clocking are shown to reduce power consumption significantly (doi: 10.1109/JETCAS.2011.2158344)
2011	Experimental demonstration of dipole-coupled nanomagnets operating near thermal equilibrium, achieving energy dissipation close to thermodynamic limits (doi: 10.1109/TNANO.2011.2152851)
2010	Development of a nanomagnetic switch for logic applications. Utilization of spin transfer torque for magnetization switching in a nanomagnetic switch (doi: 10.1109/JPROC.2010.2064150)
2010	Experimental demonstration of a nanomagnetic majority gate. Experimental verification of its functionality (doi: 10.1109/SNW.2010.5562573)
2010	Experimental demonstration of fanout for nanomagnetic logic (doi: 10.1109/TNANO.2010.2060347)
2009	A submicron-sized extraordinary Hall-effect sensor is developed to electrically probe the output states and magnetic properties of field-coupled magnetic logic gates and submicron-scale single-domain dots (doi: 10.1109/ESSDERC.2009.5331551)

circuits proposed. In 2018, magnonic majority gates were designed and analyzed. NML circuits optimized for area and power. Modeling and analysis of SOT nano-oscillators were conducted. Efficient and reliable implementations of nanomagnetic circuits using non-binary SOT-magnetic logic gates proposed, and stochastic aspects in NML performance were explored.

In 2017, power-efficient majority gate design is emphasized, along with discussions on magnetic majority gate design and implementation. Principles and trends in quantum nano-electronics and nanomagnetics were explored, examining the transmission of bits using non-uniform states in nanomagnets. In 2016, perpendicular nanomagnetic logic with an interfaced switching mechanism was proposed. A single-domain nanodot-shaped nanomagnet was suggested for high-performance nonvolatile spin transfer torque magnetic RAM. Logic-in-memory circuits using 3D-integrated nanomagnetic logic were explored, along with investigations into controlled domain wall pinning in permalloy nanowire by nanoparticle stray fields. Stochastic computing circuits using nanomagnetic logic were designed, and a systematic study on the feasibility of error-free switching of two nanomagnets coupled via their dipolar fields was discussed (cf. Figure 4.1).

In 2015, straintronics was introduced for strain-switched multiferroic nanomagnets. The exploration of nanomagnetic logic circuits with nonvolatile gates was investigated. In 2014, reliability-aware design in nanomagnetic logic circuits was considered, along with discussions on the thermal stability and reliability analysis of perpendicular nanomagnetic tunnel junctions. In 2013, all-spin logic circuits for energy-efficient computation were proposed. The magnetic switching characteristics of nanomagnets were analyzed for reliability optimization.

In 2012, dynamic logic circuits with nanomagnetic devices were introduced, along with explorations toward high-performance nonvolatile memory (NVM) based on perpendicular nanomagnetic tunnel junctions. In 2011, a comprehensive study on power dissipation in nanomagnetic logic circuits was conducted. The development of a reliable and low-power nanomagnetic logic circuit was proposed. In 2010, perpendicular nanomagnetic logic with nonvolatile latching

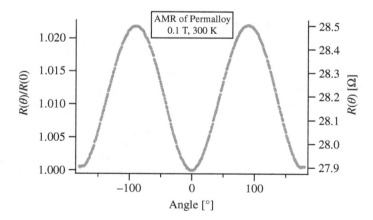

Figure 4.1 This plot shows the resistance as a function of the angle of an applied external field for a thin Permalloy film at room temperate. File: AMR of Permalloy.png https://commons.wikimedia.org/wiki/File:AMR_of_Permalloy.png Source: A13ean/Wikimedia Commons/CC BY-SA 3.0.

was discussed. In 2009, exploration into the computational complexity of high-variability low-energy barrier magnet technology was initiated.

4.1 Overview of Nanomagnetic Logic Architectures

This section talks about the overview of nanomagnetic logic architectures, briefs the introduction to architectural design in Section 4.1.1, and describes the architectural evolution in Section 4.1.2.

Challenges persist in NML implementation. Nanoscale magnetic state control is crucial. Reliable switching without interference remains tough. Techniques like spin transfer torque are explored. Reliability and robustness need careful consideration. Magnetic states are susceptible to external influences. Error-tolerant designs and coding enhance fault tolerance. NML extends beyond classical computing. Quantum computing holds potential with nanomagnetic materials. Quantum-dot cellular automata (QCA) use nanomagnetic spins for quantum information (cf. Figure 4.2). This intersection opens possibilities for computational power.

4.1.1 Introduction to Architectural Design

This section talks about the introduction to architectural design throughout and briefs the importance of architectural considerations in Section 4.1.1.1.

Nanomagnetic logic architectures offer enhanced stability and advanced components. Reduced power consumption and nonvolatility improve future logic systems. Square lattice arrangements ensure reliable logic operations, reaching 94% reliability. 3D nanomagnet logic with vertical nanopillars introduces extra magnetic degrees. Nanomagnetics and spintronics promise a new paradigm for information processing.

Circular magnetoelastic nanomagnets lead to energy-efficient logic, scalable to smaller dimensions. Improved nanomagnet design reduces error rates in nanomagnetic logic circuits. Systolic pattern matcher for nanomagnet logic

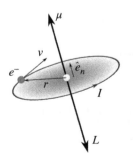

Figure 4.2 Magnetic Moment File: Momento Magnetico 02.svg https://commons.wikimedia.org/ wiki/ File: Momento_Magnetico_02.svg Source: Salviano/Wikimedia Commons/CC BY-SA 4.0.

advances large-scale devices. Biaxial anisotropy in nanomagnetic devices doubles logic density and encodes four-states. Field-coupled nanomagnets enable digital information processing with magnetostatic interaction.

Systematic design of NML circuits generates predictable solutions, considering thermal noise. Anisotropy-engineered chains optimize nanomagnetic logic speed and reliability. Experimental determination of speed and reliability in nanomagnetic logic is crucial. Reconfigurable systolic arrays efficiently manage complexity in nanomagnet logic. NML systolic architectures enable low-power, high-throughput systems for various tasks.

Nanomagnetic logic circuits achieve dissipationless, reversible computation under specific conditions. Advantages of nanomagnetic logic include non-volatility, density, low power, and radiation hardness. Novel clocking mechanism enhances nanomagnetic logic reliability, fabrication, and operating frequency. Spin Hall effect clocking reduces energy dissipation significantly in nanomagnetic logic. Hybrid CMOS-magnetic logic achieves over 95% energy reduction in adders and multipliers.

Thermoplasmonic nanomagnetic logic offers fast, energy-efficient, and reconfigurable in-memory computation. Experimental clocking with strain achieves nanomagnetic Boolean logic with energy efficiency. Magnetic domain walls clocking provides error-free antiferromagnetic ordering in nanomagnetic logic devices. pNML compact model enables fast simulations and easy integration of novel materials in 3-D logic architectures. Focused-electron-beam-induced-deposited cobalt nanopillars are critical for correct configurations in nanomagnetic logic.

Magnetic-electrical interface facilitates nanomagnet logic integration with transistor-based circuits. On-chip clocking is crucial for implementing unidirectional dataflow in nanomagnet logic devices. Majority voter characterization enhances practical knowledge of nanomagnetic logic technology. Multiferroic four-state nanomagnets double logic density, suitable for various information processing applications. Chirality-encoded domain wall logic demonstrates potential for high-speed, ultralow power computational architectures. Tuning magnetic monopole population and mobility aids in designing nanomagnet logic devices.

4.1.1.1 Importance of Architectural Considerations

NML architecture integrates nanomagnetic elements seamlessly. It requires careful design, surpassing silicon norms. A paradigm shift is crucial for execution, including: balancing advantages and practicalities in planning; optimizing non-volatility, lowering energy with consideration; ensuring the design addresses complexities, ensuring reliability; arranging nanomagnetic elements within the framework carefully; considering architectural design comprising arrangement, interaction, and manipulation of magnetic states; navigating challenges and finding equilibrium in exploiting advantages; integrating the promise of non-volatility with nanoscale control reality; shaping the future of magnetic computing with architectural considerations.

Each NML device serves as memory/logic, and design tools enhance understanding of NML circuits. Formal methodology guarantees correct nanomagnet circuit evolution, and NML is extensively studied for its implementation. ToPoliNano tool aids analysis using targeted models, while a reconfigurable systolic array executes various algorithms. The logic-in-memory structure is a step towards innovation, and nanomagnetic logic is scalable to atomic spins, operating at room temperature. Dipolar coupling enables local element construction.

Dipole field coupling drives nanomagnet-based logic functionality, and NML offers potential for low-power consumption. Graphene potential in replacing copper for clocking depends on parameters, as does the error rate in field-coupled systems. Markov random fields are central to nanoscale architectures, enabling nanomagnetism for information processing via quantum cellular automata. A VHDL model evaluates nanomagnetic circuit behavior and power, while mapping CMOS circuits to NML needs efficiency. Layer thickness and size crucial for MEI, and CMOL designs enhance speed, density, and power.

Primitive computational cell uses resistive interconnections, whereas a majority gate offers non volatile logic at room temperature. Nanomagnetic logic is clocked without a magnetic field, and 3D integration and non-volatility are key features. Clocked magnetic logic matches low-power CMOS circuits, and asynchronous logic is a solution for nano-magnet circuits. Nanomagnetic circuits stable against temperature fluctuations, and nanomagnetic devices promising for low-power computing. Power-aware design is crucial for nanoscale MOS logic gates, while nanowire crossbars show promise for post-CMOS architectures.

Magnet-shape-dependent switching aids Boolean logic, ensuring nanomagnet logic enables reduction in hardware requirements. SOTs torques aid nanomagnetic logic at room temperature, yet realistic fabrication mechanisms are essential for logical correctness. Nanomagnets' behavior aligns with Landauer's principle formulations, demonstrating nanomagnetic logic experiments' room temperature functionality. Nanomagnet logic gates are stable against thermal noise, and CMOS devices are approaching nano-transistor regime in recent times. Nanoassemblers are crucial for smart electronic nano structures, and an accumulator-based network tackles challenges in nanoelectronic circuits.

Majority gate allows non volatile logic with global clocking, whereas three-dimensional systems exhibit volumetric efficiency. Minimum-energy state design considers nearest and next-nearest neighbor couplings, while a magnetoelastic clock system promises remarkable improvements. New logic synthesis is needed for emerging nanotechnologies' evaluation, and alteration of magnets affects timing and energy consumption. Nanomagnet logic meets requirements for digital logic, and spin-wave nanochannels promise reconfigurable logic devices.

Nanomagnets offer a low-power alternative to traditional computing, and NML is extensively studied as an implementation.

4.1.2 Architectural Evolution

This section talks about the architectural evolution throughout and briefs the historical progression of nanomagnetic logic architectures in Section 4.1.2.1.

NML's architecture evolves using spintronic devices like magnetic tunnel junctions and spin valves (cf. Figure 4.3). Magnetic tunnel junctions use tunnel magnetoresistance for reading and writing magnetic states, demonstrating nanomagnetic technology.

Precision in controlling nanoscale magnetic states is crucial. Reliable switching without external influences is a challenge. Techniques like spin transfer torque and SOT address these challenges (cf. Figure 4.4). Reliability and robustness demand meticulous attention. Error-correction mechanisms counter

Figure 4.3 A spin valve in the high resistance state.
File: High-resistance.jpg https://commons.wikimedia.org/wiki/File:High-resistance.jpg Source: Stiner905/Wikimedia Commons/Public Domain.

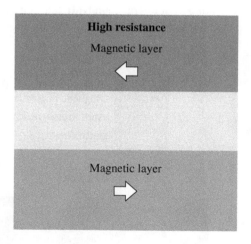

Figure 4.4 Diagram showing the possible spin angular momentum values for 1/2 spin particles.
File: Quantum projection of *S* onto *z* for spin half particles https://commons.wikimedia.org/wiki/File:Quantum_projection_of_S_onto_z_for_spin_half_particles.svg Source: Theresa knott/Wikimedia Commons/CC BY-SA 4.0.

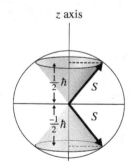

susceptibility to external factors. Researchers focus on error-tolerant designs, enhancing fault tolerance in logic circuits.

NML's architectural evolution extends beyond traditional computing. Quantum computing intersects with nanomagnetic technology. QCA leverages nanomagnetic spins for quantum information processing, promising unprecedented computational power.

4.1.2.1 Historical Progression of Nanomagnetic Logic Architectures

MTJs revolutionized nanomagnetic logic. MTJs, using tunnel magnetoresistance, enabled unparalleled magnetic state read and write precision, enhancing NML reliability (cf. Figure 4.5). Spin valves, controlling spin-polarized electrons, became vital in advanced NML structures, introducing new dimensions to magnetic state manipulation (cf. Figure 4.6). Innovations in error-correction mechanisms improved NML architecture robustness. Advanced coding schemes and error-tolerant designs address challenges tied to magnetic states susceptibility, elevating fault tolerance in NML logic circuits.

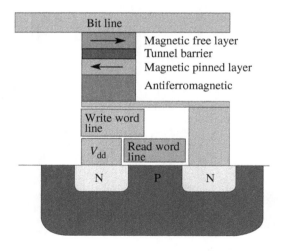

Figure 4.5 A simplified MRAM cell structure. File: MRAM-Cell-Simplified.svg https://commons.wikimedia.org/ wiki/File:MRAM-Cell-Simplified .svg Source: Cyferz/Wikimedia Commons/CC BY-SA 3.0.

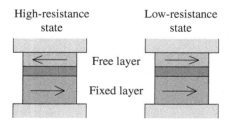

Figure 4.6 Schematic diagram of the high- and low-resistance states in a spin valve. File: Spin valve schematic.svg https://commons.wikimedia.org/wiki/ File: Spin_valve_schematic.svg Source: A13ean/Wikimedia Commons/CC BY-SA 3.0.

4.1.2.2 Trends Shaping Contemporary Architectural Design

Contemporary design trends favor enhanced miniaturization. NML's compatibility with semiconductor processes allows for smaller, (cf. Figure 4.7) energy-efficient devices. This aligns with industry needs for compact form factors and reduced power consumption.

Implementation involves single domain nanomagnets on a plane. Magneto-dynamic interaction achieves logic computation and information propagation. Circuits include wires and logic gates using nanomagnets. Cross-wire blocks enable crossing without interferences, vital for complexity. Clock mechanism uses an externally applied magnetic field. Current through wires generates the necessary magnetic field. Multiphase clock system handles thermal noise limitations. Circuits divided into clock zones for efficient switching. Clock signals with 120° phase difference applied. Experimental focus on NML due to fabrication feasibility. Clock zones prevent interference between switching and stable magnets.

Advanced circuits studied include microprocessors, decoders, and systolic arrays. Finite-element micromagnetic simulators unsuitable for complex circuits. ToPoliNano tool aids in NML circuit design. Tool uses VHDL language, follows top-down approach. Tool's flow chart involves VHDL input, circuit layout, and simulation. ToPoliNano analyzes circuits and performs logical simulation, and fault analysis. Tool offers area and power estimation for NML circuits. ToPoliNano developed by Politecnico di Torino's VLSI group. Over 100k lines of C++ code, compatible with Linux, Windows, and Mac OS X. Tool uses a top-down approach

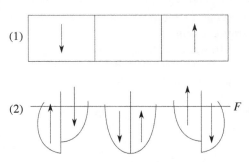

Figure 4.7 Splitting of density of electron states in ferromagnetic and nonmagnetic metals. (1) Three-layered structure with two ferromagnetic layers (arrows indicates magnetization direction) and nonmagnetic layer. (2) Splitting of density of electron states for electrons with different spin direction (arrows indicates direction of spin). F, Fermi level. Only electrons above them take part in conduction. If part of electrons states with some direction of spin above Fermi level is greater, then conductivity for them is larger. Note, that magnetization direction is antiparallel to total spin above Fermi level. File:Electron density in magnets.svg https://commons.wikimedia.org/wiki/File:Electron_density_in_magnets.svg Source: Alex-engraver/Wikimedia Commons/Public Domain.

similar to CMOS technology. Place&Route engine optimizes HDL graph for NML technology. Cross-wire minimization techniques reduce the total number of crossings. Modified-Barycenter algorithm focuses on simplicity and speed. Tool provides logical simulation, fault analysis, and area/power estimation.

4.2 Major Nanomagnetic Logic Architectures

This section talks about the major nanomagnetic logic architectures, it briefs the Combinational Logic Architectures in Section 4.2.1, and describes the sequential logic architectures in Section 4.2.2. Nanomagnetic logic transforms computation. It uses unique properties of nanomagnetic materials for logic and memory. Key components and characteristics include the following:

Spintronics Foundations: Spintronics underlies nanomagnetic logic, leveraging electron spin for information processing.

MTJs: MTJs are crucial. They use tunnel magnetoresistance for precise reading and writing.

Spin-Transfer Torque (STT): STT is vital. Spin-polarized current imparts torque for controlled switching (cf. Figure 4.8).

SOT: SOT adds dimension. It uses spin-orbit coupling for alternative magnetic state manipulation.

Skyrmion-Based Architectures: Skyrmions gain attention. Stable spin textures make them potential for storage and operations.

NML and QCA: QCA merges nanomagnetic spins with quantum computing. Interdisciplinary nature defines nanomagnetic logic architectures (cf. Figure 4.9).

Customization for Specific Applications: Nanomagnetic logic adapts. Tailoring designs ensures versatility for diverse computing needs.

Hybrid Architectures: Integration with semiconductors defines hybrids. It optimizes electronic and magnetic logic strengths for efficiency in in-plane and perpendicular-to-plane (cf. Figure 4.10).

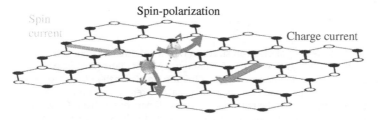

Figure 4.8 Schematic of the inverse spin Hall effect in graphene. File: InverseSpinHall.png https://commons.wikimedia.org/wiki/File:InverseSpinHall.png Source: Alex-engraver/Wikimedia Commons/Public Domain.

Figure 4.9 The top figure (a) represents a Stern Gerlach type experiment, as used in ordinary EPR experiments. The bottom part (b) represents a "space–time" EPR experiment. Details regarding this type of experiment are presented in https://arxiv.org/abs/1812 .11450 and see also [arXiv:1812.11450 [gr-qc]]. File: An Ordinary EPR exp vs a spacetime EPR exp.png https://commons.wikimedia .org/wiki/File:An_Ordinary_ EPR_exp_vs_a_spacetime_ EPR_exp.png Source: Nemirov1/Wikimedia Commons/CC BY-SA 4.0.

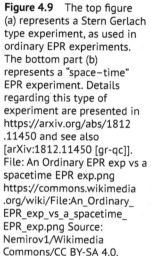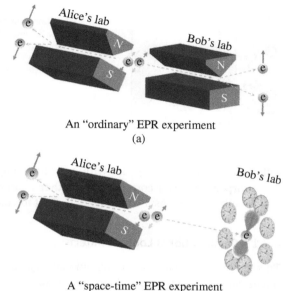

An "ordinary" EPR experiment
(a)

A "space-time" EPR experiment
(b)

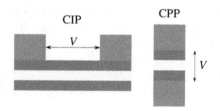

Figure 4.10 Current-in-plane (CIP) and current-perpendicular-to-plane (CPP) connections of spin-valve sensor. *V* is applied voltage between conductors – schematically ferromagnetic and non magnetic layers in spin-valve. File: Spin-valve CIP CPP.svg https://commons.wikimedia.org/wiki/ File: Spin-valve_CIP_CPP.svg Source: Alex-engraver/Wikimedia Commons/CC BY-SA 3.0.

Error-Correction Mechanisms: Robustness is key. Innovations centralize fault tolerance against external disturbances in nanomagnetic logic circuits.

Miniaturization and Compatibility: Compatibility facilitates minia-turization. Nanomagnetic logic aligns with industry standards for compact, energy-efficient devices.

Foundations rely on spintronics. Magnetic tunnel junctions and Bio Sensors with TMR and GMR enable precise manipulation (cf. Figure 4.11). Incorporating STT and SOT introduces dynamic elements. Skyrmion-based architectures offer intriguing possibilities for storage and logic. Exploration extends to quantum-dot

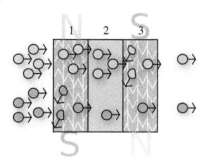

Figure 4.11 Giant magnetoresistance. File: GMR layers.svg https://commons.wikimedia .org/wiki/File:GMR_layers.svg Source: Antimoni/Wikimedia Commons/CC BY-SA 3.0.

cellular automata. Convergence with quantum computing showcases adaptability. Incorporating error-correction strengthens nanomagnetic logic. Innovations ensure reliability against disturbances for fault tolerance.

4.2.1 Combinational Logic Architectures

This section talks about the combinational logic architectures throughout, briefs the designs for logic gates and circuits in Section 4.2.1.1, describes the majority gate-based architectures in Section 4.2.1.2, and presents in detail the innovations in combinational logic in Section 4.2.1.3. Architectures crucial for information processing use nanomagnetic elements for logic gates and circuits. This exploration examines into combinational logic intricacies, covering designs, majority gate-based structures, and recent innovations.

4.2.1.1 Designs for Logic Gates and Circuits

Advanced Nanomagnetic Logic Circuitry: Principles and Implementations include the following:

Fundamental Logic Gate Principles: Utilizing nanomagnetic elements, logic gates perform fundamental operations.

Nanomagnetic XOR Gate Designs: XOR gates showcase nanomagnetic logic performance in exclusive operations.

AND and OR Gate Implementations: Nanomagnetic AND and OR gates redefine logical circuitry principles.

Parallelism in Logic Circuitry: Achieving parallelism, nanomagnetic circuits enhance information processing efficiency.

Reliability and Error Mitigation: Nanomagnetic logic ensures reliability, integrating error-mitigating techniques.

Design Considerations for Nanomagnetic Circuits: Nanomagnetic circuit designs prioritize efficiency, adaptability, and fault tolerance.

Efficient Signal Processing with Nanomagnetic Logic: Signal processing benefits from efficient nanomagnetic logic circuit architectures.

Nanomagnetic Memory Integration: Combinational logic seamlessly integrates with nanomagnetic memory architectures.

In addition to the principles above, **implementations includes the following:** Bennett proposed a nanomagnet-based computation system. Universal logic gate was experimentally demonstrated with nanomagnets. Nanomagnets interact via magnetostatic dipole fields for logic. Nanomagnets patterned into elliptical or rectangular shapes for uniformity. Logic performed by coercing magnetizations along the hard axis. External magnetic field acts as a clock for alignment. Dipole coupling mediates cascade of magnetizations predictably. Cascade success relies on hard axis magnetization stability. Biaxial anisotropy enhances nanomagnet hard axis stability. Improved universal logic gate addresses certain conditions. Uniaxial anisotropy insufficient against perturbing effects like fluctuations. Shape-induced instabilities mitigated by introducing inherent stability. Novel method reduces effects from perturbative instabilities. Elongated nanomagnets have inherent uniaxial anisotropy term. Introducing biaxial anisotropy term enhances stability along the short axis. 3-input majority logic gate fails under asynchronous inputs. Gates enable the formation of complex combinatorial logic.

4.2.1.2 Majority Gate-Based Architectures

The combinational logic architecture utilizes nanomagnets, specifically pNML devices. The architecture involves manipulating magnet properties through FIB irradiation or adjusting multilayer material composition. Anisotropy enables controlled switching along a defined edge for information propagation. Out-of-plane devices are clocked uniformly with an external magnetic field. In pNML, a single, global, and homogeneous clock signal is used. Architectural solutions explore implementing a Full-Adder, considering power, area, and timing advantages. Innovative solutions include, fixed-size nanomagnets to overcome limitations in elongated ones. The model assumes equal influence between neighboring nanomagnets and shape independence. Different implementations optimize area and throughput, balancing latency. The modified Full-Adder version to perform multiplier exhibits increased throughput but sacrifices area. Studies confirm correct behavior with increased latency for improved throughput (cf. Figure 4.12).

Robustness challenges and solutions in NML overview: NML combinational logic architecture faces challenges as follows: Logic locking challenges and potential attacks. Algorithmic attacks on NML. Side-channel attacks. NML vulnerabilities. Electrical imaging threats to NML. Material delayering and magnetic imaging attacks. Solution to these problems include the following: strain-protected NML proposal for security. Tamper-proof NML with polymorphism. Protection against delayering with strain shield. Hybrid CMOS/NML circuits for improved efficiency. NML islands in hybrid systems for security. Inputs to and outputs from NML islands. Timing considerations in hybrid CMOS/NML.

Algorithm 4.1 Nanomagnetic Logic Multiplier Design

1: **procedure** DEFINE REQUIREMENTS

2:     Specify the requirements for the nanomagnetic logic multiplier, including:

3:     Performance metrics (speed, power, area)

4:     Input/output specifications (operand size, precision)

5:     Technology constraints (compatibility with nanomagnetic tech)

6: **end procedure**

7: **procedure** LOGIC GATE SELECTION

8:     Choose appropriate nanomagnetic logic gates based on desired functionality and compatibility:

9:     Majority gate, XOR gate, NAND gate, USER DEFINED.

10: **end procedure**

11: **procedure** MULTIPLIER CIRCUIT DESIGN

12:     Arrange nanomagnetic logic gates to perform multiplication efficiently:

13:     Utilize parallel processing, bitwise operations, and optimization techniques.

14:     Implement Wallace Tree Multiplier or Booth Encoding for optimization.

15: **end procedure**

16: **procedure** MEMORY ELEMENT INTEGRATION

17:     Integrate memory elements (registers, flip-flops) for storing intermediate results and operands.

18: **end procedure**

19: **procedure** INTERCONNECT DESIGN

20:     Design interconnects to connect components:

21:     Minimize latency, power consumption, and signal integrity issues.

22:     Utilize hierarchical routing techniques for complexity management.

23: **end procedure**

24: **procedure** HIERARCHICAL LAYOUT DESIGN

25:     Utilize CleWin for hierarchical layout design:

26:     Create a hierarchical layout of the multiplier architecture.

27:     Manage routing and congestion efficiently.

28: **end procedure**

29: **procedure** SIMULATION SETUP

30:     Select simulation tools:

31:     MagCAD, Nmag, MATLAB/Simulink, SPICE, OOMMF, MuMax3, Ansys Maxwell, COMSOL Multiphysics.

32:     Set up simulation environments:

33:     Define models for magnetic behavior, logical operations, and interconnect performance.

34: **end procedure**

35: **procedure** MICROMAGNETIC SIMULATION

36:     Perform micromagnetic simulations using OOMMF and MuMax3:

37:     Analyze magnetic behavior of nanomagnetic logic gates and interconnects.

38:     Validate functionality under different magnetic field conditions.

39: **end procedure**

40: **procedure** LOGIC SIMULATION

41:     Conduct logic-level simulations:

42:     Verify correct operation of the multiplier circuit.

43:     Test multiplication operations for various input combinations.

44: **end procedure**

Algorithm 4.1 Nanomagnetic Logic Multiplier Design (Continued)

45: **procedure** CONTINUE ALGORITHM

46:    **procedure** SYSTEM LEVEL SIMULATION

47:       Perform system-level simulations:

48:       Use MATLAB/Simulink or custom environments.

49:       Evaluate overall performance, including speed, power, area, and signal integrity.

50:    **end procedure**

51:    **procedure** POWER PLANNING SIMULATION

52:       Utilize Ansys Maxwell for power planning simulations:

53:       Analyze power distribution and consumption across the multiplier architecture.

54:    **end procedure**

55:    **procedure** PHYSICS EXPLORATION

56:       Use COMSOL Multiphysics for physics exploration:

57:       Explore physics-related aspects such as thermal effects and electromagnetic interactions.

58:    **end procedure**

59:    **procedure** OPTIMIZATION AND ITERATION

60:       Identify optimization opportunities based on simulation results:

61:       Iterate on the design to meet performance goals and constraints.

62:    **end procedure**

63:    **procedure** VALIDATION AND VERIFICATION

64:       Validate the final design through testing and verification procedures:

65:       Ensure compliance with requirements and specifications, including functionality, performance, and reliability.

66:    **end procedure**

67: **end procedure**

Physical security considerations in hybrid systems. Algorithmic security of hybrid CMOS/NML. Levels of algorithmic attacks on NML. Potential for physical and algorithmic security.

NML architecture emergence and majority gate-based architecture are detailed. CMOS faces heat issues with smaller transistors. NML offers solutions to integration and computations. Cowburn et al. introduced NML. NML seen as potential addition to existing CMOS. Imre et al. demonstrated universal majority logic gate. Varga et al. implemented NML-based full adder. Shape anisotropy optimized nano-magnetic adder circuit. Positional anisotropy applied for mis-alignment free model. SP hybrid anisotropy used for 1-bit full adder. NML adder implemented with ferromagnetically coupled fixed input. Enables runtime reconfigurable approximate arithmetic computation.

Need for a reliable nano-magnetic based multiplier (cf. Algorithm 4.1). Dipole-coupled nanomagnets face high error chances. Error rates in dipole coupled nano-magnets are high. Majority gate error rates increase for nanomagnets below 200 nm. Inter-layer exchange coupling (IEC) provides stronger coupling. IEC shows quadrupled coupling strength compared to dipole. IEC demonstrates

Figure 4.12 Micromagnetic simulation showing the functionality of a combinatorial logic circuit with multiple gates. File: Figure 2 https://ieeexplore.ieee.org/document/4634504/ Source: David Carlton et al./With permission of IEEE.

higher thermal stability limits. IEC-based Universal Logic Gate Model proposed. IEC-based 2-bit multiplier model introduced for the first time. Model authentication to be achieved using OOMMF micro-magnetic simulator. Temperature variations studied for IEC-based multiplier architecture. Simulation results show temperature resilience for sub 50 nm. IEC 2-bit multiplier uses 30 nanomagnets. Top layer and base magnet exchange coupled strongly. IEC-based multiplier architecture proven stable. Temperature studies conducted up to Curie temperature. Simulation results indicate stability without bit flip. Stronger coupling in IEC provides robustness. Gilbert damping factor analyzed for multiplier logic operation. Dependability of IEC model compared with dipole coupled. NML offers potential for next-gen computing architectures.

4.2.1.3 Innovations in Combinational Logic

Key innovations includes;

Dynamic Nanomagnetic Circuit Designs: Dynamic designs introduce flexibility in nanomagnetic combinational logic.

Adaptive Logic Circuitry: Adaptive circuits tailor operations to suit dynamic computational requirements.

Nanomagnetic Circuits with Memory Elements: Integrating memory elements enhances nanomagnetic circuits' versatility and functionality.

Crossbar Array Architectures: Crossbar arrays amplify computational capabilities, fostering intricate logic operations.

Cellular automata control information flow through cell interactions. Magnetic quantum-dot cellular automata (MQCA) is energy efficient. MQCA demonstrates logic gates, interconnects, and digital arithmetic circuits. Universal logic gates use MQCA majority logic with ferromagnetic dots. Researchers implement on-chip

MQCA by replacing driver magnets with copper wire and potential studies reveal copper to be replaced by graphene.

A. MQCA-Based Adder Architecture Design Methodology:

Case Study 1: MQCA-based adders designed using shape-engineered nano-structured magnets. Methodology combines shape (S) and positional (P) anisotropy for MQCA. Full-Adder design incorporates slant edged and 45° inclined nanomagnets. Asymmetric shape anisotropy enables standalone inputs; positional anisotropy reduces signal loss. Complementary use of shape and positional anisotropy minimizes nanomagnet energy. Horizontal and vertical layouts provide multi-dimensional scalability for MQCA. OOMMF used for micromagnetic simulations of the designs. Results indicate 28% reduction in nanomagnets, enhancing efficiency. Architecture compared favorably with state-of-the-art MQCA architectures. Slant-edged magnets and 45° alignment maintain highly reliable structures. Architecture design requires only two clock cycles, enhancing reliability.

Test loop design verifies shape and positional hybrid approach advantages. Architecture paves the way for higher-dimensional nanomagnetic computing. Challenges in experimental realization may be addressed with precision tools. Further optimization could focus on circuit-level and materialistic approaches. Design methodology has potential for extension using single molecule magnets. This SP hybrid design facilitates reliable, low-power nanomagnetic computing (cf. DOI: 10.1109/TNANO.2018.2874206).

The binary Full-Adder is constructed based on the logical operations defined by Eqs. (4.1) and (4.2). In these equations, A and B represent the operands, C_i denotes the carry input, while C_0 signifies the carry output, and S represents the sum produced by the full adder

$$C_0 = AB + BC_i + AC_i \tag{4.1}$$

$$S = ABC_i + A'B'C_i + A'BC_i' + AB'C_i' \tag{4.2}$$

B. IEC-Based Multiplier Model Methodology:

Case Study 2: 2-bit IEC nano-magnetic multiplier architecture deploying the OOMMF micro-magnetic simulator for its implementation. The architecture focuses on processing two 2-bit binary numbers, $A(a_1, a_0)$ and $B(b_1, b_0)$, generating a 4-bit output $M(M_3, M_2, M_1, M_0)$. At the core of its operation lies the Majority Logic Block (MG), in the multiplier's logical functionality. The multiplier design uses multiple MGs to execute the requisite logical operations for each bit of the output. IEC-Based Multiplier Model Equations: (cf. Eqs. 4.3–4.9)

1. **Majority Logic Operation:**

$$MG(a, b, c) = ab + bc + ca \tag{4.3}$$

2. **OR Gate Operation:**

$$MG(a, b, 1) = a + b \tag{4.4}$$

3. **AND Gate Operation:**

$$MG(a, b, 0) = ab \tag{4.5}$$

4. **Output M_0 Calculation:**

$$M_0 = a_0 b_0 \tag{4.6}$$

5. **Output M_1 Calculation:**

$$M_1 = a_0 b_1 + a_1 b_0 \tag{4.7}$$

6. **Output M_2 Calculation:**

$$M_2 = a_1 b_1 a_0 b_0 + a_1 b_1 \tag{4.8}$$

7. **Output M_3 Calculation:**

$$M_3 = a_1 a_0 b_1 b_0 + a_0 b_1 + a_1 b_0 \tag{4.9}$$

These expressions encapsulate the intricate logic operations executed by the IEC-based multiplier architecture. The connection of majority logic gates results in the generation of the final 4-bit output, for the input binary numbers A and B. The simulation of the IEC-based multiplier model, by the OOMMF micromagnetic simulator, showcases a resilient and reliable performance. With varied input conditions, exhibiting correct logic functionality without the bit flip. The IEC-based multiplier architecture emerges as a robust solution for nano-magnetic logic operations. Its distinctive utilization of IEC positions emerges to mitigate the challenges confronting conventional dipole-coupled nanomagnets. As research advances in this direction, the prospect of shaping the next-generation computing architecture is high (cf. DOI: 10.1109/TNANO.2021.3115936).

4.2.2 Sequential Logic Architectures

This section talks about the sequential logic architectures throughout, briefs the flip-flop and latch designs in Section 4.2.2.1, describes the clocking, storage in nanomagnetic logic in Section 4.2.2.2.

Nanomagnetic flip-flops (FFs) and latches store and manipulate sequential information. Leveraging nanomagnetic elements, they offer reliable, low-power storage solutions. Nanomagnetic states' non-volatility ensures data retention without constant power supply, ideal for energy-efficient applications.

4.2.2.1 Flip-Flop and Latch Designs

Flip-flops are crucial in digital circuits. Magnetic elements introduce novel design concepts. Controlled switching ensures robust operation, aligning with modern computing's miniaturization trend. FFs significantly impact VLSI power consumption. MTJs enhance FFs with nonvolatility. MTJ switching mechanisms include STT, SOT, and voltage-controlled magnetic anisotropy (VCMA). STT and SOT face energy and latency disadvantages. VCMA-MTJ utilizes VCMA effect for low energy. Double-stage pre-charge sense amplifier (DPCSA) reduces sensing delay. VCMA-MTJ-based multi-bit shared magnetic nonvolatile FF (MNV-FF) designed. MNV-FF achieves high speed and energy efficiency. Magnetic non-volatile register group (MNV-RG) extends MNV-FF capabilities. SMIC 55 nm CMOS design kit used for simulation. Simulation results show improvements in backup and restore. MNV-FF uses MTJs for nonvolatile data backup. STT writing mechanism enhances MTJ efficiency. SOT and VCMA explored as alternatives to STT. VCMA effect enables low-energy MTJ switching. Proposed DPCSA accelerates sensing and reduces dynamic power. Self-adaptive write circuit minimizes redundant writing, lowers power. Four operation modes: backup, stand-by, sleep, and restore. MNV-RG is obtained by multiplexing MNV-FF units. Simulation performed on Cadence/Spectre Platform with 55 nm CMOS. Gate width of transistors plays a crucial role. Functional verification of DPCSA includes pre-charging, sensing, and stand-by modes. Results demonstrate improvements in sensing and dynamic power. VCMA-MTJ model described in Verilog-A language. Architecture aims for high efficiency and low power.

Latches are vital for maintaining state information. Nanomagnetic elements provide adaptability and fault tolerance, creating reliable latch architectures. Integrating nanomagnetic memory elements enhances storage capabilities and ensures resilience against external disturbances.

Clocking strategies, foundational in sequential logic, transform in nanomagnetic logic. Controlled magnetic state manipulation synchronizes operations, enabling precise timing. Synergy between clocking strategies and nanomagnetic elements ensures efficient data flow, reducing power consumption for energy-efficient computing. Implementation of latches using nanomagnetic logic and their role in sequential circuits is detailed with a comprehensive analysis of nanomagnetic latches. This section discusses their design principles, operational characteristics, advantages, limitations, and potential applications in sequential circuit design. As aforestated, nanomagnetic latches are fundamental building blocks in sequential circuits. They are responsible for storing and propagating information. Unlike traditional CMOS-based latches, which rely on electrical signals, nanomagnetic latches exploit the magnetization states of nanomagnets to store binary information.

Design Principles: Nanomagnetic latches consist of interconnected nano-magnets arranged in a specific topology to enable stable information storage. The basic design comprises two main components: a magnetic storage element (e.g. a nanomagnet) and a control element (e.g. a magnetic field or spin current). By manipulating the magnetization state of the storage element, nanomagnetic latches transition between two stable states, representing binary values.

Operational Characteristics: The operation of nanomagnetic latches relies on the interaction between magnetic fields, spin currents, and nanomagnet configurations. Writing data to a nanomagnetic latch involves applying external magnetic fields or spin currents to switch the magnetization orientation of the storage element. Reading the stored data typically entails detecting the resistance or magnetoresistance of the nanomagnetic structure.

Advantages: Nanomagnetic latches offer several advantages over traditional CMOS-based latches, including: (a) Ultra-low power consumption: Nanomagnetic logic operates without significant electrical currents, leading to minimal power dissipation. (b) Non-volatility: Nanomagnetic latches retain their state even in the absence of power, enabling instant-on functionality and power-off memory retention. (c) Scalability: The small size of nanomagnetic components allows for high-density integration and potential for scaling beyond the limits of CMOS technology. (d) Radiation tolerance: Nanomagnetic devices exhibit inherent radiation tolerance, making them suitable for space and high-radiation environments.

Limitations: Despite their potential, nanomagnetic latches face several challenges, including: Fabrication complexity: Fabricating nanomagnetic devices with precise dimensions and orientations requires advanced nanofabrication techniques. Limited operating temperature range: Nanomagnetic devices may exhibit variations in performance at extreme temperatures. Sensitivity to external magnetic fields: Nanomagnetic latches are susceptible to interference from external magnetic fields, requiring shielding or error correction mechanisms. **Applications:** Nanomagnetic latches find applications in various sequential circuits, including FFs. Nanomagnetic flip-flops enable the storage and sequential transfer of binary data in digital systems. Registers: Nanomagnetic registers provide temporary storage for data manipulation and processing in microprocessors and digital signal processors. Counters: Nanomagnetic counters facilitate the generation of sequential binary sequences for applications such as frequency division and timing control. Nanomagnetic latches represent a promising paradigm for realizing energy-efficient and scalable sequential circuits. Their unique combination of low-power consumption, non-volatility, and radiation tolerance makes them attractive for a wide range of applications in digital computing and beyond. Continued research and development efforts are essential to overcome existing challenges and unlock the full potential of nanomagnetic logic in future electronic systems.

Nanomagnetic logic employs magnetic properties of nanoscale materials. Flip-flops store binary data, relying on magnetic orientation. Registers store multiple bits of data. Counters tally and store values. Nanomagnetic logic FFs use MTJs. Each FF consists of two ferromagnetic layers separated by a non-magnetic layer. The magnetization of one layer acts as a reference, while the other layer's magnetization represents the stored data. FF states change when a magnetic field surpasses a threshold. Registers employ arrays of FFs to store data. They allow parallel data storage and retrieval. Counters accumulate and display binary values. They comprise interconnected FFs with specific configurations. Nanomagnetic logic counters operate similarly to conventional ones. They count pulses by changing state sequentially. MTJs facilitate high-density integration. They consume less power than traditional CMOS circuits. Their non-volatility ensures data retention without power. However, they face challenges in scalability and manufacturing consistency. Sequential circuits form the backbone of digital systems. They process data in a sequential manner. Nanomagnetic logic sequential circuits offer potential advantages in low-power applications. Research continues to enhance their performance and reliability.

4.2.2.2 Clocking, Storage in Nanomagnetic Logic

Clocking strategies ensure orderly sequential operations, leveraging nanomagnetic elements for precise timing, synchronization, and energy-efficient circuits. Spin-wave-based clocking, manipulating spin waves, introduces novel timing approaches, contributing to nanomagnetic circuit efficiency (cf. Figure 4.13). Integration of spin-transfer torque and SOT enhances dynamic control over magnetic states for precise sequential logic clocking, boosting nanomagnetic circuit adaptability (cf. Figure 4.14).

Clocking strategies address signal propagation delays, clock skew, and power consumption. Aligning clocking strategies with nanomagnetic properties minimizes energy expenditure, meeting the need for energy-efficient computing.

Moreover, clocking strategies play a crucial role in the efficient operation of architectures utilizing nanomagnetic logic. NML employs magnetic elements to perform logic functions, offering potential advantages in terms of low-power consumption and high integration density. However, effective clocking methodologies are essential to synchronize the operations of various components within NML architectures. One common clocking strategy in NML architectures is the use of a global clock signal. This approach involves distributing a single clock signal throughout the entire circuit, ensuring synchronous operation of all components. Global clocking simplifies the design and synchronization process but may lead to increased power consumption and signal propagation delays due to long interconnects.

Wavefunction for a free spin-1/2 particle

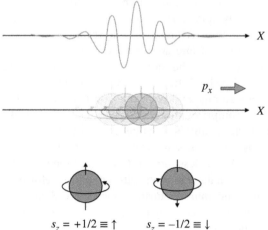

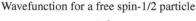

Figure 4.13 Spin-wavefunction. File: Spin-wavefunction.svg https://commons
.wikimedia.org/wiki/ File: Spin-wavefunction.svg Source: Maschen/Wikimedia
Commons/Public Domain.

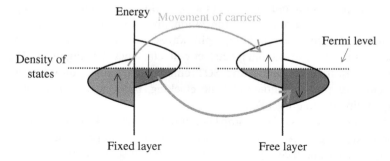

Figure 4.14 A simplified explanation of spin-transfer torque using the Stoner model of
ferromagnetism. Current flowing out of the fixed layer is spin polarized. When it reaches
the free layer the majority spins relax into a lower-energy state, applying a torque to the
free layer in the process. File: Spin Transfer Torque with Stoner model.svg
https://commons.wikimedia.org/wiki/ File: Spin_Transfer_Torque_with_Stoner_model.svg
Source: A13ean/Wikimedia Commons/CC BY-SA 3.0.

Alternatively, local clocking strategies are employed to mitigate the drawbacks
of global clocking. Local clocks are generated and distributed to specific regions
or functional units within the NML architecture. By synchronizing only relevant
components, local clocking reduces power consumption and minimizes signal
propagation delays. However, implementing local clocks necessitates additional

circuitry for clock generation and distribution, potentially increasing the complexity of the overall design.

Moreover, self-timed or asynchronous clocking techniques offer another approach to clocking in NML architectures. Unlike synchronous clocking, self-timed designs do not rely on a global or local clock signal to control operations. Instead, signals propagate through the circuit asynchronously, with each component activating based on its input conditions. Self-timed designs potentially offer improved energy efficiency and robustness against variations in operating conditions. However, designing self-timed circuits requires sophisticated control mechanisms to manage timing constraints and ensure correct operation.

In addition to traditional clocking strategies, hybrid approaches combining synchronous, asynchronous, and semi-synchronous techniques have been proposed for NML architectures. These hybrid strategies aim to leverage the benefits of each approach while mitigating their respective drawbacks. For example, a semi-synchronous design may employ a global clock signal for coarse-grained synchronization while allowing certain components to operate asynchronously for fine-grained control. Furthermore, dynamic clocking strategies adapt clock frequencies and voltages dynamically based on workload and operating conditions. Dynamic clocking techniques optimize power consumption and performance by adjusting clock frequencies and voltages in real time to meet application requirements. However, implementing dynamic clocking in NML architectures may pose challenges due to the inherent limitations of nanomagnetic devices in supporting variable clock frequencies and voltages. Clocking strategies play a critical role in the design and operation of architectures using nanomagnetic logic. Global, local, asynchronous, and hybrid clocking techniques offer different trade-offs in terms of power consumption, performance, and complexity. Selecting the appropriate clocking strategy depends on the specific requirements and constraints of the NML architecture, considering factors such as power efficiency, timing constraints, and scalability. Further research and development are needed to explore novel clocking methodologies tailored to the unique characteristics of nanomagnetic devices.

Memory and Storage Architectures:

Nanomagnetic logic revolutionizes information storage with **non volatile**, reliable, and compact **memory elements**, reshaping nanoscale data storage. Exploring nanomagnetic memory architectures, highlighting MTJs for NVM (cf. Figure 4.15). Integration with traditional semiconductor processes forms hybrid architectures, optimizing electronic and magnetic memory for efficient, scalable storage solutions.

Innovations extend to skyrmion-based memory, leveraging stable spin textures for unique information storage. Controlled generation and manipulation of

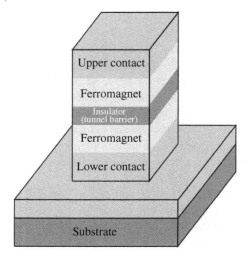

Figure 4.15 Schematic view of a magnetic tunnel junction. File: Magnetic Tunnel Junction.png https://commons.wikimedia.org/wiki/ File: Magnetic_Tunnel_Junction.png Source: Elessar911/Wikimedia Commons/CC BY-SA 3.0.

skyrmions add a new dimension to nanomagnetic memory, potentially transforming non volatile storage. Continuous evolution aligns with data-intensive demands, promising efficient, reliable, and scalable solutions for the digital era. Sequential Logic Architectures usher in a new era of nanoscale sequential operations. Addressing challenges in energy efficiency, adaptability, and data storage. NML-based memory architectures typically consist of arrays of nanomagnetic elements arranged in rows and columns. Each nanomagnet is magnetized in different directions to encode binary information. Reading from and writing to these memory cells is achieved by applying magnetic fields or currents to manipulate the magnetization state of the nanomagnets.

Storage architectures based on NML utilize similar principles but may involve more complex configurations to enable functionalities such as data encryption, error correction, and wear-leveling. These architectures may include additional circuitry and algorithms to manage data access and ensure reliability. One key advantage of NML-based architectures is their potential for high-density integration. The small size of nanomagnetic elements allows for densely packed memory arrays, enabling the storage of large amounts of data in a compact footprint. Another benefit is the non volatile nature of NML-based storage. Unlike traditional volatile memories such as DRAM, which require constant power to retain data, NML-based memories retain their state even when power is removed. This property makes them suitable for applications requiring persistent storage, such as solid-state drives (SSDs) and NVM modules.

NML-based architectures also offer the potential for energy-efficient operation. The manipulation of nanomagnetic elements typically requires lower energy compared to traditional semiconductor-based memories, leading to reduced power consumption and longer battery life in portable devices. Furthermore, NML-based architectures may exhibit inherent resistance to certain types of physical attacks, such as electromagnetic interference and radiation-induced soft errors. The use of magnetic elements for data storage makes NML-based systems more robust in harsh environments and less susceptible to external disturbances.

However, NML-based architectures also face challenges and limitations. One challenge is achieving reliable reading and writing operations at the nanoscale, where factors such as thermal fluctuations and manufacturing variations affect device performance. Overcoming these challenges requires advanced fabrication techniques and error correction mechanisms. Another limitation is the relatively slow speed of magnetic switching compared to electronic switching in conventional semiconductor devices. While NML-based memories offer high-density storage, their access times may be slower, limiting their suitability for certain high-performance applications. Memory and storage architectures based on nanomagnetic logic offer promising opportunities for high-density, energy-efficient data storage and processing. By leveraging the unique properties of nanomagnets, these architectures provide non volatile, resilient, and scalable solutions for a wide range of applications, from consumer electronics to data centers and beyond. However, overcoming technical challenges and optimizing performance will be essential for realizing the full potential of NML-based systems in practical implementations.

4.3 Fundamentals of NML Architecture

This section talks about the fundamentals of NML architecture, it briefs the QCA and its magnetic implementation in Section 4.3.1, describes the pros and cons: - mastering requires to know both aspects of any technology in Section 4.3.2, briefs the quantum aspect of NML – an indepth understanding in Section 4.3.3, describes the bench-marking concepts – food for thought: in Section 4.3.4, presents the why do we need NML? in Section 4.3.5 and details the vision in Section 4.3.6.

4.3.1 QCA and Its Magnetic Implementation

Background includes:

a) First idea was to try to make computers by using arrays of quantum dots, which are patches of semiconductor in which electrons are confined to a space.
b) Rather than using wires, neighboring dots would transmit information by the Coulomb forces, which attract opposite charges and repel like charges.

c) These physical interactions could be used in an appropriately structured array to perform logic operation.

d) It ran into technical limitations, in part because it was hard to control size variations when constructing quantum dots.

e) Realized magnetic dots, which are much more stable, easier to fabricate, and capable of operating at room temperature, might be a good alternative.

f) Prof. Cowburn, University of Cambridge, demonstrated the first magnetic implementation of QCA.

g) In MQCA, the term "Quantum" refers to the quantum-mechanical nature of magnetic exchange interactions and not to the electron-tunneling effects.

4.3.2 Pros and Cons: Mastering Requires to Know Both Aspects of Any Technology

Pros, Cons, and analysis of NML include:

a) They are inherently insensitive to radiation.

b) They switch pretty much indefinitely without degradation.

c) They are nonvolatile, requiring no energy to retain data when they're not switching.

d) At the same time, they are very slow by modern transistor standards, maxing out at about one-hundredth the speed of a traditional transistor.

e) That means nanomagnet logic will likely never reach gigahertz speeds. But with mm waves and THz waves – with varied material considerations researchers show it has a potential (a pointer for you to think upon?)

f) But, the potential energy savings still make it an attractive alternative for the many applications that don't require such speeds.

g) Much of the energy advantage comes in at the circuit level. Though the front-end and back-end CMOS/hybrid options to serve input to the computing block and retrieve output from the computing block are till date costly in terms of the required computing energy, making it almost null energy efficient as claimed (a crucial challenge to think upon?).

h) However, because of the way nanomagnets interact to perform logic operations, it take as few as five magnets to add two 1-bit numbers together.

i) For comparison, it take 20 – 30 transistors to construct a similar adder in silicon.

j) Scalability in building larger circuits is currently a challenge. However, the task of connecting or converting a first output to a second input shows promise for resolution through various interconnection techniques and signal mitigation strategies. For instance, photonic integration could be a promising avenue to explore.

4.3.3 Quanutm Aspect of NML - An Indepth Understanding

This requires a clear understanding on:

a) Electronic QCA is termed "quantum" because it uses quantum mechanical tun-nelling of charge between dots to change logic state; classical electrostatics are used thereafter to propagate the logic state.

b) The quantum mechanical interactions in the MQCA networks are exchange interactions between spins within a single dot in order to form a single giant classical spin.

c) These classical magnetostatic interactions are then used to propagate informa-tion along the chain of dots. Furthermore, this fabrication requirement is one order of magnitude less stringent than that of electronic QCA.

d) Brown's fundamental theorem states that, because of a competition between magnetostatic energy and quantum mechanical exchange energy, magnetic domain formation should be entirely suppressed in very small magnetic particles, causing nano magnets to behave as single giant spins – Single Domain Nanomagnets.

e) Magnetic QCA, commonly referred to as MQCA is based on the interaction between magnetic nanoparticles. The magnetization vector of these nanopar-ticles is analogous to the polarization vector in all other implementations.

f) In MQCA, the term "Quantum" refers to the quantum-mechanical nature of magnetic exchange interactions and not to the electron-tunneling effects. Devices constructed this way could operate at room temperature.

4.3.4 Bench-marking Concepts - Food for Thought:

Single MQCA dot analogous to a transistor

Integration density:

$$5500 \text{ million cm}^{-2} \text{ compared with } 6.6 \text{ million cm}^{-2}$$

For dots as small as 20 nm in diameter, possible integration density:

$$250,000 \text{ million cm}^{-2}$$

Maximum dot dissipation:

$$10^{-17} \text{ J per clock cycle} (10^4 \text{ times less than the power delay product of CMOS})$$

Energy required to keep thermally induced data errors below one per year:

$$\text{At least } 40 \, k_B T$$

Microprocessor based on MQCA dissipates around 1 W.

Switching time for sub micrometer magnetic particles:

Less than 1 ns

Maximum expected across-chip clock frequency of MQCA devices:

Order of 100 MHz

Entire networks are constructed on a single plane, allowing for stacking of multiple planes for three-dimensional hardware.

Priority applications include:

a) Nanomagnetic processors could be used to build wireless networks of autonomous sensors.
b) They could potentially operate on energy they scrounge out of the environment.
c) MQCA design methodology leads to area efficient computing architecture favorable to be deployed in low power portable design applications.
d) Internet of Things (IOT) and Cyber Physical Systems (CPSs) applications require advance signal processing algorithms and architectures to be processed at high-speed and ultra-low power.
e) Make it practical to run neural networks locally on smartphones or even to embed them in household appliances. "Compute on the Go."

4.3.5 Why Do We Need NML?

In today's computing landscape, energy efficiency is a pressing concern. The rise of cloud computing and the widespread use of mobile devices have heightened the demand for energy-efficient solutions. Chip manufacturers are adjusting their focus to prioritize reducing energy consumption in response to environmental and economic considerations. This includes key factors/considerations such as:

a) On a larger, industrial scale, as computing increasingly moves into "the cloud," the electricity demands of the giant cloud data centers are multiplying, collectively taking an increasing share of the country's and world's electrical grid.
b) Lowering energy use is a relatively recent shift in focus for chip manufacturers after decades of emphasis on packing greater numbers of increasingly tiny and faster transistors onto chips.
c) The only way solution is to shrink the amount of energy needed for computing.
d) The biggest challenge in designing computers and, in fact, all our electronics today is reducing their energy consumption.
e) This is critical for mobile devices, which demand powerful processors that can run for a day or more on small, lightweight batteries.
f) This is the right time to think on alternative Computing Paradigm which could potentially lead to Energy Efficient Computing!

g) **Can Nanomagnetic Computing solve this issues and provide a potential replacement or assistance to traditional computing?** NanoMagnetic computing has garnered attention due to several key advantages, lets understand in details:

A. First lets understand Why Computing with Nanomagnets is of importance?

1) **Differentiated Magnetic Bits:** Magnetic bits in nanomagnetic computing are differentiated by direction, offering a novel approach to computing architecture.

2) **Intrinsic Memory Devices:** In this emerging magnetic computing paradigm, logic devices would inherently serve as memory devices, streamlining the architecture.

3) **Low-Energy Consumption:** Nanomagnetic computing demonstrates promising energy efficiency, requiring only 15 millielectron volts of energy to flip a magnetic bit from one state to another.

4) **Nonvolatility:** Nanomagnetic computing devices exhibit nonvolatility, retaining their state even when power is off, thus eliminating the need for booting up and conserving time and energy.

5) **Energy Minimization Nature:** Collections of nanomagnetic disks tend to couple magnetically with one another to minimize the total magnetic energy of the system, contributing to energy efficiency.

B. Why not implement this technology in the chip? (WNITIC?)

Despite its potential, several challenges hinder the on-chip implementation of nanomagnetic computing:

1) **Copper-Based Implementation:** Current chip technology primarily relies on copper, posing a challenge for integrating nanomagnetic computing into existing architectures.

2) **Performance Issues:** Copper's performance limitations, including leakage current and heat generation, present significant obstacles.

3) **Heat Dissipation:** Nanomagnetic computing may exacerbate heat dissipation issues, further complicating on-chip integration.

4) **Need for Material Replacement:** The necessity to replace copper with materials better suited for NanoMagnetic computing poses a critical question for researchers and manufacturers.

B. I. WNITIC: Search for Suitable Materials?

The pursuit of materials suitable for nanomagnetic computing revolves around several key criteria:

1) **Low-Power Dissipation:** Materials must minimize power dissipation to ensure energy efficiency.

2) **High-Current Density:** High-current density capabilities are essential for optimal performance.
3) **Reduced Heat Dissipation:** Materials should exhibit minimal heat dissipation to mitigate thermal challenges.
4) **Scaling and Reliability:** Addressing scaling issues and ensuring reliability are paramount considerations in material selection.
5) **Graphene as a Potential Solution:** Graphene holds promise as a candidate material due to its unique properties, including high conductivity, low power dissipation, and potential for scalability.

B. II. WNITIC: Need for energy efficient circuitry
Graphene-based computing presents an avenue for energy-efficient circuitry:

1) **Utilization of Intrinsic Magnetism:** Graphene-based computing leverages the intrinsic magnetism and enhanced electronic transport properties of graphene, offering a novel approach to computation.
2) **Unrivaled Properties of Graphene Nanoribbons:** Studies highlight the unparalleled intrinsic magnetism and electronic transport capabilities of free-standing graphene nanoribbons, suggesting their potential in energy-efficient computing.
3) **Experimental Investigation:** Experimental investigations utilizing tools such as nano fabrication and DFT-based first principle calculations using Quantum Espresso offer insights into the feasibility and performance of graphene-based computing devices.

The 2017 Magnetism Roadmap emphasizes graphene-based magnetism's potential for advancing spin logic, novel devices, and spintronic circuits, further underscoring the importance of energy-efficient computing solutions. NML incorporating graphene-based computing represent promising avenues for addressing the energy consumption challenges inherent in traditional computing paradigms, offering potential solutions for a more sustainable and efficient future targeted toward rebooting computing.

4.3.6 Vision

The author's vision is as:

a) The need for high-performance computing data centers is rapidly growing to cater to the increasing demand for processing and storing Big Data arising from the Digital World Post-COVID-19.
b) Let us work toward a vision of realizing resource-constrained magnetic chips for ultra low-power portable artificial intelligence (AI) applications.
c) Many modern systems, such as speech and face recognition systems and IoT-enabled devices for remote health monitoring, require highly computationally and energy-intensive neural networks.

d) Hence, it is not practically affordable to perform these computations on portable handheld devices.

e) With these major limitations, all the machine learning algorithms used in these AI applications run on remote systems.

f) To address these issues, highly intensive convolutions should be performed using ultra-low-power, least energy-consuming, and area- and energy-efficient design.

g) This is the motivation to explore the magnetic quantum-based nanomagnetic architecture designs for the next-generation rebooting computing platform.

"Performing AI computing on the edge with approximate nanomagnetic logic deployed on the magnetic ICs is an envisaged attempt towards futuristic computations."

4.4 Parallel and Pipelined Architectures

This section talks about the parallel and pipelined architectures, briefs the parallel processing in nanomagnetic logic in Section 4.4.1, and describes the pipelined architectures in Section 4.4.2.

Let's explore parallel and pipelined architectures in nanomagnetic logic. Emphasize their roles in computing, data handling, and efficiency.

Parallel Architectures in Nanomagnetic Logic:

Utilizing magnetic elements for simultaneous data manipulation redefines parallel processing. Magnetic states control enables multiple simultaneous operations, aligning with modern computing needs. Analysis includes, design considerations: magnetic coupling effects, spin interactions (cf. Figure 4.16), and integrating parallel processing with semiconductor techniques. The result optimizes computational capabilities while ensuring energy efficiency.

Pipelined Architectures in Nanomagnetic Logic:

Nanomagnetic-pipelined architectures enhance data flow and streamline processes. Sequential stages of pipelining, with nanomagnetic elements, create a sophisticated information processing framework. Exploring pipelined architectures involves examining nanomagnetic registers, memory elements, and integrating magnetic states within stages. Controlled data progression minimizes latency, optimizing overall system performance. Nanomagnetic pipelining applies to diverse tasks: signal processing, image recognition, and complex computations. Adaptability to specific computational requirements highlights their versatility.

Technical Advancements and Challenges:

Integration of parallel and pipelined architectures faces signal integrity, magnetic interference, and scalability challenges. Exploration of advanced materials,

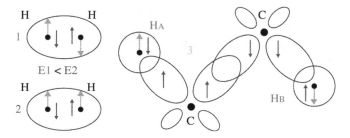

Figure 4.16 Fermi interaction of NMR: Arrows represents nuclear spin ($\pm\frac{1}{2}$), and electron spin ($\pm\frac{1}{2}$). In the context of direct J-coupling in hydrogen gas, the 1H nucleus polarizes the nearby electron to the opposite spin state, leading to the coexistence of antiparallel spins within the same orbital, following the Pauli exclusion principle. The minus $\frac{1}{2}$ electron subsequently polarizes another σ-bond electron to the $+\frac{1}{2}$ state, and the $+\frac{1}{2}$ electron polarizes the 1H nucleus to the minus $\frac{1}{2}$ state. This results in coupled and antiparallel 1H nuclei, with a positive coupling constant. The gyromagnetic ratio ($\gamma > 0$) for 1H nuclei leads to antiparallel magnetic moment vectors (μ). This antiparallel alignment is energetically favorable. The second scenario involves 1H nuclei in a parallel configuration, forming a less stable state with higher energy (E_2) compared to the previous case (E_1). Finally, positive vicinal J-coupling of 1H nuclei via 12C or 13C nuclei is facilitated by parallel electron spins on p-orbitals, following Hund's rules. File: J-coupling Fermi contact mechanism.svg https://commons.wikimedia.org/wiki/ File: J-coupling_Fermi_contact_mechanism.svg Source: Keministi/Wikimedia Commons/CC BY-1.0.

including novel nanomagnetic materials and MTJs, aims to overcome challenges. These materials offer improved reliability, signal-to-noise ratios, and magnetic state control for nanomagnetic architectures' success.

4.4.1 Parallel Processing in Nanomagnetic Logic

This section talks about the parallel processing in nanomagnetic logic throughout and briefs the enhancing computational throughput in Section 4.4.1.1.

Parallel processing in nanomagnetic logic marks a pivotal advancement in computational methodologies. This section examines into the core concepts, implementations, computational throughput enhancements, and application-s/limitations of parallelism in nanomagnetic logic.

Nanomagnetic parallel processing leverages pipelining stages to enhance throughput. Pipelining involves breaking down computational tasks into smaller stages, enabling simultaneous execution of multiple tasks. Each stage performs a specific operation on the input data, passing it to the next stage without waiting for the entire task to complete. This parallelism minimizes idle time, maximizing overall throughput.

At the core of nanomagnetic parallel processing are nanomagnetic elements. These elements, such as nanomagnets or magnetic tunnel junctions, serve as the building blocks for computational units. They possess distinct magnetic states representing binary values (0s and 1s) and can be controlled using external magnetic fields. Nanomagnetic elements enable parallel manipulation and processing of multiple bits of information simultaneously, enhancing computational efficiency.

Integration of magnetic states is crucial for optimizing information flow in nanomagnetic parallel processing systems. By coordinating the magnetic states of nanomagnetic elements across pipeline stages, seamless data propagation is ensured. This integration enables the synchronized execution of computational tasks, minimizing delays and maximizing system throughput.

Nanomagnetic registers play a vital role in storing intermediate results and facilitating data transfer between pipeline stages. These registers utilize nanomagnetic elements to store binary information temporarily. By incorporating nanomagnetic registers into the pipeline architecture, data dependencies are managed efficiently, enabling continuous data processing and minimizing latency.

Memory units based on nanomagnetic technology further enhance the performance of parallel processing systems. These memory units utilize arrays of nanomagnetic elements to store and retrieve data. By leveraging the parallelism inherent in nanomagnetic memory, simultaneous access to multiple memory locations is achieved, boosting overall system throughput.

Integrated states within pipeline stages optimize system performance by coordinating controlled progression. By synchronizing the activation and deactivation of pipeline stages, data movement is regulated, minimizing latency and maximizing computational efficiency. This controlled progression ensures that computational tasks progress smoothly through the pipeline, without bottlenecks or delays.

4.4.1.1 Enhancing Computational Throughput

This section focuses on nanomagnetic parallel processing for enhanced throughput. Pipelining stages, nanomagnetic elements, and magnetic states integration optimize information flow (cf. Figure 4.17). Nanomagnetic registers, memory, and integrated states in pipeline stages are key. Controlled progression minimizes latency, optimizing system performance for diverse computing tasks.

Applications and Limitations:

Nanomagnetic parallelism has limitations, including scalability and heat dissipation. Fabrication complexities require continual exploration and innovation. Nanomagnetic parallelism in neuromorphic computing mimics human brain parallel processing. Controlled magnetic state manipulation creates intricate neural networks for advanced AI applications.

Overall, nanomagnetic parallel processing offers a promising approach to enhancing system throughput in diverse computing tasks. By utilizing pipelining

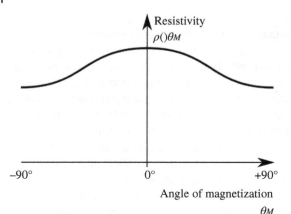

Figure 4.17 Angular dependence of the AMR. File: AMR-angular.svg https:// commons.wikimedia.org/wiki/ File: AMR-angular.svg Source: Zeptomoon/Wikimedia Commons/Public Domain.

stages, integrating nanomagnetic elements, and optimizing magnetic states, efficient information flow is achieved. Nanomagnetic registers, memory, and integrated states within pipeline stages play critical roles in optimizing system performance. Through controlled progression, latency is minimized, ensuring efficient execution of computational tasks.

4.4.2 Pipelined Architectures

This section talks about the pipelined architectures throughout and briefs the introduction to pipelining in Section 4.4.2.1.

4.4.2.1 Introduction to Pipelining

NML pipelining shifts computational design, enabling sequential instruction processing. Data flows continuously through various stages. This section technically introduces foundational pipelining concepts in NML. Controlled data progression uses nanomagnetic registers and memory elements (cf. Figure 4.18). Magnetic states ensure efficient information transfer between stages. Pipelining principles in NML enhance computational throughput.

Foundational pipelining concepts and architecture in nanomagnetic logic revolve around controlled data progression utilizing nanomagnetic registers and memory elements. Pipelining is a fundamental technique employed in nanomagnetic logic architectures to enhance computational throughput and efficiency. At its core, pipelining involves breaking down the execution of instructions into discrete stages, allowing multiple instructions to be processed concurrently. In nanomagnetic logic, this is achieved through the utilization of nanomagnetic registers and memory elements, which enable the storage and manipulation of data at the nanoscale.

Figure 4.18 Usage of spin valve in magnetic random access memory (MRAM). 1. Spin-valve sensor as memory cell (arrows indicates ferromagnetic layers occurrence). 2. Row line. 3. Column line. Ellipses with arrows indicates magnetic field lines when current flow in row and column lines. File: Spin-valve in MRAM.svg https://commons.wikimedia.org/ wiki/File:Spin-valve_in_MRAM.svg Source: Alex-engraver/Wikimedia Commons/CC BY-SA 3.0.

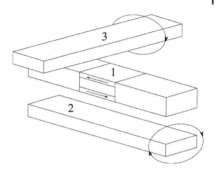

The pipeline architecture typically consists of several stages, each performing a specific operation on the data. These stages include instruction fetch, decode, execute, memory access, and write-back. By dividing the execution process into these stages, the overall throughput of the system is increased as multiple instructions can be in various stages of execution simultaneously.

Nanomagnetic registers play a crucial role in the pipelining process by providing storage elements for intermediate data values and instructions. These registers are implemented using nanomagnetic materials, which exhibit unique properties such as non volatility and low power consumption, making them suitable for use in low-power computing systems. Memory elements based on nanomagnetic technology are utilized for storing larger data sets and program instructions. These memory elements offer advantages such as high density, fast access times, and low-power consumption compared to traditional memory technologies.

Controlled data progression is achieved by synchronizing the movement of data through the pipeline stages using clock signals and control logic. This ensures that each stage of the pipeline operates in a coordinated manner, allowing for efficient execution of instructions. In nanomagnetic logic architectures, careful design considerations are required to minimize data hazards and ensure proper synchronization between pipeline stages. Data hazards occur when the result of one instruction depends on the outcome of a previous instruction that has not yet completed its execution. Techniques such as forwarding and stalling are employed to mitigate these hazards and maintain the correct order of execution.

Overall, foundational pipelining concepts and architecture in nanomagnetic logic are centered around leveraging nanomagnetic registers and memory elements to enable controlled data progression through the pipeline stages. By exploiting the unique properties of nanomagnetic materials, such as non-volatility and low power consumption, nanomagnetic logic architectures offer promising opportunities for high-performance and energy-efficient computing systems.

Achieving Efficient Data Flow

Nanomagnetic logic pipelines demand efficient data flow. Seamless data transfer through magnetic state manipulation and optimized pipeline design holds prominence. Technical advancements and challenges include to address: scalability, heat dissipation, and fabrication complexities through materials science and innovative design. Continuous pursuit of technical advancements keeps NML pipelines cutting-edge.

Magnetic state manipulation facilitates efficient data transfer by utilizing the inherent properties of magnetic materials. This approach minimizes energy consumption and enhances the speed of data transmission.

Optimized pipeline design ensures streamlined data flow within the pipelined architecture for nanomagnetic logic PANML framework. By carefully orchestrating the sequence of operations, potential bottlenecks are mitigated, leading to improved overall performance. Through meticulous planning and implementation, latency is minimized, and throughput is maximized.

The synergy between magnetic state manipulation and optimized pipeline design enables seamless data transfer in PANML systems. Leveraging magnetic properties for data manipulation, coupled with a well-designed pipeline architecture, results in efficient and reliable data transmission. This approach holds promise for advancing the field of nanomagnetic logic by enhancing its practical applicability and scalability.

4.5 Reconfigurable Nanomagnetic Architectures

This section talks about the reconfigurable nanomagnetic architectures, briefs the dynamic reconfiguration concepts in Section 4.5.1, and describes the case studies of reconfigurable architectures in Section 4.5.2.

This section explores reconfigurable NML, a computational paradigm shift. It covers architectures, applications, challenges, and evolution in nanomagnetic logic. Challenges involve signal integrity, interference reduction, and reconfiguration scalability. Advanced materials and hybrid integration offer solutions.

4.5.1 Dynamic Reconfiguration Concepts

This section talks about the dynamic reconfiguration concepts throughout and describes the applications in dynamic environments in Section 4.5.1.1.

Understanding Dynamic Reconfiguration:

Dynamic reconfiguration adapts a system's structure and functionality on-the-fly. In NML, magnetic states are manipulated for dynamic computational pathway changes.

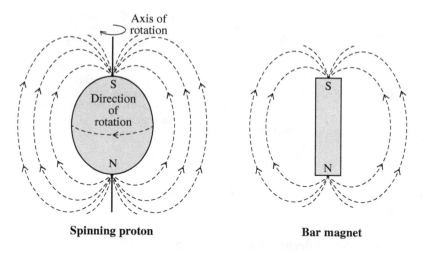

The magnetic field lines (dashed lines) of a spinning rotating proton are comparable to those of a tiny bar magnet.

Figure 4.19 A spinning proton producing a magnetic field which is comparable to a tiny bar magnet's magnetic field. File: Spinning proton magnetic field lines.jpg https:// commons.wikimedia.org/wiki/File:Spinning_proton_magnetic_field_lines.jpg Source: Mgianino/Wikimedia Commons/CC BY-1.0.

Technical Underpinnings:

Technical aspects involve nanomagnetic materials, spin dynamics, and magnetic fields for state manipulation (cf. Figure 4.19). MTJs play a pivotal role, enabling precise control over magnetic states.

Challenges and Solutions:

Challenges include maintaining signal integrity, minimizing energy consumption, and ensuring reliability. Integrating topological insulators with nanomagnetic elements enhances adaptability. Adaptability in NML transforms computing systems. It refers to the system modifying pathways, configurations, and functionalities based on dynamic environmental conditions. Nanomagnetic elements facilitate swift adjustments, optimizing performance.

4.5.1.1 Applications in Dynamic Environments

In dynamic environments, NML's adaptability applies in diverse domains like IoT and autonomous systems, enhancing real-time decision-making. **Trade-offs and Challenges in Reconfigurability**: While adaptability offers potential, trade-offs include the balance between adaptability and energy efficiency. Challenges involve precise control over magnetic states and maintaining signal integrity during transitions. **Technical Insights:** Achieving adaptability involves a deep

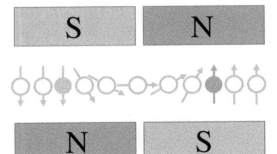

Figure 4.20 Spin orientation in the direction of material magnetization. File: Spintronic poles and electons.png https://commons.wikimedia.org/wiki/File:Spintronic_poles_and_electons.png Source: MykhailoZ/Wikimedia Commons/CC BY-SA 3.0.

understanding of MTJs, spin (cf. Figure 4.20) dynamics, and exploring topological effects like skyrmions.

4.5.2 Case Studies of Reconfigurable Architectures

This section talks about the case studies of reconfigurable architectures throughout and describes the benefits and limitations in specific applications in Section 4.5.2.1.

This part explores practical applications and case studies of reconfigurable architectures in NML. Case studies include adaptive signal processing, autonomous robotics, scientific simulations, adaptive cybersecurity, and dynamic resource allocation in cloud computing.

Technical Insights: Reconfigurable architectures leverage unique properties of nanomagnetic elements, like MTJs. Challenges include efficient error correction mechanisms and balancing adaptability with energy efficiency.

Real-World Examples of Reconfigurable Designs

Real-world examples highlight NML's reconfigurable designs in satellite communications, autonomous exploration, edge computing, cybersecurity, and industrial automation.

Technical Insights: NML's reconfigurability relies on nanomagnetic elements and topological effects like skyrmions. Challenges include error correction mechanisms and achieving a balance between adaptability and energy efficiency.

4.5.2.1 Benefits and Limitations in Specific Applications

Examining benefits and limitations in specific applications reveals insights:

- In satellite communications, NML enhances adaptive signal processing, but challenges include error correction and power consumption.
- In autonomous exploration, NML enables dynamic path planning, but challenges include computational complexity and energy consumption.

- In edge computing, NML optimizes processing efficiency, but challenges include synchronization and adaptation latency.
- In cybersecurity, NML enables dynamic security protocols, but challenges include algorithmic complexity and key management.
- In industrial automation, NML allows task-specific adaptation, but challenges include programming overheads and mechanical constraints.

Technical Insights and Challenges:
Implementing reconfigurable designs showcases NML's technical depth. Challenges include error correction, power consumption, and computational complexity.

4.6 Conclusion

This section talks about the conclusion, briefs the summarizing key nanomagnetic logic architectures in Section 4.6.1, and describes the Chapter End Quiz in Section 4.6.2.

Concluding the exploration of NML necessitates a comprehensive understanding of its architectural intricacies. This chapter examined into various aspects, providing a foundation for the subsequent material on material design considerations.

4.6.1 Summarizing Key Nanomagnetic Logic Architectures

This section talks about the summarizing key nanomagnetic logic architectures throughout and describes the material design considerations in Section 4.6.1.1.

NML architectures include parallel processing, pipelined architectures, reconfigurable designs, and hybrid architectures. Understanding these structures is crucial for harnessing NML's full potential. Summarizing architectural designs reinforces the unique features of each NML paradigm. Parallel processing offers enhanced computational speed, pipelined architectures ensure efficient data flow, reconfigurable designs provide adaptability, and hybrid architectures integrate diverse computing paradigms. Connecting architectural designs to the fundamentals of NML provides a holistic view of NML's evolution.

4.6.1.1 Material Design Considerations
The exploration of NML architectures showcases possibilities. Each design offers unique advantages. It contributes to the collective potential of NML in rebooting computational paradigm. The fundamentals of NML serve as the key for architectural considerations. This connection emphasizes the importance of a holistic

understanding. It also bridge theoretical foundations with practical implementations. Recapping architectures reinforces the idea that NML is adaptable to diverse computing needs. It is crucial in real-world scenarios where requirements changes time to time. The next step involves material considerations. Success lies on careful material design. Analyzing nanoscale material properties becomes essential for overcoming challenges and optimizing performance. Material design considerations form a base in the exploration of NML. The shift toward material science is crucial to address practical challenges in NML implementation. The next chapter examines into the critical role materials play in optimizing NML performance. The architectural journey in NML is a proof to its potential in revolutionizing computing. This multi-inter disciplinary connection among architectures, fundamentals, and materials underscores the need for a comprehensive approach. As we get into the material-centric chapter. The stage is set for a deeper exploration into the details of NML. It lays the groundwork for practical applications and innovations.

4.6.2 Chapter End Quiz

An online Book Companion Site is available with this fundamental text book in the following link: www.wiley.com/go/sivasubramani/nanoscalecomputing1
   For Further Reading:
   Bhoi et al. (2021), Garlando et al. (2018), Csaba et al. (2015), Liu and Kuhn (2015), Turvani et al. (2014, 2017), Yilmaz and Mazumder (2012), Causapruno et al.(2016a, 2016b), Rajaram (2014), Dalbouchi et al. (2023), O'connor et al. (2011), Gartside et al. (2020), Zhou et al. (2021), Adamatzky (2018), Wang et al. (2006), and Strukov and Likharev (2006)

References

Andrew Adamatzky. *Unconventional Computing: A Volume in the Encyclopedia of Complexity and Systems Science*. Springer Publishing Company, Incorporated, 2018.

Bandan Kumar Bhoi, Nirupma Pathak, Santosh Kumar, and Neeraj Kumar Misra. Designing digital circuits using 3D nanomagnetic logic architectures. *Journal of Computational Electronics*, 20(3):1310–1325, 2021.

Giovanni Causapruno, Umberto Garlando, Fabrizio Cairo, et al. A reconfigurable array architecture for NML. In *2016 IEEE Computer Society Annual Symposium on VLSI (ISVLSI)*, pages 99–104. IEEE, 2016a. https://ieeexplore.ieee.org/abstract/document/7560180.

Giovanni Causapruno. *Architectural Solutions for Nanomagnet Logic*. Doctoral Thesis, PORTO @ Research Institutional Repository 2016b. https://hdl.handle.net/11583/2643285.

Gyorgy Csaba, Gary Hirshon Bernstein, Alexei Orlov, et al. 12 nanomagnetic logic: from magnetic ordering to magnetic computing. *CMOS and Beyond: Logic Switches for Terascale Integrated Circuits*, page 301, 2015.

Roukaya Dalbouchi, Chiraz Trabelsi, Majdi Elhajji, and Abdelkrim Zitouni. A model-driven platform for dynamic partially reconfigurable architectures: a case study of a watermarking system. *Micromachines*, 14(2):481, 2023.

Umberto Garlando, Fabrizio Riente, Giovanna Turvani, et al. Architectural exploration of perpendicular nano magnetic logic based circuits. *Integration*, 63:275–282, 2018.

Jack C Gartside, Son Gyo Jung, Seung Yeun Yoo, et al. Current-controlled nanomagnetic writing for reconfigurable magnonic crystals. *Communications Physics*, 3(1):219, 2020.

Tsu-Jae King Liu and Kelin Kuhn. *CMOS and Beyond: Logic Switches for Terascale Integrated Circuits*. Cambridge University Press, 2015.

Ian O'connor, Junchen Liu, Jabeur Kotb, et al. Emerging technologies and nanoscale computing fabrics. In *VLSI-SoC: Technologies for Systems Integration: 17th IFIP WG 10.5/IEEE International Conference on Very Large Scale Integration, VLSI-SoC 2009, Florianópolis, Brazil, October 12–14, 2009, Revised Selected Papers 17*, pages 1–20. Springer, 2011.

Srinath Rajaram. *Multilayer Nanomagnetic Systems for Information Processing*. University of South Florida, 2014.

Dmitri B Strukov and Konstantin K Likharev. A reconfigurable architecture for hybrid CMOS/nanodevice circuits. In *Proceedings of the 2006 ACM/SIGDA 14th International Symposium on Field Programmable Gate Arrays*, pages 131–140, 2006.

Giovanna Turvani, A Tohti, Matteo Bollo, et al. Physical design and testing of nano magnetic architectures. In *2014 9th IEEE International Conference on Design & Technology of Integrated Systems in Nanoscale Era (DTIS)*, pages 1–6. IEEE, 2014.

Giovanna Turvani, Fabrizio Riente, Fabrizio Cairo, et al. Efficient and reliable fault analysis methodology for nanomagnetic circuits. *International Journal of Circuit Theory and Applications*, 45(5):660–680, 2017.

Gang Wang, Wenrui Gong, and Ryan Kastner. On the use of bloom filters for defect maps in nanocomputing. In *Proceedings of the 2006 IEEE/ACM International Conference on Computer-aided Design*, pages 743–746, 2006.

Yalcin Yilmaz and Pinaki Mazumder. Nonvolatile nanopipelining logic using multiferroic single-domain nanomagnets. *IEEE Transactions on Very Large Scale Integration (VLSI) Systems*, 21(7):1181–1188, 2012.

Huaijuan Zhou, Carmen C Mayorga-Martinez, Salvador Pané, et al. Magnetically driven micro and nanorobots. *Chemical Reviews*, 121(8):4999–5041, 2021.

5

Material Design for Nanoscale Computing

Specific, Measurable, Achievable, Relevant, and Time-bound (SMART) learning objectives for: material design for nanoscale computing

- Grasp the foundational role of materials in nanoscale computing.
  Explain the importance of material diversity in nanoscale computing.
- Identify and evaluate the significance of ferromagnetic materials.
  Analyze the properties and applications of antiferromagnetic and ferrimagnetic Materials.
- Classify dielectric and insulating materials in nanoscale computing.
  Differentiate between conductive and semiconductive materials.
- Assess the integration of multiferroic materials in nanomagnetic logic.
  Explore the utilization of spintronics in material design.
- Examine practical examples illustrating material selection principles.
  Engage in a tutorial on the effective design of nanomagnetic materials.
- Summarize key considerations in material design for nanoscale computing.
  Evaluate the implications of various material choices on computational performance.

After completing this chapter, you will possess a comprehensive understanding of the foundational and diverse aspects of material selection for nanoscale computing. You will be able to critically analyze and apply this knowledge to practical examples, demonstrating proficiency in material design principles. This chapter talks about the importance of material selection in nanoscale computing in Section 5.1, briefs the magnetic materials for nanomagnetic logic in Section 5.2, describes the nonmagnetic materials in nanoscale computing in Section 5.3, and presents in detail the multiferroic and spintronic materials in Section 5.4.

Nanoscale Computing: The Journey Beyond CMOS with Nanomagnetic Logic,
First Edition. Santhosh Sivasubramani.
© 2025 The Institute of Electrical and Electronics Engineers, Inc. Published 2025 by John Wiley & Sons, Inc.
Companion website: www.wiley.com/go/sivasubramani/nanoscalecomputing1

5.1   Importance of Material Selection in Nanoscale Computing

This section talks about the importance of material selection in nanoscale computing, briefs the foundational role of materials in Section 5.1.1, and describes the material diversity in nanoscale computing in Section 5.1.2. Nanoscale computing relies on advanced materials. Precision crucial for nanoscale device performance enhancement. The significance of material choice cannot be overstated. Materials impact functionality, efficiency, and overall device success. The five major considerations are as follow:

Importance in Material Choice: Selection affects electronic properties in nanoscale components. Optimal materials enhance signal processing and data storage. Silicon, graphene, and carbon nanotubes popular choices today (cf. Figure 5.1). Diverse materials cater to specific nanoscale computing requirements.

Performance Considerations: Silicon's widespread use due to semiconductor properties. Graphene's exceptional conductivity suits nanoscale electronic applications. Carbon nanotubes' structural strength supports nanodevice durability.

Efficiency and Energy Consumption: Materials influence energy consumption in nanoscale devices. Efficient materials promote low-power consumption and sustainability. Improvements in energy efficiency essential for nanoscale computing progress.

Challenges in Material Selection: Challenges include compatibility, scalability, and manufacturability. Identifying materials meeting these criteria crucial for nanoscale success. Advances in nanoscale fabrication techniques contribute to material compatibility.

Emerging Materials: Researchers exploring novel materials for nanoscale computing. Quantum dots, 2D materials, and topological insulators gaining attention. Emerging Materials promise improved performance and unique functionalities.

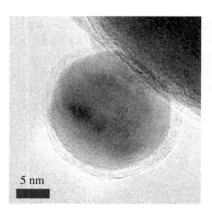

Figure 5.1   A cobalt nanoparticle coated with layers of graphene. File: Cobalt-graphene-nanoparticle.tif https://commons.wikimedia .org/wiki/File:Cobalt-graphene-nanoparticle .tif Source: Supermaster2011/Wikimedia Commons/CC BY SA 3.0.

Figure 5.2 The image illustrates the entanglement of two qubits prepared in the plus state, which is the eigenstate of the X Pauli operator. When a measurement is performed over the first qubit, it affects the second qubit, causing it to assume a different outcome state. File: Measurement Based (two Qubits).jpg https://commons.wikimedia.org/wiki/File:Measurement_Based_(two_qubits).jpg Source: Sebastiano Corli/Wikimedia Commons/CC BY-SA 4.0.

Quantum computing relies on specialized materials for its operation. Quantum bits (qubits) require materials with quantum coherence. Superconductors, topological insulators, and trapped ions are key in quantum applications. Quantum materials face environmental interference. Material sensitivity complicates maintaining quantum states. Research aims to address these challenges. Neuromorphic computing mimics human brain functions. Material properties impact synaptic connections and neural network efficiency. Phase-change materials, memristors, and spintronic devices are pivotal in neuromorphic computing. Advanced memory technologies are crucial in nanoscale computing. Non-volatile memories like resistive random-access memory (RRAM) and phase change memory (PCM) are essential. Reversible phase transition materials support high-density and rapid datastorage.

Sensing and actuation in nanoscale devices depend on material properties. Piezoelectric and thermoelectric materials are ideal for sensors. Smart materials enhance actuation efficiency in nanoscale systems. Material stability is vital for sensing devices. Functionality must be maintained across varying environmental conditions. Robust materials ensure reliable and accurate sensing performance. Material selection significantly impacts nanoscale computing performance and energy efficiency. Silicon, graphene, and carbon nanotubes are common choices. Research explores emerging materials like quantum dots and topological insulators. The challenges in material selection, especially in quantum computing and sensing devices, necessitate continuous innovation. Material breakthroughs will shape the future of nanoscale computing. The image illustrates qubit entanglement, crucial for quantum computing operations. Measurement on one qubit affects the state of the entangled partner, highlighting material sensitivity challenges (cf. Figure 5.2).

5.1.1 Foundational Role of Materials

This section talks about the foundational role of materials throughout. The foundational role of materials in nanomagnetic logic is pivotal for the progression of nanoscale computation technologies. This section explores the significance of magnetic materials and their properties, which are essential for enabling advanced computational methods at the nanoscale.

Importance of Material Selection: The selection of appropriate materials is crucial for the development and efficiency of nanomagnetic logic. The choice of materials directly influences the performance, reliability, and stability of nanoscale devices. Materials must exhibit specific properties to support the fundamental operations required in nanomagnetic logic systems. These properties include high magnetic anisotropy, coercivity, temperature stability, and the ability to sustain distinct magnetic states.

Foundations of Nanomagnetic Logic: Permanent magnets and magnetizable materials serve as the fundamental building blocks for nanomagnetic logic. These materials are critical for creating the magnetic elements manipulated for logical operations at the nanoscale. The magnetic states of these materials are used to represent binary information, making them integral to the functionality of nanomagnetic logic devices.

High Magnetic Anisotropy: High magnetic anisotropy is essential for maintaining stable magnetic states, which are crucial for the reliable operation of nanomagnetic logic systems. Magnetic anisotropy refers to the directional dependence of a material's magnetic properties, which ensures that the magnetic moments are aligned in a specific direction. This alignment provides stability against thermal fluctuations and external perturbations, thereby enhancing the robustness of the logical operations performed by nanomagnetic devices.

Coercivity: Materials with high coercivity are crucial for ensuring reliable computation in nanomagnetic logic. Coercivity is the resistance of a magnetic material to changes in its magnetization. High coercivity materials require significant energy to alter their magnetic states, thereby preventing unintended switching due to external magnetic fields or thermal noise. This property is vital for maintaining the integrity of stored information and ensuring the accuracy of logical operations.

Temperature Stability: Temperature stability is another critical factor for the consistent operation of nanomagnetic logic devices. Magnetic materials must maintain their properties over a wide range of temperatures to ensure reliable performance. Variations in temperature can lead to changes in magnetic states, causing errors in computation and data storage. Therefore, materials with stable magnetic properties across varying temperatures are essential for the effective deployment of nanomagnetic logic systems. The parallel and antiparallel states in a spintronic sandwich illustrate how magnetic materials enable spin state manipulation (cf. Figure 5.3). This mechanism is crucial in spintronics applications, where spin valves and magnetic tunnel junctions (MTJs) are employed. Efficient control of these states is essential for developing advanced nanomagnetic logic systems.

Figure 5.3 Spintronic sandwich with two possible states – parallel and antiparallel. File: Spintronic sandwich P AP.png https://commons.wikimedia.org/wiki/File:Spintronic_sandwich_P_AP.png Source: MykhailoZ/Wikimedia Commons/CC BY-SA 3.0s.

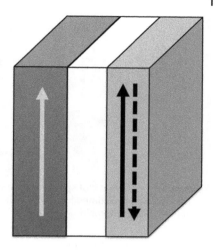

Magnetic Materials for Information Storage: Magnetic materials play an integral role in non volatile information storage technologies. Examples include magnetic hard disks and magnetic tapes, which have been widely used for decades. Advances in magnetic materials have led to increased data storage capacities and improved performance of these storage devices.

Magnetic Random Access Memory (MRAM): MRAM is a type of non volatile memory that utilizes magnetic properties for data storage and retrieval. MRAM devices rely on MTJs, which are critical components in their operation. These junctions consist of two ferromagnetic layers separated by a thin insulating layer. The relative orientation of the magnetic moments in the ferromagnetic layers determines the resistance of the junction, which can be used to represent binary data.

MTJs: MTJs are essential in MRAM technology due to their ability to exhibit distinct magnetic states. The resistance of the junction changes depending on the alignment of the magnetic moments, allowing for the reliable storage and retrieval of data. This mechanism provides a high degree of data integrity and ensures that information remains stable over time.

Material Challenges in MRAM Technology: There are several material challenges associated with MRAM technology, including achieving low-power consumption and fast access times. Identifying materials that balance stability with dynamic switching capabilities is essential for the widespread adoption of MRAM. Research is focused on overcoming these challenges by developing new materials and optimizing existing ones to improve the performance and efficiency of MRAM devices.

Nanomagnetic Logic in Quantum Computing: Nanomagnetic logic also plays a role in quantum computing, where magnetic materials are used for

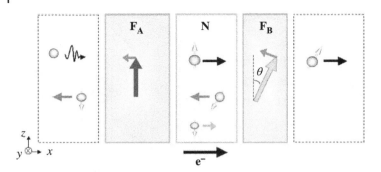

Figure 5.4 Schematic of spin transfer torque in metallic junctions. File: ST twolayers.JPG https://commons.wikimedia.org/wiki/File:ST_twolayers.JPG Source: Montigny/Wikimedia Commons/Public Domain.

quantum information processing. Quantum dots with magnetic properties are explored for their potential in quantum logic operations. These quantum dots can represent quantum bits (qubits), which are the fundamental units of information in quantum computing. Spin transfer torque in metallic junctions demonstrates the dynamic interaction between magnetic materials and electron spins (cf. Figure 5.4). This effect is fundamental to MRAM technology, where MTJs play a critical role. Understanding spin transfer torque aids in addressing challenges related to low-power consumption and fast access times in MRAM devices.

Quantum Dots with Magnetic Properties: Quantum dots with magnetic properties are investigated for their application in quantum logic. These nanoscale structures can confine electrons and exhibit discrete energy levels, which can be manipulated using magnetic fields. The magnetic properties of quantum dots enable the control of their quantum states, making them suitable for quantum computing applications.

Magnetic Resonance-Based Quantum Computing: Magnetic resonance-based quantum computing relies on the magnetic properties of materials to perform quantum operations. Techniques such as electron spin resonance (ESR) and nuclear magnetic resonance (NMR) are used to manipulate the spin states of electrons and nuclei, respectively. These techniques enable precise control of quantum states, which is essential for the implementation of quantum algorithms.

Material Challenges in Quantum Nanomagnetic Logic: Maintaining quantum coherence in magnetic states poses significant challenges for quantum nanomagnetic logic. Quantum coherence is the preservation of the phase relationships between quantum states, which is essential for the correct functioning of quantum operations. Magnetic materials used in quantum applications are susceptible to environmental interference, such as thermal noise and external magnetic fields, which can disrupt quantum coherence. Ongoing research is focused on addressing these material challenges to make quantum nanomagnetic logic practical for real-world applications.

Role in Spintronics: Spintronics, or spin electronics, leverages the spin of electrons for information processing. Magnetic materials are crucial in spintronics because they enable the manipulation of electron spin states. This technology has the potential to provide faster and more efficient electronic devices compared to traditional charge-based electronics. The four possible domain states in ferroelectric ferromagnetic materials highlight the complex interplay of polarization and magnetization (cf. Figure 5.5). These domain interactions are vital for applications in nanomagnetic logic and quantum computing. Understanding the domain walls and their classifications can lead to improved material stability and performance in various nanotechnologies.

Spin Valves: Spin valves and MTJs are essential components in spintronic devices. A spin valve is a layered structure consisting of alternating ferromagnetic and non-magnetic layers. The resistance of the spin valve depends on the relative orientation of the magnetic moments in the ferromagnetic layers, which can be used to encode information. MTJs, as previously discussed, also play a critical role in spintronics by allowing for the control of electron spin states through the relative alignment of magneticmoments.

Long Spin Lifetimes: Materials with long spin lifetimes are crucial for efficient spintronics. Spin lifetime refers to the duration over which an electron retains its spin orientation. Long spin lifetimes are essential for reliable spin-based operations, as they ensure that the spin states can be manipulated and read without significant loss of information. Magnetic semiconductors and ferromagnetic materials are explored for their potential to provide long spin lifetimes and enhance the performance of spintronic devices.

Figure 5.5 Schematic picture of the four possible domain states of a ferroelectric ferromagnetic material in which both the polarization (electric dipole indicated by charges) and the magnetization (red arrow) have two opposite orientations. The domains are separated by different types of domain walls, classified by the order parameters that change across the wall. File: Multiferroic domains.svg https://commons .wikimedia.org/wiki/File:Multiferroic_domains.svg Source: Tim Hoffmann/Wikimedia Commons/CC BY-SA 4.0.

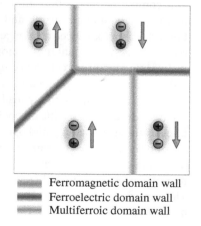

Ferromagnetic domain wall
Ferroelectric domain wall
Multiferroic domain wall

Energy Efficiency in Nanomagnetic Logic: Energy efficiency is a critical consideration in the design and implementation of nanomagnetic logic systems. Magnetic materials impact the overall energy consumption of these systems, as efficient materials enable low-power magnetic switching operations. Energy-efficient nanomagnetic logic is vital for sustainable computing, as it reduces the power requirements and operational costs of nanoscale devices.

Efficient Materials for Low-Power Operation: Identifying efficient materials that support low-power operation without compromising reliability is essential for the development of energy-efficient nanomagnetic logic. These materials must exhibit properties such as low switching energy, high stability, and fast response times. Research focuses on material innovations that improve the energy efficiency of nanomagnetic logic systems, thereby contributing to the development of sustainable computing technologies.

Balancing Energy Efficiency and Stability: Balancing energy efficiency with material stability poses significant challenges in nanomagnetic logic. Materials must provide stable magnetic states while minimizing energy consumption during switching operations. Achieving this balance requires a deep understanding of the material properties and their interactions with external factors such as temperature and magnetic fields. Ongoing research aims to optimize material properties to enhance both energy efficiency and stability in nanomagnetic logic systems.

Magnetic materials play a foundational role in the development and implementation of nanomagnetic logic-based computation. Permanent magnets and magnetizable materials are essential building blocks for nanoscale devices, providing the necessary magnetic properties for reliable and efficient logical operations. The selection of appropriate materials is crucial for achieving stable, energy-efficient, and high-performance nanomagnetic logic systems. Magnetic materials are integral to various applications, including information storage, quantum computing, and spintronics. Each application presents unique challenges that require ongoing research and innovation in material science. For instance, MRAM technology faces challenges related to low power consumption and fast access times, while quantum nanomagnetic logic must address issues related to maintaining quantum coherence in magnetic states. The role of magnetic materials in spintronics highlights their importance in enabling the manipulation of electron spin states, which can lead to faster and more efficient electronic devices. Energy efficiency is a critical consideration across all applications, as it directly impacts the sustainability and operational costs of nanomagnetic logic systems.

Continuous advancements in material science are essential for the future development of nanomagnetic logic. Researchers must focus on optimizing material properties to enhance stability, energy efficiency, and performance. The future of nanomagnetic logic depends on the successful integration of innovative materials that meet the stringent requirements of nanoscale computation technologies. The foundational role of materials in nanomagnetic logic-based computation is undeniable. Magnetic materials, including permanent magnets and magnetizable materials, serve as the building blocks for nanoscale devices. Addressing the challenges in material selection, particularly in achieving low-power consumption and maintaining stability in quantum states, drives ongoing research. The future of nanomagnetic logic hinges on continuous material innovations, ensuring efficient, reliable, and sustainable computing solutions.

5.1.1.1 Impact on Performance and Reliability

Magnetic anisotropy remains essential for stable magnetic states. High anisotropy ensures that these states persist under various conditions. Coercivity defines the resistance of materials to magnetic field changes. High coercivity materials enhance computation reliability. Temperature stability is vital for consistent performance. Materials must withstand diverse operating temperatures without degradation.

Nanomagnetic logic devices depend on specific materials. These include permanent magnets and magnetizable substances. Each material has unique properties suited to different logical operations. Their manipulation allows for precise nanoscale computational processes. In information storage, magnetic materials are crucial. Non volatile storage relies on these materials for data retention. Hard disks utilize magnetic properties for reliable data storage. Magnetic tapes offer another storage medium based on these principles. MRAM also relies heavily on specific materials. MTJs are fundamental to MRAM's operation. Distinct magnetic states enable effective data storage and retrieval. Dynamic switching in MRAM requires materials balancing stability and speed.

Quantum nanomagnetic logic presents unique material challenges. Quantum coherence must be maintained despite external influences. Magnetic properties in quantum dots facilitate quantum logic operations. Resonance-based computing in quantum computing depends on these materials. Environmental susceptibility impacts quantum nanomagnetic logic materials. Research continues to address these vulnerabilities. Maintaining coherence in these materials remains a critical challenge. Spintronics benefits significantly from magnetic materials. Electron spin manipulation underpins spintronic devices. Spin valves rely on specific magnetic materials for operation. Long spin lifetimes in these materials enhance spintronic efficiency. Energy efficiency in nanomagnetic logic is another critical aspect. Magnetic materials directly affect energy consumption. Efficient materials enable low-power magnetic switching. This efficiency is essential for sustainable computingpractices.

Balancing energy efficiency with material stability poses a challenge. Identifying materials that support reliable low-power operation is crucial. Research

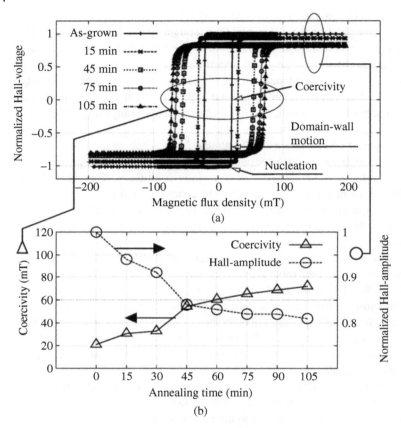

Figure 5.6 (a) The Hall-voltage hysteresis loops of the as-grown sample and after consecutive thermal annealing steps. (b) A decrease in the Hall-voltage amplitude (right axis) and an increase in the Coercivity field (left axis) is plotted versus the annealing time. File: Figure 1 https://ieeexplore.ieee.org/document/5331551/ Source: M. Becherer et al., /With permission of IEEE.

focuses on innovations for improved energy-efficient nanomagnetic logic. The Hall-voltage hysteresis loops illustrate (cf. Figure 5.6) the impact of thermal annealing. The changes in Hall-voltage amplitude and coercivity highlight material stability and reliability. These observations are crucial for understanding material behavior under different conditions. Magnetic materials form the foundation of nanomagnetic logic. Permanent magnets and magnetizable materials serve as building blocks. Continuous innovation in these materials drives progress. Future developments in this field depend on material advancements.

5.1.2 Material Diversity in Nanoscale Computing

This section talks about the material diversity in nanoscale computing throughout, briefs the range of materials suitable for nanomagnetic logic in Section 5.1.2.1, and describes the magnetic and nonmagnetic material contributions in Section 5.1.2.2.

Equations representing material properties: cf. Eqs. (5.1)–(5.5)

$$\text{High Magnetic Anisotropy} \Rightarrow \text{Stable Magnetic States} \tag{5.1}$$

$$\text{High Coercivity} \Rightarrow \text{Reliable Dynamic Switching} \tag{5.2}$$

$$\text{Temperature Stability} \Rightarrow \text{Consistent Operation} \tag{5.3}$$

$$\text{Quantum Coherence} \Rightarrow \text{Successful Quantum Applications} \tag{5.4}$$

$$\text{Long Spin Lifetimes} \Rightarrow \text{Enhanced Spintronic Functionalities} \tag{5.5}$$

Material diversity is the key of nanoscale computing, particularly in the realm of nanomagnetic logic devices. From magnetic anisotropy to quantum considerations, each material property contributes to the overall success of these devices (cf. Figure 5.7). Ongoing research continues to address challenges, ensuring that material innovations drive the future of nanoscale computing.

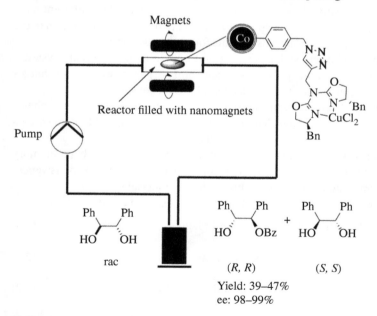

Figure 5.7 Continuous flow through magnetic-assisted reaction. File: Magnetic-chemistry-3.tif https://commons.wikimedia.org/wiki/File:Magnetic-chemistry-3 .tif Source: Supermaster2011/Wikimedia Commons/CC BY-SA 3.0.

5.1.2.1 Range of Materials Suitable for Nanomagnetic Logic

A diverse range of materials is suitable for nanomagnetic logic, each contributing unique properties essential for the functionality of these devices. From permanent magnets to quantum dots, the selection of materials plays a pivotal role in shaping the landscape of nanoscale computing.

Permanent Magnets: Permanent magnets are foundational materials in nanomagnetic logic. Their ability to maintain a stable magnetic state forms the basis for reliable computation.

Magnetizable Substances: Beyond permanent magnets, magnetizable substances contribute to the manipulation of logical operations in nanoscale devices. Their adaptability adds versatility to the range of materials.

Quantum Dots: Quantum dots, with their quantum properties, are explored for quantum nanomagnetic logic. Their unique characteristics contribute to the advancement of quantum computing applications.

MTJs: MTJs are crucial components in MRAM. These materials enable distinct magnetic states, facilitating reliable data storage and retrieval.

Magnetic Semiconductors: In the realm of spintronics, magnetic semiconductors play a vital role. These materials contribute to efficient manipulation of electron spin states, enhancing the functionalities of spintronic devices.

Ferromagnetic Materials: Ferromagnetic materials, with their inherent magnetic properties, are explored for spintronic applications. Their contribution to the field adds diversity to the materials suitable for nanoscale computing.

Piezoelectric Materials: For sensing and actuation in nanoscale systems, piezoelectric materials find application. Their ability to generate electrical charges in response to mechanical stress makes them ideal for sensors.

Phase-Change Materials: In the context of neuromorphic computing, phase-change materials are significant. These materials impact synapse-like connections and contribute to the efficiency of neural networks.

Smart Materials: Materials with smart properties, such as shape-memory alloys, contribute to efficient actuation in nanoscale devices. Their responsiveness to external stimuli adds a layer of sophistication to material selection.

Equations representing material properties: cf. Eqs. (5.6)–(5.10).

$$\text{Stable Magnetic State} \Rightarrow \text{Permanent Magnets} \tag{5.6}$$

$$\text{Quantum Properties} \Rightarrow \text{Quantum Dots} \tag{5.7}$$

$$\text{Distinct Magnetic States} \Rightarrow \text{Magnetic Tunnel Junctions} \tag{5.8}$$

$$\text{Efficient Spin Manipulation} \Rightarrow \text{Magnetic Semiconductors} \tag{5.9}$$

$$\text{Responsive to Mechanical Stress} \Rightarrow \text{Piezoelectric Materials} \tag{5.10}$$

The range of materials suitable for nanomagnetic logic devices is diverse and tailored to specific functionalities. From the foundational stability of permanent magnets to the quantum properties of quantum dots, each material brings a unique contribution. Ongoing research continues to explore new materials, expanding the possibilities for nanoscale computing and ensuring a rich and varied landscape of materials for the future.

5.1.2.2 Magnetic and Non magnetic Material Contributions

Magnetic and non magnetic materials play distinctive roles in nanomagnetic logic computing. Understanding their contributions is fundamental to harnessing their unique properties for efficient and reliable computation.

Magnetic Materials in Nanomagnetic Logic: Magnetic materials, like permanent magnets and ferromagnetic substances, provide the essential foundation for nanomagnetic logic. Their inherent magnetic properties contribute to the stability of magnetic states, ensuring the reliability of logical operations.

Contribution of Ferromagnetic Materials: Ferromagnetic materials, with their aligned magnetic domains, facilitate dynamic switching in nanomagnetic logic devices. The coercivity of these materials enables rapid transitions between magnetic states, a key aspect of efficient computation.

5.2 Magnetic Materials for Nanomagnetic Logic

This section talks about the magnetic materials for nanomagnetic logic, briefs the ferromagnetic materials in Section 5.2.1, and describes the antiferromagnetic and ferrimagnetic materials in Section 5.2.2.

5.2.1 Ferromagnetic Materials

This section talks about the ferromagnetic materials throughout and briefs the challenges and strategies for optimization in Section 5.2.1.1.

5.2.1.1 Challenges and Strategies for Optimization

The journey toward optimizing materials for nanomagnetic logic computing is not without its challenges. As we examine into the details of material selection, numerous concerns arise, demanding strategic solutions. To navigate this complexity, we must explore fundamental and applied equations and adhere to established laws, in order to establish robust theories and concepts.

Challenges in Material Optimization: Several challenges obstruct the seamless optimization of materials for nanomagnetic logic computing. These challenges include issues related to stability, energy efficiency, and compatibility with existing technologies.

Equation representing stability challenges: cf. Eqs. (5.11)–(5.13).

$$\text{Material Instability} \Rightarrow \text{Data Corruption} \qquad (5.11)$$

Strategies for Stability Improvement: To address stability challenges, strategies involve optimizing the magnetic anisotropy and reducing susceptibility to external disturbances. These strategies enhance the reliability of nanomagnetic logic devices.

Representing Magnetic Anisotropy Optimization:

$$\text{Optimized Magnetic Anisotropy} \Rightarrow \text{Enhanced Stability} \qquad (5.12)$$

Energy Efficiency as a Key Concern: Energy efficiency is a paramount concern in nanomagnetic logic computing. Materials that contribute to low power consumption are pivotal for developing sustainable and energy-efficient computing architectures.

Representing Energy-Efficient Materials:

$$\text{Low Power Consumption} \Rightarrow \text{Sustainable Computing} \qquad (5.13)$$

Compatibility Challenges with Existing Technologies: Integrating new materials with existing technologies poses compatibility challenges. Ensuring seamless integration requires a comprehensive understanding of the interplay between different material systems.

Representing Compatibility Challenges:

$$\text{Compatibility Issues} \Rightarrow \text{Integration Challenges} \qquad (5.14)$$

Strategies for Compatibility Enhancement: Strategies for enhancing compatibility involve the development of interface materials that facilitate smooth transitions between different material components. This promotes harmonious integration within nanomagnetic logic devices.

Interface Material Development:

$$\text{Optimized Interface Materials} \Rightarrow \text{Smooth Transitions} \qquad (5.15)$$

$$\text{Magnetization Reversal} \Rightarrow \text{Ferromagnetic Optimization} \qquad (5.16)$$

Fundamental Laws Governing Material Optimization: Adherence to fundamental laws is crucial for successful material optimization. The Stoner–Wohlfarth Model, describing the magnetization reversal in ferromagnetic materials, is one such fundamental law guiding the optimization process.

Stoner–Wohlfarth Model:

$$\theta_H = \arccos\left(\frac{H}{H_k}\right)$$

$$\Phi = \frac{1}{2}\sin^2\theta_H$$

$$\beta = \frac{1}{2}\cos^2\theta_H$$

$$M_s = M_s^0\left[1 - \exp\left(-\frac{K_u V}{k_B T}\right)\right]$$

$$H_k = \frac{2K_u}{\mu_0 M_s^0}$$

where θ_H is the angle between the magnetization and the applied field

Φ and β are parameters

M_s is the saturation magnetization

H_k is the anisotropy field

K_u is the magnetic anisotropy constant

V is the particle volume

k_B is the Boltzmann constant

T is the temperature

μ_0 is the vacuum permeability

Significance in Real-Time and Analytical Explanation:

The Stoner–Wohlfarth Model provides insights into the behavior of magnetic particles in external magnetic fields, focusing on the transition from single-domain to multi-domain states. Crucial for various real-time applications, particularly in magnetic data storage technologies.

- θ_H determines the alignment of the magnetic moments with respect to the external magnetic field (H). This angle is critical for understanding the stability of the magnetic state of nanoparticles.
- Φ and β are parameters derived from θ_H that characterize the energy landscape of the magnetic particle. They provide information about the relative stability of different magnetic orientations.
- M_s represents the saturation magnetization of the material, indicating its maximum achievable magnetization. This equation accounts for thermal effects, influencing the reliability and performance of magnetic devices.
- H_k calculates the anisotropy field, which is crucial for determining the stability of the magnetic state against thermal fluctuations and external perturbations. In other words, it is the magnetic field strength required to overcome the magnetic anisotropy energy barrier and change the orientation of the magnetic moments within the material. It is essential for designing magnetic materials with desired properties.

Overall, these equations illustrate the physical mechanisms governing the behavior of magnetic nanoparticles, facilitating the development of efficient and reliable magnetic storage and emerging computing technologies. The significance of the Stoner–Wohlfarth model lies in its ability to predict and understand the magnetic behavior. By considering the balance between magnetic anisotropy energy and thermal energy, the model helps to determine the critical size below which a magnetic particle will exhibit single-domain behavior, crucial for achieving high-density magnetic storage. Physically, the model elucidates how the interplay between external magnetic fields, thermal fluctuations, and material properties influences the magnetic state of nanoparticles.

Advanced Theories for Material Optimization: Establishing advanced theories for material optimization involves considering quantum mechanical effects in nanomagnetic logic devices. Quantum tunneling effects, described by the Wentzel Kramers Brillouin (WKB) approximation, influence the behavior of materials at the nanoscale.

$$\text{Quantum Tunneling Effects} \Rightarrow \text{Nanoscale Behavior} \tag{5.17}$$

Representing WKB Approximation:

The WKB approximation is a method used to approximate solutions to differential equations that vary rapidly compared to the length scale over which they are defined. It is particularly useful for solving problems in quantum mechanics and wave propagation, where oscillatory behavior is prevalent.

The basic idea behind the WKB approximation is to assume that the solution to the differential equation can be written as a rapidly oscillating function multiplied by a slowly varying envelope function. This allows us to separate the rapid oscillations from the slow variations, making the problem more manageable.

In the context of nanomagnetic logic and material optimization, nanomagnetic logic devices operate at the nanoscale, where quantum mechanical effects become increasingly significant. These devices rely on the manipulation of magnetic moments of individual nanoparticles for computation and data storage. However, as the size of nanoparticles decreases, classical models become inadequate for describing their behavior, and quantum effects such as tunneling start to dominate.

Quantum tunneling, described by the WKB approximation, is a phenomenon where particles penetrate through potential barriers that they classically would not be able to overcome. In nanomagnetic logic devices, quantum tunneling can lead to unintended flipping of magnetic states, which can degrade device performance and reliability.

Therefore, the WKB approximation is essential for understanding and predicting the behavior of nanomagnetic logic devices at the quantum level. By considering quantum effects such as tunneling, researchers optimize material properties and device designs to minimize quantum-induced errors and enhance device performance.

Mathematically, consider a second-order differential equation of the form:

$$\frac{d^2y}{dx^2} + k^2(x)y = 0$$

where $k(x)$ is a slowly varying function of x. The WKB approximation seeks a solution of the form:

$$y(x) = A(x)e^{iS(x)/\hbar}$$

where $A(x)$ is the slowly varying envelope function, $S(x)$ is a slowly varying phase function, and $\hbar$ is the reduced Planck's constant.

Substituting this ansatz into the differential equation, we obtain:

$$\frac{d^2A}{dx^2}e^{iS/\hbar} + 2i\frac{dA}{dx}\frac{dS}{dx}e^{iS/\hbar} + iA\frac{d^2S}{dx^2}e^{iS/\hbar} + k^2Ae^{iS/\hbar} = 0$$

Dividing through by $e^{iS/\hbar}$ and simplifying, we get:

$$\frac{1}{A}\frac{d^2A}{dx^2} + 2i\frac{1}{\hbar}\frac{dS}{dx}\frac{1}{A}\frac{dA}{dx} + i\left(\frac{1}{\hbar}\frac{dS}{dx}\right)^2 + k^2 = 0$$

This equation can be separated into real and imaginary parts, leading to two equations: one governing the amplitude (A) and the other governing the phase (S): 1. Amplitude Equation:

$$\frac{1}{A}\frac{d^2A}{dx^2} + \left(\frac{1}{\hbar}\frac{dS}{dx}\right)^2 - k^2 = 0$$

2. Phase Equation:

$$\frac{dS}{dx} = \pm\sqrt{\hbar^2k^2 - \left(\frac{d^2A}{dx^2}\right)A^{-1}}$$

These equations are then solved iteratively, typically using boundary conditions to determine the integration constants and ensure the solutions match the physical behavior of the system. The WKB approximation provides a semi-classical approach to understanding quantum phenomena in nanomagnetic logic devices and has applications in various fields, including solid-state physics, optics, and quantum chemistry.

In the context of magnetism, the WKB approximation can be used to analyze the behavior of magnetic moments in materials with complex magnetic structures or in the presence of external fields. Here is how the connection between these two (though not direct but to be used in conjunction):

Stoner–Wohlfarth Model: The Stoner–Wohlfarth model describes the behavior of magnetic particles in the presence of an external magnetic field, particularly focusing on the transition from single-domain to multi-domain states. While the WKB approximation complements the understanding of how quantum effects, such as tunneling, influence the stability and switching behavior of magnetic moments in nanomagnetic structures.

Material Optimization in Nanomagnetic Logic: In the context of material optimization for nanomagnetic logic devices, understanding quantum effects is crucial for designing materials with desirable magnetic properties, such as high stability, low-energy consumption, and fast switching times. By incorporating the WKB approximation into the analysis of established magnetism models, researchers gain insights into how quantum effects influence the behavior of magnetic materials at the nanoscale and optimize material properties accordingly. By considering quantum effects alongside classical magnetism models, researchers will be able to develop a comprehensive understanding of how to optimize magnetic materials for various applications in nanotechnology.

The challenges and strategies for optimizing materials in nanomagnetic logic computing demand a multifaceted approach. By addressing stability, energy efficiency, and compatibility concerns, and incorporating fundamental laws and advanced theories, we pave the way for the development of robust computing technologies.

5.2.2 Antiferromagnetic and Ferrimagnetic Materials

This section talks about the antiferromagnetic and ferrimagnetic materials throughout, briefs the unique characteristics and applications in Section 5.2.2.1, describes the enhancing stability and performance in Section 5.2.2.2 and presents in detail the role in advanced nanomagnetic logic architectures in Section 5.2.2.3.

The domain of nanomagnetic logic computing expands as we examine into the unique properties of antiferromagnetic and ferrimagnetic materials. Understanding the fundamental and applied equations, along with Governing Laws, is imperative in establishing theories and concepts that pave the way for advancements in this field.

Antiferromagnetic Materials: Antiferromagnetic materials exhibit unique magnetic properties, where adjacent magnetic moments align in opposite directions. This inherent property results in a cancelation of macroscopic magnetization. The Néel Temperature (T_N) is a critical parameter that governs the transition between ferromagnetic and antiferromagnetic states. Cf. Eqs. (5.18)–(5.22)

Néel Temperature: The Néel temperature is often described by the Weiss molecular field theory as:

$$T_N = \frac{C}{3k_B} \cdot S(S+1) \tag{5.18}$$

where T_N is the Néel temperature, C is the exchange coupling constant, k_B is the Boltzmann constant, and S is the spin quantum number.

The Néel temperature (T_N) is the temperature at which a material undergoes a transition from a ferromagnetic to an antiferromagnetic state. The Néel temperature represents the critical temperature at which thermal energy overcomes the magnetic exchange interaction in a material. Above the Néel temperature, thermal fluctuations disrupt the ordered alignment of magnetic moments, leading to a transition from a ferromagnetic to an antiferromagnetic state.

Significance:

Understanding the Néel temperature is crucial for studying the magnetic properties of materials, particularly in the context of magnetically ordered systems. The transition at the Néel temperature signifies a change in the magnetic behavior of the material, impacting its magnetic susceptibility, coercivity, and other magnetic properties. Additionally, the Néel temperature provides insight into the stability of magnetic ordering in a material and is utilized in the design and optimization of magnetic materials for various applications, such as data storage, spintronics, and magnetic sensors.

Applied Aspects of Antiferromagnetic Materials: In nanomagnetic logic computing, antiferromagnetic materials find applications in spintronics and memory devices. The manipulation of antiferromagnetic domains holds promise for enhancing information storage and processing capabilities.

Antiferromagnetic Domain Manipulation:

$$\text{Domain Manipulation} \Rightarrow \text{Enhanced Processing} \qquad (5.19)$$

Antiferromagnetic domain manipulation refers to the controlled manipulation of the domains within an antiferromagnetic material. Antiferromagnetic materials consist of sublattices with antiparallel alignment of magnetic moments, resulting in a net magnetization of zero. Each sublattice is known as an antiferromagnetic domain.

Enhanced processing indicates that by manipulating these antiferromagnetic domains, we potentially enhance the processing capabilities or performance of a device or system. This manipulation involve techniques such as applying external magnetic fields, temperature control, or using spin-polarized currents to influence the orientation and arrangement of antiferromagnetic domains.

Equation (5.19) symbolically represents the relationship between domain manipulation and enhanced processing. It suggests that by effectively manipulating the antiferromagnetic domains, we achieve improvements in processing capabilities or performance. In summary, antiferromagnetic domain manipulation holds promise for enhancing processing capabilities in various technological applications.

Ferrimagnetic Materials: Ferrimagnetic materials, characterized by opposing magnetic moments that are unequal, exhibit a net magnetization. This unique property makes them valuable for various applications in nanomagnetic logic computing.

Ferrimagnetic net magnetization:

Net Magnetization $\Rightarrow$ Ferrimagnetic Materials $\qquad$ (5.20)

Ferrimagnetic net magnetization refers to the overall magnetic moment exhibited by a ferrimagnetic material, which arises from the unequal alignment of magnetic moments in its sublattices. Ferrimagnetic materials consist of two or more sublattices with magnetic moments aligned in opposite directions, but with different magnitudes, resulting in a net magnetization.

Ferrimagnetic materials are materials that exhibit ferrimagnetism, characterized by the presence of two or more sublattices with antiparallel magnetic moments. Examples include magnetite (Fe_3O_4) and ferrites, commonly used in magnetic recording media and microwave devices.

Equation (5.20) symbolically represents the relationship between net magnetization and the presence of ferrimagnetic materials. It suggests that the observation of net magnetization indicates the presence of ferrimagnetic properties in a material.

In technical terms, the net magnetization of a ferrimagnetic material arises from the partial cancellation of magnetic moments within its sublattices, leading to a residual magnetization. This behavior is fundamental to the magnetic properties of ferrimagnetic materials and is crucial for understanding their applications in various technologies, such as data storage, magnetic sensors, and microwave devices. Equation (5.20) highlights the association between net magnetization and the presence of ferrimagnetic properties in materials, providing insight into their magnetic behavior and applications.

Governing Laws: The magnetic properties of antiferromagnetic and ferrimagnetic materials are governed by the exchange interaction, which influences the alignment of magnetic moments. The Heisenberg Exchange Interaction (J) is a fundamental parameter dictating the stability of these materials.

Heisenberg Exchange Interaction:

Stability governed by $J \Rightarrow$ Exchange Interaction

$$\mathcal{H} = -J\sum_{\langle i,j \rangle} \mathbf{S}_i \cdot \mathbf{S}_j \qquad (5.21)$$

Where:

- $\mathcal{H}$ represents the Hamiltonian of the system,
- J is the Heisenberg Exchange Interaction parameter,
- $\sum_{\langle i,j \rangle}$ denotes a sum over nearest-neighbors pairs of magnetic moments $\mathbf{S}_i$ and $\mathbf{S}_j$.

The Heisenberg Exchange Interaction (J) parameterizes the strength of the interaction between neighboring magnetic moments in a material. It arises from the exchange of virtual particles between neighboring atoms or ions and is a fundamental parameter governing the stability, magnetic ordering, and behavior of magnetic materials.

In the context of nanomagnetic logic-based computing material optimization, the Heisenberg Exchange Interaction plays a crucial role in determining the stability and functionality of magnetic devices. Understanding and controlling J is essential for designing materials with desirable magnetic properties, such as high stability, low energy consumption, and efficient information processing capabilities. Optimization of J allows for the development of magnetic materials tailored for specific applications in nanomagnetic logic-based computing, enabling advancements in computing speed, energy efficiency, and data storage density. Equation (5.21) highlights the significance of the Heisenberg Exchange Interaction in governing the stability and behavior of materials, emphasizing its relevance to nanomagnetic logic-based computing material optimization.

Theories and Concepts: Establishing theories and concepts in nanomagnetic logic computing involving antiferromagnetic and ferrimagnetic materials requires considering phenomena like spin waves and magnetic coupling. Spin wave excitations, described by the Landau–Lifshitz–Gilbert (LLG) equation, play a pivotal role in understanding dynamic magnetic behaviors.

Modified LLG equation considering antiferromagnetic and ferrimagnetic materials, as well as spin wave excitations:

$$\frac{d\mathbf{M}}{dt} = -\gamma \mathbf{M} \times \mathbf{H}_{\text{eff}} + \alpha \mathbf{M} \times \frac{d\mathbf{M}}{dt} + \mathbf{T}_{\text{ext}} + \mathbf{T}_{\text{exchange}} + \mathbf{T}_{\text{anisotropy}} + \mathbf{T}_{\text{spin-wave}}$$

$$(5.22)$$

where M is the magnetization vector, t is the time, γ is the gyromagnetic ratio, H_{eff} is the effective magnetic field acting on the material, α is the Gilbert damping parameter, $\mathbf{T}_{\text{ext}}$ represents any external torque terms, $\mathbf{T}_{\text{exchange}}$ accounts for exchange coupling terms specific to antiferromagnetic and ferrimagnetic materials, $\mathbf{T}_{\text{anisotropy}}$ includes anisotropy field contributions arising from the magnetic interactions within these materials, and $\mathbf{T}_{\text{spin-wave}}$ captures additional torque contributions associated with spin wave excitations in the material.

The first term $-\gamma \mathbf{M} \times \mathbf{H}_{\text{eff}}$ represents the precession of the magnetization vector around the effective magnetic field ($\mathbf{H}_{\text{eff}}$), where γ is the gyromagnetic ratio (cf. Eq. 5.22).

The second term $\alpha \mathbf{M} \times \frac{d\mathbf{M}}{dt}$ is the damping term, where α is the Gilbert damping parameter, accounting for energy dissipation and relaxation of the magnetization.

The additional terms $\mathbf{T}_{\text{ext}}$, $\mathbf{T}_{\text{exchange}}$, $\mathbf{T}_{\text{anisotropy}}$, and $\mathbf{T}_{\text{spin-wave}}$ represent external torques, exchange coupling terms specific to antiferromagnetic and ferrimagnetic materials, anisotropy field contributions, and torque contributions associated with spin wave excitations, respectively.

Significance and Relevance:

Antiferromagnetic and Ferrimagnetic Materials: The inclusion of terms specific to antiferromagnetic and ferrimagnetic materials (e.g. exchange coupling and anisotropy fields) allows for a more accurate description of the magnetic behavior in these systems. This is significant as antiferromagnetic and ferrimagnetic materials play crucial roles in various technological applications, including spintronics, data storage, and magnetic sensors.

Spin Wave Excitations: Spin waves are collective excitations of the magnetization in a material, and their dynamics are essential for understanding magnetic phenomena such as magnonics and spin transport. By incorporating terms related to spin wave excitations, the modified LLG equation provides insights into the propagation and behavior of spin waves in magnetic materials, which is relevant for designing spin-based devices and optimizing their performance.

In summary, the modified LLG equation provides a comprehensive framework for studying the dynamics of magnetization in magnetic materials, accounting for the behavior of antiferromagnetic/ferrimagnetic materials and spin wave excitations. Its significance lies in its ability to accurately describe and predict magnetic phenomena relevant to various technological applications. Exploring the realm of antiferromagnetic and ferrimagnetic materials in nanomagnetic logic computing unveils exciting possibilities. By getting into fundamental and applied equations, Governing Laws, and advanced theories, we chart a course towards harnessing the full potential of these materials for future technological innovations.

5.2.2.1 Unique Characteristics and Applications

Exploring the materials used in nanomagnetic logic computing reveals various characteristics and applications. This exploration leads to studying fundamental and applied symbolic equations, as well as governing laws, to establish theories and concepts cf. Eqs. (5.23)–(5.27).

Diversity of Materials: Materials for nanomagnetic logic computing come in various forms, each possessing distinct characteristics. Understanding their unique properties is crucial for tailoring materials to specific applications and optimizing performance.

Representation for Material Diversity:

$$\text{Diversity} \Rightarrow \text{Tailoring to Applications} \Rightarrow \text{Optimization} \tag{5.23}$$

Applications in Nanomagnetic Logic: These materials find applications in spintronics, memory devices, and logic gates. The integration of magnetic properties enhances information storage, processing, and computational capabilities in nanoscale computing.

Representation for Magnetic Integration:

$$\text{Integration} \Rightarrow \text{Enhanced Computational Capabilities} \tag{5.24}$$

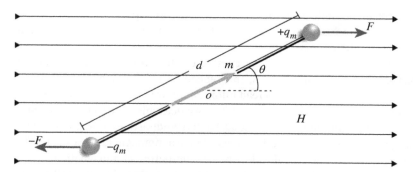

Figure 5.8 In the pole model of a dipole, an *H* field (to the right) causes equal but opposite forces on a N pole (+*q*) and a S pole (−*q*) creating a torque. File: Dipole in uniform H field.svg https://commons.wikimedia.org/wiki/File:Dipole_in_uniform_H_field .svg Source: Fred the Oyster/Wikimedia Commons/CC BY-0.

Characteristics Defining Materials: The characteristics of these materials include magnetic susceptibility, coercivity, and remanence. Understanding these properties is essential for predicting the behavior of materials in various magnetic environments.

Governing Laws: Governing the behavior of materials are fundamental laws such as Ampere's Law and Faraday's Law. These laws brief the relationship between magnetic fields, currents, and induced electromotive forces (cf. Figure 5.8).

Representation for Ampere's Law and Faraday's Law:

Ampere's Law is like a rulebook that tells us how electric currents create magnetic fields. It says that when electricity flows through a wire, it creates a magnetic field around it, just like how a magnet creates a magnetic field around itself.

Mathematically, Ampere's Law states that the magnetic field ($\vec{B}$) around a closed loop is directly proportional to the electric current (I) passing through the loop. This relationship is described by the equation:

$$\oint \vec{B} \cdot d\vec{l} = \mu_0 \cdot I \tag{5.25}$$

Faraday's Law tells us how changing magnetic fields can create electricity. It is like saying if you move a magnet near a wire, it can generate electricity in that wire.

Mathematically, Faraday's Law states that the electromotive force ($\mathscr{E}$) induced in a closed circuit is equal to the rate of change of magnetic flux (Φ_B) threading the circuit. This is represented as equation:

$$\mathscr{E} = -\frac{d\Phi_B}{dt} \tag{5.26}$$

Connection with NML: Nanomagnetic logic utilizes tiny magnets to represent bits of information (0s and 1s). These magnets are manipulated by applying external magnetic fields/electric currents. The principles of Ampere's and Faraday's laws are essential in understanding how these magnets interact and how we control their behavior.

For instance, in nanomagnetic logic circuits, electric currents are used to switch the magnetization of nanomagnets, thus encoding information. Faraday's Law comes into play when we want to read information from these nanomagnetic circuits. By detecting changes in magnetic flux, we determine the state of the nanomagnets, whether they are in a 0 or 1 state. Understanding these fundamental laws helps engineers and scientists design and optimize nanomagnetic logic devices for various applications, such as ultra low-power computing and magnetic memory storage, paving the way for smaller, faster, and more energy-efficient electronics.

Theories and Concepts: Theories in nanomagnetic logic computing involve understanding spin dynamics, magnetic domains, and quantum effects. Concepts like spin waves and domain-wall motion contribute to the development of advanced magnetic logic devices.

Representation for magnetostatic energy density (U):
The equation describing magnetic domains is typically represented using the concept of magnetization ($\vec{M}$) and magnetic field ($\vec{H}$). In a simplified form, the equation governing magnetic domains can be described by the magnetostatic energy density (U) as:

$$U = -\frac{1}{2}\vec{M} \cdot \vec{H} \tag{5.27}$$

where U is the magnetostatic energy density, which is the energy per unit volume associated with the magnetic state of the material. $\vec{M}$ is the magnetization vector, which represents the magnetic moment per unit volume of the material. $\vec{H}$ is the magnetic field, which is the external magnetic field applied to the material. This equation explains the interaction between the magnetization of the material and the external magnetic field, influencing the formation and behavior of magnetic domains within the material.

The dot product $\vec{M} \cdot \vec{H}$ represents the alignment or misalignment between the magnetization and the applied magnetic field. When the magnetization is aligned with the applied field, the dot product is positive, indicating a lower-energy state. Conversely, when the magnetization is opposed to the applied field, the dot product is negative, indicating a higher energy state. The factor $-\frac{1}{2}$ is a convention that arises from the derivation of the magnetostatic energy density and ensures that the energy associated with the magnetic state of the material is properly accounted for. This equation helps in understanding and predicting the behavior of magnetic materials, such as the formation of magnetic domains and the response of materials to applied magnetic fields.

Optimizing Material Performance: Optimizing the performance of materials requires tailoring them to specific applications. Considerations include magnetic anisotropy, exchange interactions, and the thermal stability of materials in nanoscale environments. The significance of materials in nanomagnetic logic computing lies in their unique characteristics and diverse applications.

5.2.2.2 Enhancing Stability and Performance

In the realm of nanomagnetic logic computing, materials play a crucial role in enhancing stability and performance. cf. Equations (5.28)–(5.34).

Stability in Nanomagnetic Logic: The stability of materials plays a pivotal role in ensuring the reliability of nanomagnetic logic devices. Stability, denoted by S, is defined as the resistance to external perturbations in the magnetic field.

Fundamental Representation for Stability:

$$S = \frac{\text{Resistance to Perturbations}}{\text{Time}} \tag{5.28}$$

Applied Equation for Stability Improvement: To enhance stability, optimizing the magnetic anisotropy, denoted by K, and minimizing the impact of thermal fluctuations, characterized by $k_B T$, becomes crucial.

$$S = \frac{K}{k_B T} \tag{5.29}$$

Performance Optimization: Achieving optimal performance involves a delicate balance of material parameters. Magnetic exchange interactions, spin dynamics, and coercivity must be tuned to maximize computational speed while minimizing energy consumption.

Fundamental Equation for Performance:

$$\text{Performance} = \frac{\text{Computational Speed}}{\text{Energy Consumption}} \tag{5.30}$$

Enhancing Coercivity: Coercivity, represented by H_c, is fundamental to stability and performance. Increasing coercivity improves the resistance of materials to demagnetization, resulting in more reliable nanomagnetic logic devices (cf. Figure 5.9).

Applied Equation for Coercivity Enhancement:

$$H_c = \frac{\text{Magnetic Anisotropy}}{\text{Exchange Interactions}} \tag{5.31}$$

Governing Law for Thermal Stability: The laws of thermodynamics, specifically the relationship between temperature (T), entropy (S), and internal energy (U), govern the thermal stability of materials.

$$\Delta U = T \Delta S \tag{5.32}$$

Concept of Spin Dynamics: Understanding spin dynamics is paramount to stability and performance. Spin waves, characterized by magnons, influence the behavior of spins in materials (cf. Figure 5.10).

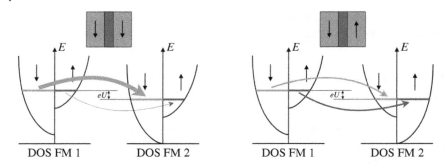

Figure 5.9 Sketch of the tunnel current for tunnel magnetoresistance (TMR) File: TunnelSchema TMR.png https://commons.wikimedia.org/wiki/File:TunnelSchema_TMR .png Source: Markus Meinert/Elessar911/Wikimedia Commons/CC BY-SA 3.0.

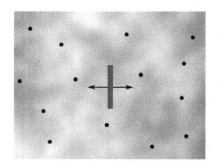

Figure 5.10 In 1909, Einstein proposed a thought experiment to demonstrate wave–particle duality. He envisioned a mirror placed in a cavity containing particles of an ideal gas and black body radiation, with the entire system in thermal equilibrium. The mirror was constrained in its motions to a direction perpendicular to its surface. Due to the gas molecules' Brownian motion, the mirror experienced jiggling. Since the mirror was in a radiation field, it transferred some of its kinetic energy to the radiation field. Einstein concluded that radiation exhibited simultaneous wave and particle aspects. File: Einstein's Mirror in Radiation Field.svg https://commons.wikimedia.org/wiki/File:Einstein %27s_Mirror_in_Radiation_Field.svg Source: Prokaryotic Caspase Homolog/Wikimedia Commons/CC BY SA 4.0.

Fundamental Equation for Spin Dynamics:

$$\text{Spin Dynamics} = \frac{\text{Interaction between Spins}}{\text{External Magnetic Field}} \tag{5.33}$$

Magnetic Anisotropy Tuning: Optimizing the magnetic anisotropy, often denoted by K, allows tailoring materials to specific applications. It influences the preferred direction of magnetization, impacting stability and reliability.

Applied Equation for Anisotropy Tuning:

$$K = \frac{\text{Material Specific Characteristics}}{\text{External Magnetic Field}} \tag{5.34}$$

5.2.2.3 Role in Advanced Nanomagnetic Logic Architectures

Exploring nanomagnetic logic computing unveils the essential role of materials in shaping advanced architectures. As we investigate the intricate world of nanomagnetic logic, we encounter equations and laws that define the fundamental principles guiding the establishment of theories and concepts. cf. Equations (5.35)–(5.39)

Architectural Impact of Materials: Materials serve as the backbone of advanced nanomagnetic logic architectures, influencing the design, functionality, and efficiency of computing devices. Their role extends beyond conventional expectations, paving the way for unprecedented advancements.

Fundamental Representation for Architectural Impact:

$$\text{Architectural Impact} = \frac{\text{Material Properties}}{\text{Computing Device Efficiency}} \tag{5.35}$$

Applied Equation for Device Design Optimization: Optimizing device design involves a delicate balance between magnetic anisotropy (K), exchange interactions (J), and spin dynamics. The following equation encapsulates this optimization process:

$$\text{Device Efficiency} = \frac{K}{J} \tag{5.36}$$

Role of Magnetic Anisotropy: Magnetic anisotropy (K) emerges as a key player in shaping advanced architectures. Tailoring K allows engineers to steer the magnetic orientation, offering control over device behavior.

Applied Equation for Magnetic Anisotropy Tailoring:

$$K = \frac{\text{External Magnetic Field}}{\text{Material-Specific Characteristics}} \tag{5.37}$$

Exchange Interactions in Architecture: Exchange interactions (J) predicts the strength of magnetic coupling between adjacent spins. Harnessing the potential of J enables the creation of intricate architectures with enhanced computational capabilities.

Fundamental Representation for Exchange Interactions:

$$J = \frac{\text{Magnetic Coupling Strength}}{\text{Distance Between Spins}} \tag{5.38}$$

Spin Dynamics in Advanced Architectures: Understanding spin dynamics is necessary to realize the full potential of nanomagnetic logic devices. Spin waves, represented by magnons, introduce a dynamic element crucial for advanced computing architectures.

Fundamental Representation for Spin Dynamics:

$$\text{Spin Dynamics} = \frac{\text{Interaction between Spins}}{\text{External Magnetic Field}} \tag{5.39}$$

Materials play a critical role in advanced nanomagnetic logic architectures. Fundamental equations and laws govern their impact on device efficiency, stability, and computational capabilities. As we navigate this intricate landscape, we pave the way for the next generation of computing advancements.

5.3 Nonmagnetic Materials in Nanoscale Computing

This section talks about the nonmagnetic materials in nanoscale computing, briefs the dielectric and insulating materials in Section 5.3.1, and describes the conductive and semiconductive materials in Section 5.3.2.

Exploring nanoscale computing reveals the significant role of nonmagnetic materials in nanomagnetic logic. In this intricate domain, nonmagnetic materials influence performance and innovation, shaping the landscape of nanomagnetic logic computing.

Nonmagnetic Materials and Nanomagnetic Logic: Nonmagnetic materials bring a unique perspective to nanomagnetic logic computing. Unlike their magnetic counterparts, these materials abstain from participating directly in magnetic interactions. This distinction sparks a series of considerations and potentials adding complexity to nanoscale computing.

Performance Implications of Nonmagnetic Materials: The integration of non-magnetic materials into nanomagnetic logic devices introduces a complex interplay of forces. Their presence affects not only the immediate magnetic surroundings but also resonates with the broader computing architecture.

Considerations for Nonmagnetic Material Integration: Incorporating nonmagnetic materials demands a careful approach. Parameters such as thermal conductivity, electron mobility, and lattice structure come to the forefront, necessitating a fine balance for optimal performance. cf. Equations (5.40)–(5.42).

Fundamental Equation for Thermal Conductivity:

$$\text{Thermal Conductivity} = \frac{\text{Heat Flow}}{\text{Temperature Gradient}} \tag{5.40}$$

Applied Equation for Electron Mobility: Electron mobility, a critical factor, contributes to the overall speed and efficiency of nanomagnetic logic devices. Its calculation involves the drift velocity (v_d) and the applied electric field (E):

$$\text{Electron Mobility} = \frac{v_d}{E} \tag{5.41}$$

Lattice Structure Impact on Device Stability: The lattice structure of non-magnetic materials plays a pivotal role in determining the stability of nanomagnetic logic devices. Deviations from ideal structures introduce defects that can impact the overall device performance.

Fundamental Representation for Lattice Energy:

$$\text{Lattice Energy} = \frac{\text{Force}}{\text{Distance}} \tag{5.42}$$

Innovations with Nonmagnetic Materials: The integration of nonmagnetic materials fosters innovations in nanoscale computing. From enhancing thermal management to introducing new pathways for information processing. These materials redefine the possibilities of nanomagnetic logic.

Challenges in Nonmagnetic Material Utilization: Despite their promise, nonmagnetic materials bring forth challenges. Achieving compatibility with existing magnetic components, ensuring uniformity in device properties, and addressing potential degradation over time are hurdles that researchers must navigate.

Strategies for Optimization: Optimizing the utilization of nonmagnetic materials involves a strategic approach. Balancing thermal considerations, electron mobility, and lattice stability forms the key of strategies aimed at enhancing overall device performance.

5.3.1 Dielectric and Insulating Materials

This section talks about the dielectric and insulating materials throughout, briefs the isolation and insulation considerations in Section 5.3.1.1, describes the impact on signal propagation and noise reduction in Section 5.3.1.2, and presents in detail the selection criteria for optimal performance in Section 5.3.1.3.

In nanomagnetic logic computing, dielectric and insulating materials shape device foundations with their unique electrical properties.

Dielectric Materials in Nanomagnetic Logic: Dielectric materials play a crucial role in nanomagnetic logic devices, contributing to the isolation of magnetic components and ensuring controlled interactions. Their ability to store electrical energy without dissipation introduces a dynamic element to the computing architecture.

Insulating Materials and Electron Flow Control: Insulating materials act as barriers, controlling the flow of electrons within nanomagnetic logic circuits (cf. Figure 5.11). This control is fundamental for preventing unintended interactions and maintaining the integrity of information processing.

Dielectric Constant and Material Selection: The dielectric constant, a key property, identifies the ability of a material to store electrical energy. Researchers carefully select dielectric materials based on their constant to achieve desired levels of capacitance cf. Equations (5.43) and (5.44).

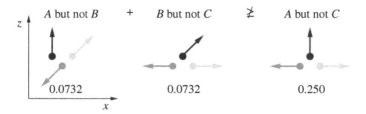

Figure 5.11 Spins of electrons in a Bell's theorem experiment File: Bell's Theorem QM JCB.jpg https://commons.wikimedia.org/wiki/File:Bell%27s_Theorem_QM_JCB.jpg Source: Joseph C Boone/Wikimedia Commons/CC BY-SA 4.0.

Fundamental Representation for Capacitance:

$$\text{Capacitance} = \frac{\text{Dielectric Constant} \times \text{Area}}{\text{Distance between Plates}} \tag{5.43}$$

Insulating Materials as Magnetic Field Modulators: Beyond their insulating role, certain materials exhibit the capability to modulate magnetic fields. This dual functionality opens avenues for designing devices with enhanced functionalities, introducing a level of versatility in nanomagnetic logic.

Applied Equation for Magnetic Modulation: The modulation of a magnetic field *(B)* can be expressed as:

$$\text{Magnetic Modulation} = \frac{\text{Change in Magnetic Field}}{\text{Time}} \tag{5.44}$$

Challenges in Dielectric and Insulating Material Integration: While dielectric and insulating materials offer immense potential, their integration presents challenges. Achieving optimal thickness, ensuring uniformity, and addressing issues related to heat dissipation require attention.

Strategies for Overcoming Integration Challenges: Strategies involve a balance between material selection and device design. Optimizing dielectric thickness, incorporating effective insulating layers, and mitigating heat-related issues become essential considerations.

Future Directions and Innovations: The evolution of dielectric and insulating materials in nanomagnetic logic computing is an ongoing journey. Exploring novel materials, refining integration techniques, and pushing the boundaries of performance open new frontiers for future innovations.

Dielectric and Insulating Materials: In nanomagnetic logic, dielectric and insulating materials play a crucial role. They influence electron and magnetic field behavior, defining nanoscale computing.

5.3.1.1 Isolation and Insulation Considerations

Navigating the complex areas of nanomagnetic logic computing involves attention to materials isolation and insulation considerations. In this paradigm, the subtle interactions between materials define the success and efficiency of nanoscale computing devices.

Fundamentals of Materials Isolation: Materials isolation forms the basic of nanomagnetic logic computing, preventing unintended interactions that can compromise device functionality. The careful selection and integration of materials aim to create an environment where magnetic elements can perform without interference.

Insulation for Controlled Electron Flow: Insulation, a key in the nanomagnetic logic architecture, directs the controlled flow of electrons. This controlled flow is vital for maintaining the stability and reliability of information processing within these advanced computing systems.

Balancing Thickness and Uniformity: One of the primary challenges lies in balancing the thickness of isolation layers and ensuring uniformity across the device. Achieving these goals is crucial for consistent performance and preventing variations in electron transport.

Optimizing Heat Dissipation Strategies: The integration of materials for isolation must also consider the thermal aspects. Optimizing heat dissipation strategies becomes imperative to avoid overheating, which could compromise the stability and longevity of nanomagnetic logic devices.

Materials Selection for Isolation Layers: Careful consideration goes into selecting materials for isolation layers. The dielectric constants, magnetic properties, and compatibility with the overall device architecture influence these choices. The goal is to find a balanced combination that facilitates isolation without sacrificing performance.

Ensuring Dielectric Compatibility: In the complex interplay of materials, dielectric compatibility is essential. Ensuring that dielectric materials work seamlessly with the magnetic elements contributes to the overall efficiency of nanomagnetic logic computing.

Striking a Balance with Insulating Materials: Insulating materials must strike a balance between preventing electron leakage and allowing controlled movement. This thin-line equilibrium ensures the reliability of information processing while minimizing energy loss.

Equations Governing Materials Interaction: In the realm of nanomagnetic logic and material interaction, the flow of electrons plays a pivotal role, identifying the dynamics of information processing and material behavior. Captured by the equation cf. Eq. (5.45):

$$\text{Electron Flow} = \frac{\text{Voltage}}{\text{Resistance}} \tag{5.45}$$

This fundamental relationship underscores the essence of electron movement within nanoscale systems, where voltage gradients propel electrons through materials characterized by varying degrees of resistance. Nanomagnetic logic leverages the manipulation of magnetic domains to encode and process information. By controlling the orientation of magnetic domains at the nanoscale, logical operations are implemented without relying on traditional electron flow, offering a promising avenue for next-generation computing architectures. Additionally, material interactions within nanomagnetic systems influence the behavior of these magnetic domains, and understanding these interactions is crucial for optimizing performance. Surface properties, magnetic domains, and material composition collectively shape the resistance encountered by electron streams, thereby influencing the efficiency and reliability of information processing.

It enables new computing paradigms and facilitates breakthroughs in material science. Through interdisciplinary efforts, researchers continue to study the complex interplay between electron dynamics, magnetic phenomena, and material properties. This effort drives innovation toward a future where nanomagnetic logic integrates smoothly with material interactions, unlocking significant capabilities.

5.3.1.2 Impact on Signal Propagation and Noise Reduction
Understanding the relationship between materials and signal propagation in nanomagnetic logic computing reveals how the choice of materials affects the efficiency of information transfer. It also addresses the challenge of noise reduction.

Signal Propagation Mechanisms: Signal propagation in nanomagnetic logic depends on the interplay between magnetic elements and the facilitating materials. This interaction defines the speed and reliability of information transfer within these computing systems.

Magnetic Modulation Equations: The speed of signal propagation can be quantified through magnetic modulation equations, linking the change in magnetic field to time. These equations guide the optimization process, aiming for swift and controlled signal propagation.

Noise Reduction Strategies: Noise, an inherent challenge in nanoscale computing, can disrupt the delicate balance of signals. Strategies for noise reduction involve material choices that minimize interference, shielding the signals from external disturbances.

Material-Induced Noise Reduction Equations: Noise reduction is expressed through equations incorporating material parameters. The goal is to create materials that actively contribute to a noise-free environment for signal propagation (cf. Eq. (5.46)).

$$\text{Noise Reduction} = \frac{\text{Signal Power}}{\text{Total Power}} \tag{5.46}$$

Role of Dielectric Constants: Dielectric constants of materials play a pivotal role in signal propagation. Their influence on the capacitance between magnetic elements affects the overall efficiency of information transfer.

Dielectric Constant Equations: Quantifying the impact of dielectric constants involves equation that link capacitance, charge, and voltage. This equation guide the selection of materials for optimal signal propagation (cf. Eq. (5.47)).

$$\text{Capacitance} = \frac{\text{Charge}}{\text{Voltage}} \tag{5.47}$$

5.3.1.3 Selection Criteria for Optimal Performance

Material selection criteria for optimal performance – nanomagnetic logic computing: optimal performance in nanomagnetic logic computing depends on careful material selection, where every choice contributes to the functioning of the system. A set of criteria governs these choices, guiding engineers and researchers toward materials that align with the goals of nanoscale computing (cf. Eqs. (5.48)–(5.50)).

Magnetic Stability Criteria: Materials must exhibit robust magnetic stability to ensure consistent performance. Criteria involve coercivity, anisotropy, and saturation magnetization, all of which contribute to the resilience of magnetic elements.

Material Stability Equations: Quantifying magnetic stability involves equation incorporating coercivity, anisotropy energy, and saturation magnetization. This equation serve as benchmarks for selecting materials that endure the demands of nanomagnetic logic.

$$\text{Coercivity} = -\frac{\text{Change in Magnetization}}{\text{Change in Applied Field}} \tag{5.48}$$

Compatibility with Integration Processes: Materials should seamlessly integrate into the fabrication processes of nanomagnetic logic devices. Compatibility criteria address issues such as deposition methods, patterning techniques, and overall manufacturability.

Compatibility Equations: Ensuring compatibility involves equations assessing parameters like deposition rates and material adhesion. These equations guide the selection of materials that align with manufacturing processes.

$$\text{Deposition Rate} = \frac{\text{Change in Film Thickness}}{\text{Change in Time}} \tag{5.49}$$

Thermal Conductivity Considerations: The impact of thermal conductivity on material selection cannot be understated. Materials should dissipate heat efficiently to prevent overheating, a critical factor in maintaining optimal performance.

Thermal Conductivity Equations: Quantifying thermal conductivity involves equation with parameters such as temperature gradient and heat transfer. This equation guides the selection of materials that actively contribute to thermal management (cf. Figure 5.12).

$$\text{Heat Transfer} = \text{Thermal Conductivity} \times \text{Area} \times \text{Temperature Gradient} \tag{5.50}$$

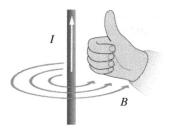

I

B

Figure 5.12 Illustration of the right-hand rule a process for determining the direction of the magnetic field created by an electric current passing through a wire. If the conductor is grasped by the right hand so the thumb points in the direction of the conventional current (flow of positive charge) in the wire, the fingers circling the wire will point in the direction of the magnetic field lines. File: Manoderecha.svg https://commons.wikimedia .org/wiki/File:Manoderecha.svg Source: Jfmelero/Wikimedia Commons/CC BY-SA 4.0.

5.3.2 Conductive and Semiconductive Materials

This section talks about the conductive and semiconductive materials throughout, briefs the role in nanomagnetic logic interconnects in Section 5.3.2.1, describes the enhancing signal transmission efficiency in Section 5.3.2.2.

In the emerging landscape of nanomagnetic logic computing, the role of conductive and semiconductive materials is pivotal. These materials form the foundation upon which the logic and functionality of nanoscale devices are built.

Conductive Materials in Nanomagnetic Logic: Conductive materials play a crucial role in facilitating the flow of electric current. The choice of conductive materials significantly influences the overall performance and efficiency of these circuits. cf. Equations (5.51)–(5.55).

Ohm's Law for Conductive Materials: Ohm's Law is a fundamental equation governing the behavior of conductive materials in the context of electrical circuits. It establishes the relationship between voltage, current, and resistance.

$$V = I \times R \tag{5.51}$$

Semiconductive Materials and Logic Operations: Semiconductive materials offer a unique advantage in nanomagnetic logic computing due to their ability to switch between conducting and insulating states. This property is harnessed for logic operations in nanoscale devices.

Semiconductor Equation for Carrier Mobility: Carrier mobility is a critical parameter in semiconductive materials, influencing their ability to transport charge carriers. The equation governing carrier mobility helps in understanding and optimizing semiconductor behavior.

$$\mu = q \times \tau \tag{5.52}$$

Integration of Conductive and Semiconductive Elements: The synergy between conductive and semiconductive materials is evident in the integration of these elements within nanomagnetic logic devices. Their collaboration allows for the creation of complex circuits and logic gates.

Charge Conservation Representation: Ensuring charge conservation is essential when integrating conductive and semiconductive elements. The charge conservation equation serves as a guide to maintaining a balanced flow of charge within the nanoscale circuitry.

$$\text{Incoming Charge} = \text{Outgoing Charge} \tag{5.53}$$

Material Selection Criteria for Conductive Elements: Choosing the right conductive materials involves considering factors such as conductivity, resistivity, and compatibility with the nanofabrication process. Equations related to these parameters aid in material selection.

Material Selection Criteria Representation: An equation encompassing conductivity, resistivity, and other relevant parameters guides engineers in selecting conductive materials that align with the requirements of nanomagnetic logic devices.

$$\text{Material Score} = \frac{\text{Conductivity}}{\text{Resistivity}} \tag{5.54}$$

Semiconductor Bandgap Considerations: Semiconductor bandgap, a defining characteristic, directly influences the logic operations achievable with semiconductive materials. The bandgap equation aids in understanding and manipulating this crucial property.

Bandgap Equation for Semiconductors: The energy–bandgap equation relates the bandgap to temperature, providing insights into the semiconductive material's behavior under varying conditions.

$$E_g(T) = E_{g0} - \alpha \cdot T^2 \tag{5.55}$$

E_g**(T)**: This represents the bandgap energy of a material as a function of temperature T. Bandgap energy is a key property of materials in semiconductor physics, representing the energy difference between the top of the valence band and the bottom of the conduction band. In this equation, the bandgap energy is assumed to vary with temperature. E_{g0}: This is the bandgap energy at absolute zero temperature ($T = 0$). It represents the bandgap energy when the material is at its lowest possible temperature. α: This is a coefficient that represents the rate of change of the bandgap energy with temperature. It is typically a material-specific constant. T: This represents the temperature at which bandgap energy is evaluated. This equation expresses the bandgap energy of a material as a function of temperature, where the bandgap energy decreases quadratically with increasing temperature, with α determining the rate of decrease.

5.3.2.1 Role in Nanomagnetic Logic Interconnects

The interplay of conductive and semiconductive materials in the realm of interconnects is crucial for nanomagnetic logic computing. cf. Equations (5.56)–(5.59).

Interconnects: The Neural Pathways of Nanomagnetic Logic: Interconnects play a vital role in connecting different components in nanomagnetic logic circuits. Conductive and semiconductive materials manage the signals that traverse these pathways.

Equations Governing Signal Propagation in Interconnects: The propagation of signals through interconnects involves electrical parameters. Equation detailing signal velocity, resistance, and capacitance guides engineers in optimizing interconnect performance.

$$\text{Signal Velocity} = \frac{1}{\sqrt{LC}} \tag{5.56}$$

Unique Challenges in Nanomagnetic Logic Interconnects: Nanomagnetic logic poses unique challenges due to the nanoscale dimensions. Conductive and semiconductive materials must navigate these challenges, including resistance-induced heating and quantum effects.

Quantum Conductance Equations: In the nanoscale regime, quantum effects become significant. Conductance in quantum conductors is quantized, and equation governing quantum conductance shed light on the conductive behavior at this scale.

$$G = n \cdot \frac{2e^2}{h} \tag{5.57}$$

G: represents the conductance of the quantum conductor. Conductance is a measure of a material's ability to conduct electric current. n: denotes the number of conducting channels available for electron flow in the quantum conductor. These channels are typically determined by the geometry and quantum properties of the conductor. e: represents the elementary charge, the fundamental unit of electric charge. h: represents Planck's constant, a fundamental physical constant that relates the energy of a photon to its frequency.

So, the equation expresses the conductance of a quantum conductor. It reveals how conductance becomes quantized at the nanoscale due to the discrete nature of conducting channels and fundamental physical constants.

Role of Semiconductors in Reducing Interconnect Power Consumption: Semiconductive materials play a vital role in reducing power consumption in interconnects. By leveraging the ability to switch between conducting and insulating states, semiconductors enhance the efficiency of signal transmission.

Power Dissipation Equation: Power dissipation is a critical consideration in interconnect design. The equation relating power dissipation, resistance, and current provides insights into the energy efficiency of nanomagnetic logic interconnects.

$$P = I^2 R \tag{5.58}$$

Material Considerations for Nanomagnetic Logic Interconnects: Selecting materials for interconnects involves a careful balance between conductivity, thermal stability, and manufacturability.

Future Prospects: The Quantum Frontier of Interconnects: As nanomagnetic logic evolves, the quantum realm becomes increasingly relevant. Conductive and semiconductive materials face new frontiers as quantum interconnects come to the forefront of research.

Quantum Interconnect Equations: Equation describing quantum phenomena in interconnects, such as quantum tunneling and entanglement, provides the theoretical foundation for the next generation of nanomagnetic logic interconnects (cf. Figure 5.13).

$$\text{Quantum Tunneling Probability} \propto e^{-\frac{2d\sqrt{2m(V-E)}}{\hbar}} \tag{5.59}$$

d: the thickness of the potential barrier. m: effective mass of the particle tunneling through the barrier. V: height of the potential barrier. E: energy of the particle. $\hbar$: reduced Planck constant.

5.3.2.2 Enhancing Signal Transmission Efficiency

The seamless flow of information in nanomagnetic logic computing relies on the capability of conductive and semiconductive materials to enhance signal transmission efficiency.

The Conduction Ensemble: Conductive materials play a crucial role in signal transmission, providing low-resistance pathways for electrons. In the nanoscale ensemble, they facilitate the transmission of electrical signals accurately.

Semiconductors: The Quantum Commanders: Semiconductors, with their unique ability to toggle between conducting and insulating states, play a pivotal role in the quantum command and control of signals. They provide the nuanced control required for efficient signal processing.

Equations Guiding Conductivity in Nanomagnetic Logic: The efficiency of signal transmission is quantified by equations governing conductivity. These equations encapsulate the material-specific parameters influencing the conduction of electrons (cf. Eqs. (5.60)–(5.62)).

$$\sigma = ne\mu \tag{5.60}$$

σ: conductivity. n: charge carrier density. e: elementary charge. μ: charge carrier mobility.

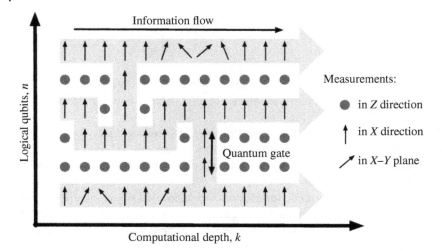

Figure 5.13 Measurement-based techniques involve entangling a cluster of Qubits and performing a set of measurements. The correlation between the entangled Qubits enables the flow of information from left to right through the measurements in the cluster on the physical Qubits. Each row in the cluster represents a logic gate to be implemented on a logical qubit. The dot and arrow symbols indicate the basis for the measurement. File: Notation-as-it-relates-to-a-one-way-quantum-computation-3-copyright-2001-by-the-APS.png https://commons.wikimedia.org/wiki/File:Notation-as-it-relates-to-a-one-way-quantum-computation-3-copyright-2001-by-the-APS.png Source: Mercedes Gimeno-Segovia, Pete Shadbolt, Dan E. Browne, Terry Rudolph/Wikimedia Commons/Public Domain.

Quantum Efficiency Considerations: In the nanomagnetic logic realm, quantum effects become prominent. Semiconductive materials harness quantum efficiency to optimize signal transmission, accounting for tunneling and quantum confinement.

Quantum Efficiency Equation: The quantum efficiency equation examines into the probabilistic nature of electron transmission, considering factors like tunneling probability and energy states.

$$\text{Quantum Efficiency} = 1 - \text{Tunneling Probability} \tag{5.61}$$

Reducing Signal Degradation: Signal degradation is a challenge in nanoscale systems. Conductive and semiconductive materials collaborate to minimize signal degradation, employing equations that describe the impact of resistance and capacitance.

$$\text{Signal Degradation} \propto RC \tag{5.62}$$

5.4   Multiferroic and Spintronic Materials

This section talks about the multiferroic and spintronic materials, briefs the integration of multiferroic materials in Section 5.4.1, describes the utilizing spintronics in material design in Section 5.4.2, presents the case studies in nanomagnetic logic computing in Section 5.4.3, discusses the challenges and future directions in Section 5.4.4.

Multiferroic Materials: Multiferroic materials possess both ferroelectric and ferromagnetic properties, crucial for nanomagnetic logic. They control magnetic states through electric fields, enabling novel computing paradigms (cf. Figure 5.14).

Spintronics Unveiled: Spintronics, combining spin and electronics, explores electron spin. In nanomagnetic logic, spintronic materials redefine information storage and processing by emphasizing spin orientation.

Multiferroic-Spintronic Synergy: Multiferroic and spintronic materials collaborate, influencing spin orientations and enhancing functionality, paving the way for robust nanomagnetic logic circuits.

Challenges in Integration: While promising, collaboration presents challenges, including efficient coupling and mitigating energy losses. Material selection, considering compatibility, stability, and response to stimuli, is paramount.

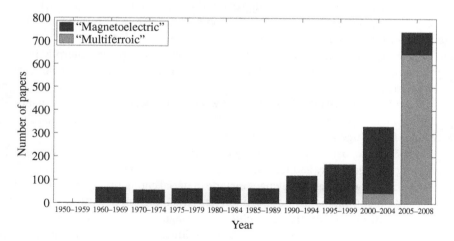

Figure 5.14   Bar chart showing number of papers using the keyword terms magnetoelectric and multiferroic. Data collected via standard Web of Science search for magneto-electric, magnetoelectric, multi-ferroic, and multiferroic. File: Multiferroics history use of terms magnetoelectric and multiferroic.png https://commons.wikimedia .org/wiki/File:Multiferroics_history_use_of_terms_magnetoelectric_and_multiferroic.png Source: Tom Forrest and Sean Muir/Wikimedia Commons/Public Domain.

Emerging Trends: Advancements in multiferroic and spintronic technologies drive trends such as non volatile memory, low-power computing, and spin-based information processing.

Applications: Beyond nanomagnetic logic, these materials find applications in magnetic sensors, memory devices, and medical devices.

Spin Hall Effect and Multiferroic Devices: Exploring spin Hall effect in multiferroic materials aids in designing devices with lower power consumption and enhanced functionality.

Materials' Role in Signal Manipulation: Multiferroic and spintronic materials enable precise signal routing and processing, utilizing intrinsic properties.

Challenges in Commercialization: Despite scientific strides, commercialization poses challenges such as economic viability and scalability.

Future Prospects: The future promises quantum leaps in processing speed, energy efficiency, and compact design through multiferroic and spintronic advancements. The integration of multiferroic and spintronic materials in nanomagnetic logic computing holds transformative potential, reshaping information processing fundamentally.

5.4.1 Integration of Multiferroic Materials

This section talks about the integration of multiferroic materials throughout, briefs the enhancing device functionality in Section 5.4.1.1, and describes the challenges and opportunities in integration in Section 5.4.1.2.

Nanomagnetic Logic Revolution: In nanomagnetic logic, multiferroic integration introduces multistate capabilities, propelling computation beyond binary constraints.

Synergistic Potential: The synergy between multiferroic materials and nanomagnetic logic focuses on energy-efficient and non volatile computing, opening avenues for next-gen processors and memory devices.

Data Storage Innovations: Multiferroic integration enhances data storage, coupling ferroelectricity and ferromagnetism for efficient writing and reading.

Material Challenges: Challenges include material compatibility, stability, and scalability, requiring meticulous attention for harnessing full potential.

Beyond Binary Computing: Multiferroic integration propels nanomagnetic logic beyond binary, offering multistate capabilities for refined computational models.

Energy-Efficient Computing: Integration promises energy-efficient computing through nanoscale control of spin and charge, reducing power consumption.

Spintronics-Multiferroic Interplay: Spintronics and multiferroics interact dynamically, showcasing intricate relationships between material properties.

Applications Across Disciplines: Multiferroic integration extends beyond computation, impacting sensors, medical devices, and various technologies.

Strategies for Optimization: Optimizing multiferroic integration involves tailoring material properties, enhancing compatibility, and minimizing energy losses.

Industrial Adoption Challenges: Industrial adoption faces challenges in scaling up production, cost-effectiveness, and manufacturing complexities.

Emerging Trends: Hybrid architectures and interdisciplinary collaborations are emerging trends in multiferroic integration.

Future Horizons: Advancements in material science and computational innovations redefine the landscape of information processing.

5.4.1.1 Enhancing Device Functionality

Multiferroic materials revolutionize nanomagnetic logic, offering enhanced device functionality. Dynamic material characteristics, coupled electric and magnetic orders, and innovative data storage solutions redefine computation.

Dynamic Material Characteristics: Multiferroic materials exhibit dynamic ferroelectric and ferromagnetic traits, laying the foundation for enhanced device functionality.

Coupling of Electric and Magnetic Orders: Multiferroics intertwine electric and magnetic orders, enabling manipulation of charge and spin for advanced information processing.

Next-Level Information Processing: Multiferroic integration elevates information processing capabilities beyond binary constraints, introducing multistate possibilities.

Efficient Energy Utilization: Multiferroic nanomagnetic logic promises energy-efficient computing by controlling spin and charge at the nanoscale.

Navigating Material Challenges: Challenges include material compatibility, stability, and scalability, requiring solutions for optimal multiferroic integration.

Spin Control Advancements: Multiferroics contribute to advancements in spin control, expanding the toolkit for spintronics.

5.4.1.2 Challenges and Opportunities in Integration

Multiferroic materials present challenges and opportunities in nanomagnetic logic integration. Compatibility, scalability, stability, and solutions for commercialization are discussed alongside emerging trends and future prospects.

Challenges Paving the Path: Challenges include material compatibility, scalability, and stability, requiring meticulous attention for successful integration.

Material Compatibility Concerns: Seamless interaction between multiferroic and nanomagnetic logic systems poses compatibility challenges.

Scalability Dilemma: Scaling up device production with multiferroic materials demands feasibility and cost-effectiveness.

Stability Challenges: Ensuring stability under varying conditions is crucial for sustained device performance.

Opportunities on the Horizon: Enhanced device functionality, Efficient energy utilization, and innovations in information storage are opportunities presented by multiferroic integration.

Material Compatibility Strategies: Strategies include surface modifications, interface engineering, and hybrid materials for seamless integration.

Scalability Solutions: Innovative approaches, including academia-industry collaborations, address scalability challenges. Multiferroic materials, with their capacity to enhance device functionality in nanomagnetic logic computing, promise transformative advancements.

5.4.2   Utilizing Spintronics in Material Design

This section talks about the utilizing spintronics in material design throughout, briefs the spin-dependent transport mechanisms in Section 5.4.2.1, and describes the spintronic devices in nanomagnetic logic in Section 5.4.2.2.

Spintronics, leveraging electron spin for information processing, is explored in material design for nanomagnetic logic computing. Challenges, opportunities, and future directions in spintronics material design are discussed.

Unveiling Spintronics Dynamics: Spintronics explores electron spin for new functionalities, impacting material design for nanomagnetic logic computing.

Navigating Material Design Challenges: Efficient spin injection, prolonged spin lifetime, and temperature sensitivity are challenges in spintronics material design.

Spin Injection Efficiency: Enhancing spin injection efficiency is crucial for seamless spin information flow in the system.

Spin Lifetime Enhancement: Prolonging spin lifetime addresses challenges of spin relaxation, ensuring sustained and reliable spin states.

Temperature Sensitivity: Material design must account for stability and functionality over a range of temperatures.

Opportunities in Material Design: Optimizing spin transport, exploiting spin Hall effect, and innovating MTJs are opportunities in spintronics material design.

Spin Transport Optimization: Material design focuses on optimizing spin transport properties for efficient nanomagnetic logic device performance.

Spin Hall Effect Exploitation: Exploring spin Hall effect in material design opens avenues for spin manipulation and efficient information processing.

Magnetic Tunnel Junction Innovation: Material innovations revolve around MTJs, crucial for low-power and high-performance spintronic devices.

Challenges and Innovation in Spintronics: Materials with long spin coherence, integration with existing technologies, and reliability in real-world conditions are challenges addressed by material design.

Material Design Solutions: Advanced alloy compositions, surface engineering techniques, and innovative layered structures are material design solutions. **Advanced Alloy Compositions**: Crafting advanced alloy compositions is a strategy to address spintronics challenges. **Surface Engineering Techniques**: Surface modification enhances spin transport efficiency, with coating and modification as key techniques. **Innovative Layered Structures**: Designing innovative layered structures allows precise control of spin transport, facilitating diverse functionalities.

Prospects and Emerging Trends: Topological insulators integration, quantum dot utilization, and interdisciplinary collaborations are emerging trends in spintronics material design. **Topological Insulators Integration**: Integrating topological insulators enhances spin transport, contributing to robust spintronic devices. **Quantum Dot Utilization**: Quantum dots find utility in spintronics material design, offering controlled environments for spins. **Collaborations Across Disciplines**: Interdisciplinary collaborations explore novel materials and design strategies, fostering innovation in spintronics.

5.4.2.1 Spin-Dependent Transport Mechanisms

Spin-dependent transport mechanisms are crucial in nanomagnetic logic computing, revolutionizing information processing. This exploration examines into the intricate dynamics of spin-dependent transport and its role in nanomagnetic logic devices.

Introduction to Spin-Dependent Transport: Spin-dependent transport manipulates electron spins for conveying information, crucial for nanomagnetic logic computing.

Crucial Aspects of Spin-Dependent Transport: Spin injection and detection, spin relaxation mechanisms, and spin transport in nanomagnetic structures are essential aspects.

Spin Injection and Detection: Initiating spin injection and detecting spin states are crucial for efficient information transfer in nanomagnetic logic devices. **Spin Relaxation Mechanisms**: Challenges include spin relaxation, where diverse mechanisms contribute to minimizing relaxation for prolonged coherence. **Spin Transport in Nanomagnetic Structures**: Nanomagnetic structures facilitate spin transport, with material design optimizing spin transport within these structures.

Challenges in Spin-Dependent Transport: Temperature sensitivity, external interference, and intrinsic material limitations are challenges in spin-dependent transport.

Temperature Sensitivity: Sensitivity to temperature fluctuations requires material designs for stability across temperature ranges. **External Interference**: External magnetic fields can disrupt spin states, leading to material design strategies for minimizing susceptibility. **Intrinsic Material Limitations**: Inherent material properties impact spin transport, demanding material engineering to overcome limitations.

Opportunities in Spin-Dependent Transport: Spintronic logic gates, spin filters and polarizers, quantum dot integration, and innovations in material design are opportunities.

Spintronic Logic Gates: Utilizing spin-dependent transport for logic gates offers alternatives to traditional electronic logic gates. **Spin Filters and Polarizers**: Designing materials as effective spin filters enhances the purity of spintronic devices. **Quantum Dot Integration**: Integrating quantum dots into spin-dependent transport offers precision in spin manipulation for enhanced functionality.

Innovations in Material Design for Spin-Dependent Transport: Two-dimensional materials, hybrid structures and alloys, and surface modification techniques are innovations in material design.

Two-Dimensional Materials: Exploring two-dimensional materials, such as graphene, for spin transport involves leveraging unique spin properties. **Hybrid Structures and Alloys**: Designing hybrid structures and alloys optimizes spin transport, with material scientists experimenting for desired functionalities. **Surface Modification Techniques**: Surface modification enhances spin transport efficiency through coating and engineering techniques.

Prospects and Emerging Trends: Topological Insulators integration, spin wave technologies, and artificial intelligence (AI) integration are emerging trends.

Topological Insulators Integration: Exploring topological insulators for enhanced spin transport involves harnessing their properties for robust devices. **Spin Wave Technologies**: Investigating spin waves for information transfer and utilizing collective spin excitations are material design strategies. **AI Integration**: Merging spin-dependent transport with AI and exploring spintronic neuromorphic computing are emerging trends.

5.4.2.2 Spintronic Devices in Nanomagnetic Logic

The integration of spintronic devices into nanomagnetic logic computing marks a paradigm shift. This exploration examines into the roles, functionalities, and contributions of spintronic devices to nanomagnetic logic.

Crucial Roles of Spintronic Devices: Spintronic memory devices, logic gates, and spin transport devices play crucial roles in nanomagnetic logic computing.

Spintronic Memory Devices: Utilizing spin as the storage unit, these devices offer non-volatile memory with faster access times, exemplified by MRAM.

Spintronic Logic Gates: Based on spintronic principles, these gates provide alternatives to traditional electronic logic gates, contributing to energy-efficient computing. **Spin Transport Devices**: Facilitating efficient spin transport, these devices are key for nanomagnetic logic interconnects, with materials designed for enhanced efficiency.

Functionality of Spintronic Devices: Spin polarization and manipulation, spin injection and detection, and spin filtering mechanisms constitute the functionality of these devices.

Contributions to Nanomagnetic Logic Computing: Spintronic devices contribute to energy-efficient information processing, enhanced processing speeds, and synergistic integration with nanomagnetic logic.

Energy-Efficient Information Processing: Reducing energy consumption and heat dissipation, spintronic devices align with the quest for sustainable and green computing solutions. **Enhanced Processing Speeds**: Leveraging spin for computations accelerates processing, enabling faster data transfer and logic operations. **Integration with Nanomagnetic Logic**: Synergistic integration complements nanomagnetic architectures, creating a cohesive and efficient computing environment.

Challenges in Spintronic Device Implementation: Temperature sensitivity, external interference, and material compatibility pose challenges in implementing spintronic devices.

Temperature Sensitivity: Devices sensitive to temperature variations require material designs ensuring stability across temperature ranges. **External Interference**: Susceptibility to external magnetic fields necessitates shielding techniques for protection and stability. **Material Compatibility**: Compatibility challenges between materials in devices demand ongoing research in material engineering for seamlessintegration.

5.4.3 Case Studies in Nanomagnetic Logic Computing

This section talks about the case studies in nanomagnetic logic computing throughout, describes the computational processing using nanomagnetic logic in Section 5.4.3.1, and presents in detail the nanomagnetic logic in biomedical applications in Section 5.4.3.2.

Case studies provide real-world insights into nanomagnetic logic computing, exploring applications, challenges, and innovations. Specific instances showcase the practical implementation of nanomagnetic logic in diversescenarios.

Magnetic Memory Applications In-memory computing, coupled with advancements in nanomagnetic logic and magnetic memory applications, presents a potent combination for data-intensive tasks. By storing and processing data directly within memory rather than transferring it back and forth between memory and processing units, in-memory computing drastically reduces latency and energy consumption. Nanomagnetic logic contributes to this by enhancing the efficiency and speed of memory operations, particularly in MRAM and Magnetic Cache Memory. This synergy enables real-time analytics, complex simulations, and AI-driven insights with unprecedented efficiency and speed.

However, challenges like temperature sensitivity, compatibility issues, and manufacturing complexities persist, necessitating continuous innovation. Strategies such as hybrid architectures and collaborative industry efforts are crucial for overcoming these hurdles, facilitating the seamless integration of in-memory computing with nanomagnetic logic and magnetic memory applications. Together, they promise to revolutionize computing paradigms, unlocking new possibilities for data-driven decision-making and computational tasks across various domains.

5.4.3.1 Computational Processing Using Nanomagnetic Logic
Nanomagnetic logic extends beyond memory applications, influencing computational processing. Case studies explore practical implementations, benefits, and challenges in utilizing nanomagnetic logic for computational tasks.

Expanding Nanomagnetic Logic to Computational Processing: Beyond memory, nanomagnetic logic contributes to computational processing, offering energy-efficient alternatives.

Case Study: Nanomagnetic Logic in Cryptographic Processing: Implementing nanomagnetic logic in cryptographic tasks ensures secure data processing with reduced power consumption.

Energy-Efficient Cryptographic Processing: Nanomagnetic logic's inherent energy efficiency aligns with cryptographic processing needs, enhancing security without compromising power consumption.

Security and Reliability Gains: Reduced vulnerability to certain cyber threats and enhanced reliability are outcomes of nanomagnetic logic integration in cryptographic applications.

Challenges in Cryptographic Processing Adoption: Standards compliance, cryptographic algorithm adaptations, and secure key management pose challenges in widespread adoption.

Strategies for Addressing Cryptographic Processing Challenges: Collaboration with cryptographic experts, continuous research on algorithm adaptations, and development of secure key management protocols address challenges in cryptographic processing adoption.

Nanomagnetic logic utilizes magnetic elements for computation. Cryptographic processing involves encoding and decoding information. Nanomagnetic logic enables cryptographic operations due to its properties. Magnetic elements store and process data in nanoscale. These elements interact with each other magnetically. Cryptographic algorithms can be implemented using these interactions. Nanomagnets represent bits in binary form. Logic gates are constructed using nanomagnets. These gates perform Boolean operations on binary data. Cryptographic algorithms rely on these operations. Nanomagnetic logic offers potential for secure processing. Magnetic interactions enhance security against attacks. Magnetic elements exhibit nonvolatility and low energy consumption. These characteristics are advantageous for cryptographic applications. Nanomagnetic logic can resist certain types of attacks. It operates in extreme environments with stability. The technology is still in development stages. Research aims to optimize performance and scalability. Integration with conventional cryptographic methods is necessary. Nanomagnetic logic holds promise for future cryptographic processing. Further advancements may enhance security and efficiency.

Case Study: Nanomagnetic Logic in Image Processing: Utilizing nanomagnetic logic in image processing tasks enhances processing speeds and reduces energy consumption.

High-Speed Image Processing: Nanomagnetic logic accelerates image processing tasks, contributing to real-time applications like medical imaging and augmented reality.

Energy-Efficient Image Processing: Reduced power consumption in nanomagnetic logic benefits portable devices, extending battery life in image processing applications.

Challenges in Image Processing Integration: Compatibility with existing image processing architectures, optimization for specific applications, and adaptation to diverse image formats pose challenges.

Innovative Approaches for Overcoming Image Processing Challenges: Hybrid architectures combining nanomagnetic and traditional processing, software optimizations, and standardization efforts address challenges in image processing integration.

Nanomagnetic logic is used for image processing. It employs nanomagnets to store and process data. Each nanomagnet represents a pixel in the image. Data is encoded in the orientation of nanomagnets. Information processing occurs through magnetic interactions. Nanomagnets can be manipulated by external magnetic fields. Logic operations are performed by altering magnet orientations. This enables computation without electrical currents. Image processing tasks can be executed efficiently. Nanomagnetic logic offers low-power consumption. It is promising for future computing technologies. Magnetic interactions enable parallel processing of data. This enhances computational speed and efficiency. Nanomagnetic logic devices are compact and scalable. They can be integrated into existing chip architectures. Image processing algorithms can be implemented effectively. Nanomagnetic logic shows potential for high-performance computing. It could revolutionize image processing applications. Research in this field continues to advance rapidly. Practical applications of nanomagnetic logic are emerging. It offers new opportunities for image processing innovations. In summary, nanomagnetic logic enables efficient image processing tasks.

5.4.3.2 Nanomagnetic Logic in Biomedical Applications

The versatility of nanomagnetic logic finds applications in the biomedical field, showcasing innovative solutions for diagnostics and treatment. Case studies explore nanomagnetic logic's role in biomedical applications, highlighting advantages and addressing challenges.

Biomedical Applications of Nanomagnetic Logic: Nanomagnetic logic contributes to biomedical applications, offering precise and efficient solutions for diagnostics and treatment.

Case Study: Nanomagnetic Logic in Magnetic Resonance Imaging (MRI): Integrating nanomagnetic logic in MRI enhances imaging capabilities, providing high-resolution and contrast-enhanced diagnostic images.

Enhanced Imaging Precision: Nanomagnetic logic improves imaging precision, enabling detailed visualization for accurate diagnostics in MRI.

Contrast-Enhanced Imaging: Utilizing nanomagnetic contrast agents enhances imaging contrast, aiding in the detection of subtle abnormalities in MRI scans.

Challenges in MRI Integration: Compatibility with existing MRI systems, biocompatibility of nanomagnetic materials, and regulatory approvals pose challenges.

Strategies for Addressing MRI Integration Challenges: Collaboration with medical imaging experts, extensive biocompatibility testing, and regulatory engagement address challenges in MRI Integration.

Nanomagnetic logic enhances drug delivery systems, enabling precise drug release and targeted therapy. Challenges like biocompatibility and scalability hinder adoption, but innovative strategies such as biocompatible material design and interdisciplinary collaborations, aim to overcome them, improving treatment outcomes.

5.4.4 Challenges and Future Directions

This section talks about the challenges and future directions throughout, briefs the challenges in nanomagnetic logic computing in Section 5.4.4.1, and describes the future directions in nanomagnetic logic computing in Section 5.4.4.2.

While nanomagnetic logic holds transformative potential, challenges persist. Addressing these challenges is crucial for the widespread adoption and evolution of nanomagnetic logic computing. Additionally, exploring future directions guides research and development efforts toward new frontiers.

5.4.4.1 Challenges in Nanomagnetic Logic Computing

Nanomagnetic logic faces challenges that require focused attention for successful implementation. Identifying and addressing these challenges are essential for harnessing the full potential of this computing paradigm.

Temperature Sensitivity and Stability: Nanomagnetic logic devices exhibit sensitivity to temperature fluctuations, necessitating robust material engineering for stability across varying conditions.

Scalability Challenges: Scaling up production of nanomagnetic logic devices poses challenges in terms of feasibility, cost-effectiveness, and integration with existing technologies.

Material Compatibility Concerns: Ensuring seamless compatibility between diverse nanomagnetic materials remains a challenge for realizing integrated and functional systems.

Energy Efficiency Optimization: While inherently energy-efficient, continuous efforts are needed to optimize nanomagnetic logic devices for even lower power consumption, contributing to sustainable computing.

Manufacturing Complexities: The intricate fabrication processes involved in nanomagnetic logic devices demand simplification and streamlining for widespread manufacturing.

Integration with Existing Technologies: Synchronizing nanomagnetic logic with existing computing architectures and technologies requires overcoming compatibility and interoperability challenges.

Reliability and Robustness: Ensuring the reliability and robustness of nanomagnetic logic devices under real-world conditions is crucial for their acceptance in practical applications.

Security Concerns: As nanomagnetic logic advances, addressing potential security vulnerabilities and developing secure implementations become paramount.

Biocompatibility in Medical Applications: In biomedical applications, ensuring the biocompatibility of nanomagnetic materials is essential for safe and effective integration into diagnostic and therapeutic systems.

5.4.4.2 Future Directions in Nanomagnetic Logic Computing

Looking ahead, the evolution of nanomagnetic logic computing opens up exciting possibilities. Future directions encompass innovative research areas, interdisciplinary collaborations, and the exploration of new applications.

Hybrid Architectures and Multifunctionality: Exploring hybrid architectures that combine nanomagnetic logic with other emerging technologies allows for multifunctional devices with diverse capabilities.

Quantum Computing Synergy: Investigating the synergy between nanomagnetic logic and quantum computing may unlock unprecedented computational power and efficiency.

Biocompatible Nanomagnetic Materials: Advancing research to develop biocompatible nanomagnetic materials enhances the applicability of nanomagnetic logic in biomedical fields.

Neuromorphic Computing Integration: Integrating nanomagnetic logic with neuromorphic computing principles opens avenues for brain-inspired computing with enhanced learning and adaptability.

Edge Computing Paradigms: Exploring nanomagnetic logic for edge computing applications aligns with the growing demand for decentralized and real-time processing capabilities.

Sustainable and Green Computing: Continuing efforts to enhance the energy efficiency of nanomagnetic logic contribute to sustainable and green computing solutions.

Standardization and Interoperability: Developing standards and ensuring interoperability facilitate the seamless integration of nanomagnetic logic into diverse computing environments.

Educational and Ethical Considerations: Promoting education and ethical practices in nanomagnetic logic research and development ensures responsible innovation and technology use.

Space Exploration and Harsh Environments: Exploring the application of nanomagnetic logic in space exploration and harsh environments showcases its resilience and adaptability.

Social Impacts and Inclusivity: Considering the social impacts of nanomagnetic logic, including accessibility and inclusivity, fosters responsible and equitable technology deployment.

The journey of nanomagnetic logic computing is dynamic, with ongoing challenges and promising future directions. Continuous exploration, collaboration, and innovation are vital for realizing its transformative potential in diverse domains. Nanomagnetic logic computing represents a paradigm shift in the world of information processing. From the exploration of spintronics and spin-dependent transport mechanisms to the integration of spintronic devices in nanomagnetic logic, this comprehensive overview has examined into the detailed dynamics of this emerging field.

The challenges and opportunities presented by multiferroic materials, spintronics in material design, and spin-dependent transport mechanisms underscore the complexity and depth of nanomagnetic logic computing. As researchers navigate challenges related to temperature sensitivity, scalability, and material compatibility, they simultaneously uncover opportunities for enhanced functionality, energy efficiency, and innovations in information storage.

Case studies have provided tangible insights into the practical applications of nanomagnetic logic in magnetic memory, computational processing, and biomedical fields. Whether revolutionizing data storage in MRAM, enhancing cryptographic processing, or advancing biomedical imaging and drug delivery, nanomagnetic logic will demonstrate its versatility and transformative potential. As we contemplate the future of nanomagnetic logic computing, the challenges of temperature sensitivity, scalability, material compatibility, and more challenges, researchers to address them with precision and creativity. The envisioned future directions, ranging from hybrid architectures and quantum computing synergy to biocompatible materials and ethical considerations, offer a roadmap for exploration and innovation. Nanomagnetic logic computing stands at the forefront of technological advancement, inviting researchers, engineers, and innovators to embark on a journey of discovery. The dynamic interplay of challenges and opportunities, coupled with real-world case studies, paints a vivid picture of the present landscape and the vast potential that lies ahead. As the quest for more efficient, sustainable, and powerful computing solutions continues, nanomagnetic logic unfolds as a promising chapter in the ever-evolving story of technology.

5.4.5   Chapter End Quiz

This section presents the Chapter End Quiz throughout.

An online Book Companion Site is available with this fundamental text book in the following link: www.wiley.com/go/sivasubramani/nanoscalecomputing1

For Further Reading:-

National Research Council (2002), Xia et al. (2022), Gavagnin et al. (2014), Cowburn (2003), Breitkreutz von Gamm (2015), Han (2021), Barman et al. (2020), Bernstein et al. (2014), Ahn (2020), Telegin and Sukhorukov (2022), Bandyopadhyay (2021), Liu and Fang (2021), and Ardesi et al. (2022).

References

Ethan Chiyui Ahn. 2D materials for spintronic devices. *npj 2D Materials and Applications*, 4(1):17, 2020.

Yuri Ardesi, Umberto Garlando, Fabrizio Riente et al. Taming molecular field-coupling for nanocomputing design. *ACM Journal on Emerging Technologies in Computing Systems*, 19(1):1–24, 2022.

Supriyo Bandyopadhyay. Nanomagnetic boolean logic-the tempered (and realistic) vision. *IEEE Access*, 9:7743–7750, 2021.

Anjan Barman, Sucheta Mondal, Sourav Sahoo, and Anulekha De. Magnetization dynamics of nanoscale magnetic materials: a perspective. *Journal of Applied Physics*, 128(17:170901), 2020.

G. H. Bernstein, K. Butler, P. Li, F. Shah, M. Siddiq, G. Csaba, X. S. Hu, M. Niemier, and W. Porod. Nanomagnet logic - from concept to prototype. In *NSTI Nanotech*,

volume 2, pages 61–64, 2014. ISBN 978–1-4822-5827-1. https://briefs.techconnect
.org/wp-content/volumes/Nanotech2014v2/pdf/1240.pdf

Stephan Breitkreutz von Gamm. *Perpendicular Nanomagnetic Logic: Digital Logic Circuits from Field-coupled Magnets.* PhD thesis, Technische Universität München, 2015.

National Research Council. *Small Wonders, Endless Frontiers: A Review of the National Nanotechnology Initiative.* The National Academies Press, Washington, DC, 2002. ISBN 978–0-309-08454-3. doi: 10.17226/10395. URL https://nap
.nationalacademies.org/catalog/10395/small-wonders-endless-frontiers-a-review-
of-the-national-nanotechnology.

Russell P Cowburn. Digital nanomagnetic logic. In *61st Device Research Conference. Conference Digest (Cat. No. 03TH8663)*, pages 111–114. IEEE, 2003.
https://scholar.google.com/citations?user=Z0hyDlYAAAAJ&hl=de

Marco Gavagnin, Heinz D Wanzenboeck, Stefan Wachter et al. Free standing magnetic nanopillars for 3D nanomagnet logic. *ACS Applied Materials & Interfaces*, 6(22):20254–20260, 2014.

Wei Han. Magnetic memory and logic. In *Handbook of Magnetism and Magnetic Materials*, pages 1553–1592. Springer, 2021. https://scholar.google.com/citations?
user=Z0hyDlYAAAAJ&hl=de

Jiahao Liu and Liang Fang. Electric field-induced magnetization reversal of multiferroic nanomagnet. In *Magnetic Materials and Magnetic Levitation*, page 17. IntechOpen, 2021.

Andrei Telegin and Yurii Sukhorukov. Magnetic semiconductors as materials for spintronics. *Magnetochemistry*, 8(12):173, 2022.

Xiaoxing Xia, Christopher M Spadaccini, and Julia R Greer. Responsive materials architected in space and time. *Nature Reviews Materials*, 7(9):683–701, 2022.
https://www.nature.com/articles/s41578-022-00450-z

6

Nanoscale Computing at the Edge: AI Devices and Applications

Specific, Measurable, Achievable, Relevant, and Time-bound (SMART) learning objectives and goals for Chapter 6: after reading this chapter you should be able to:

- **Grasp Fundamentals of Edge Computing:**
  - Define the concept of Edge Computing.
  - Describe the motivations driving Edge Computing adoption.
  - Understand the role of Edge Computing in nanoscale computing.
- **Explore the Intersection of Nanoscale Computing and Edge AI:**
  - Identify how nanoscale computing intersects with edge AI.
  - Examine the specific contributions of nanoscale computing to edge devices.
  - Understand the integration of AI in nanoscale computing for edge devices.
- **Investigate Applications of Edge AI in Nanoscale Computing:**
  - Explore applications of edge AI in Smart IoT Devices.
  - Examine Healthcare and Wearable Devices leveraging edge AI.
  - Understand the role of Edge AI in Robotics and Autonomous Systems.
- **Analyze Case Studies and Develop Edge AI Applications:**
  - Analyze real-world examples of edge AI devices.
  - Engage in a tutorial on developing applications for edge AI.
  - Explore educational emphasis on accessible explanations for undergraduates.
- **Emphasize Educational Strategies in Edge AI:**
  - Provide accessible explanations suitable for undergraduates.
  - Encourage student projects in the field of edge AI.
- **Summarize Key Concepts and Conclude:**
  - Summarize key concepts in edge AI discussed in the chapter.
  - Conclude the exploration of nanoscale computing at the edge.

Nanoscale Computing: The Journey Beyond CMOS with Nanomagnetic Logic,
First Edition. Santhosh Sivasubramani.
© 2025 The Institute of Electrical and Electronics Engineers, Inc. Published 2025 by John Wiley & Sons, Inc.
Companion website: www.wiley.com/go/sivasubramani/nanoscalecomputing1

This chapter talks about the introduction to edge computing in Section 6.1, briefs the intersection of nanoscale computing and edge AI in Section 6.2, describes the applications of edge AI in nanoscale computing in Section 6.3, presents in detail the Edge AI in robotics and autonomous systems in Section 6.4, and elaborates the conclusion in Section 6.5.

6.1 Introduction to Edge Computing

This section talks about the introduction to Edge computing. This section briefs the defining edge computing in Section 6.1.1 and describes the motivation for edge computing in nanoscale computing in Section 6.1.2.

Edge Computing: Decentralized Data Processing is closer to devices, bringing computation near data source and reducing latency. Devices process data locally, enhancing efficiency. This computing paradigm transforms data handling strategies. Less reliance on centralized cloud infrastructure, improves performance. Edge Computing minimizes data travel and boosts responsiveness. Critical for real-time applications, like IoT devices. Empowers devices to function independently, enhancing autonomy. Benefits include reduced latency, **Bandwidth Usage** optimization. Facilitates faster decision-making, crucial for time-sensitive operations. Edge computing adapts to various industries, revolutionizing operations. Seamless integration into existing networks, simplifying infrastructure management. As technology advances, edge computing's significance amplifies. Practical applications expand across diverse sectors rapidly. A versatile solution catering to modern computing demands.

6.1.1 Defining Edge Computing

This section talks about the defining edge computing throughout and briefs the distinct characteristics and objectives in Section 6.1.1.1.

Edge computing processes data at network edges. It reduces latency by moving computation closer to data sources. This architecture supports real-time processing. Data from IoT devices are processed locally. It minimizes bandwidth usage by local data analysis. Edge nodes handle tasks like filtering, aggregation, and analysis. This reduces the load on central servers. They use lightweight protocols like MQTT. Low-power devices perform specific functions. Edge devices can operate autonomously. They interact with local sensors and actuators. Computation offloading occurs from edge to cloud. This optimizes resource usage. Containerization tools manage application deployment. Kubernetes is often used for orchestration. Network function virtualization (NFV) enables flexible edge deployments. Latency-sensitive applications benefit significantly. Examples include autonomous vehicles and industrial automation. Edge infrastructure includes micro data centers. These are geographically distributed. They provide localized computing resources. Data privacy is enhanced by local processing. This is critical in healthcare and finance. Security measures include encryption and trusted execution environments. Real-time analytics enables immediate insights. Machine learning models can run at the edge. This supports predictive maintenance. Edge computing scales with demand. It is essential for 5G networks. Edge servers have specialized hardware accelerators. These include graphics processing units (GPUs) and tensor processing units (TPUs) for intensive tasks.

6.1.1.1 Distinct Characteristics and Objectives

Edge computing shifts processing closer, optimizing data flow. A decentralized approach reduces dependency on central servers, enhancing efficiency through computational tasks at network peripheries. Real-time data analysis, crucial for **Time-Sensitive Applications**, is of emphasis. Transforming data handling localizes processing minimizes delays. Devices operate independently, empowering intelligent and autonomous functioning. Edge computing offers a **Distributed**, **Scalable Architecture** enhancing **IoT Capabilities**. Critical for quick decision-making, it minimizes **Latency**, enabling faster processing. Seamlessly adapting to diverse industry requirements, crucial for rapid data analysis. Effectively resolving bandwidth limitations, an innovative paradigm with practical implications. Aligning with modern technological demands, edge computing's importance grows exponentially (cf. Figure 6.1). Identified as a versatile solution catering to various Industry Demands.

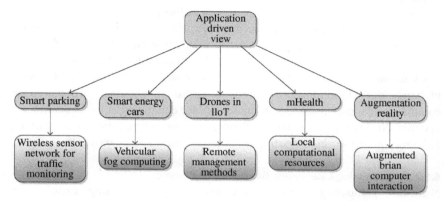

Figure 6.1 Taxonomy of edge computing applications. File:Figure 7 https://doi.org/10 .1016/j.iotcps.2023.02.004 Source: Raghubir Singh et al., 2023 Elsevier/CC BY 4.0/https:// www.sciencedirect.com/science/article/pii/S2667345223000196.

6.1.1.2 Evolution and Emergence in Modern Computing

Computing evolves, adapting to modern demands. Emergence of new technologies shapes computing landscape. Constant innovation drives transformation in computational methods. **Modern Computing** witnesses rapid technological advancements. Progression occurs across hardware, software, and algorithms. Adaptability is key in the evolving computing environment. New challenges require continual refinement of computing approaches. **Strategic Integration** of emerging technologies demands computational evolution. Shifts in paradigms influence the trajectory of computing. As computing evolves, so do its applications. Interconnected systems become integral in contemporary computing. From mainframes to edge computing, transitions define eras. As computing matures, it becomes more accessible. Emerging technologies foster collaborations and connectivity. Evolution in computing ensures **Relevance and Efficiency** (cf. Figure 6.2).

6.1.2 Motivation for Edge Computing in Nanoscale Computing

This section talks about the motivation for edge computing in nanoscale computing throughout and briefs the Addressing latency and bandwidth challenges in Section 6.1.2.1.

Nanoscale computing motivates exploration of edge computing. **Microscopic Dimensions** drive need for decentralized computation. Edge computing aligns with nanoscale's distributed nature. Motivation arises from nanoscale's unique **Computational Challenges**. Nanoscale computing demands efficient processing close by. Edge computing addresses nanoscale constraints, **Enhancing Performance**. Proximity minimizes data travel, crucial for nanoscale. Decentralized approach empowers nanoscale devices independently. Edge computing optimizes nanoscale capabilities for diverse applications. Nanoscale edge computing revolutionizes data handling in microenvironments. Efficiency and autonomy in nanoscale devices are crucial. Critical for applications requiring quick decision-making in nanoscale. Nanoscale edge computing resolves issues of bandwidth limitations. An innovative paradigm aligning with nanoscale computing demands. Strategic integration of edge computing enhances nanoscale systems. Navigating nanoscale challenges requires adaptability in computing. Efficiency in nanoscale computing relies on decentralized, adaptive strategies.

6.1.2.1 Addressing Latency and Bandwidth Challenges

Nanoscale computing demands efficiency – edge computing answers (cf. Figure 6.3). Localized processing thrives, meeting real-time demands seamlessly. Nanomagnetic logic architecture integrates seamlessly, optimizing energy efficiency significantly. Departing from traditional silicon, it encodes

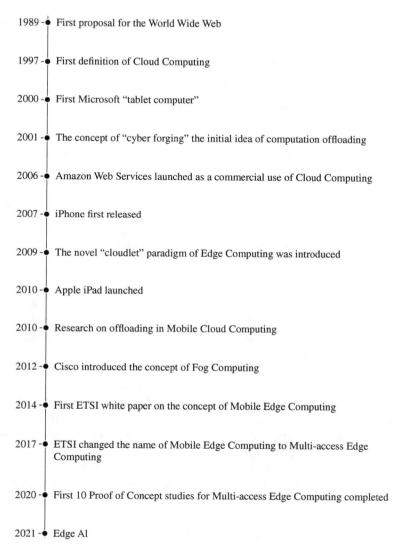

1989 – First proposal for the World Wide Web

1997 – First definition of Cloud Computing

2000 – First Microsoft "tablet computer"

2001 – The concept of "cyber forging" the initial idea of computation offloading

2006 – Amazon Web Services launched as a commercial use of Cloud Computing

2007 – iPhone first released

2009 – The novel "cloudlet" paradigm of Edge Computing was introduced

2010 – Apple iPad launched

2010 – Research on offloading in Mobile Cloud Computing

2012 – Cisco introduced the concept of Fog Computing

2014 – First ETSI white paper on the concept of Mobile Edge Computing

2017 – ETSI changed the name of Mobile Edge Computing to Multi-access Edge Computing

2020 – First 10 Proof of Concept studies for Multi-access Edge Computing completed

2021 – Edge AI

Figure 6.2 Evolution of edge computing File:Figure 1 https://doi.org/10.1016/j.iotcps .2023.02.004 Source: Raghubir Singh et al., 2023/Elsevier/CC BY 4.0/https://www. sciencedirect.com/science/article/pii/S2667345223000196.

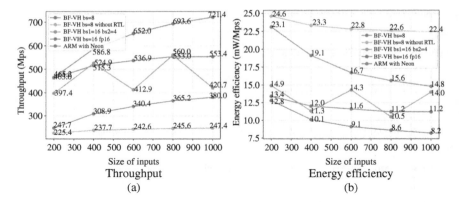

Figure 6.3 Throughput and energy efficiency with input size changes. File:Figure 14 https://doi.org/10.1016/j.micpro.2023.104824 Source: Rafael GadeaGirones, 2023/ Elsevier/CC BY 4.0/https://www.sciencedirect.com/science/article/pii/ S0141933123000704.

information magnetically, reducing energy consumption. **Material Design** refines **Magnetic Properties**, maximizing computational efficiency. Efficiency paramount in navigating data transmission delays. Optimizing data flow enhances system responsiveness significantly. Decentralized processing minimizes delays, ensuring swift responsiveness. Crucial for applications requiring real-time decision-making capabilities. Innovative strategies crucial for overcoming bandwidth constraints. Balancing Latency and bandwidth essential for optimal performance. Navigating challenges demands adaptability and innovative problem-solving. Strategic integration of technologies ensures efficient data handling. Addressing latency and bandwidth challenges requires systematic approaches. Effective solutions align with modern computing demands. An adaptive paradigm crucial for continual improvement. Efficiency in **Data Transmission** fosters overall system performance. Strategic approaches enhance system responsiveness, overcoming challenges. Balancing latency and bandwidth remains central to performance optimization. Innovative solutions for these are crucial in addressing contemporary computing challenges.

6.2 Intersection of Nanoscale Computing and Edge AI

This section talks about the intersection of nanoscale computing and edge AI, briefs the role of nanoscale computing in edge devices in Section 6.2.1, and describes the AI integration in edge devices in Section 6.2.2.

Nanoscale computing intersects with emerging edge AI. Edge AI unleashes potential at the network edge. Complementary integration of AI and edge computing, addresses issues like data storage structure effectively. Nanoscale memristor designs reduce energy in smart cameras. Biological image concepts translated into convolutional neural network (CNN)-based systems. This enables target recognition, image stabilization, and tracking. Nanowire junction arrays create integrated **Field-effect Transistor** Arrays. Basic computation is achieved through nanoscale metal/oxide/metal switches. **Nonvolatile Memory** market is transformed with memristor behavior. **Nanowire-axon** junction **arrays** enable spatially resolved detection. Stimulation and inhibition of neuronal signal propagation allows nanoscale spintronic oscillators achieve spoken-digit recognition accuracy. Indium phosphide nanowires used as building blocks for nanoscale electronics and optoelectronics, enhancing performance in electronics, computing, medical instruments. Intersection explores new functions and enhanced performance.

6.2.1 Role of Nanoscale Computing in Edge Devices

This section talks about the role of nanoscale computing in edge devices throughout, briefs the advantages of nomagnetic logic in edge computing in Section 6.2.1.1, describes the enabling efficient AI processing at the edge in Section 6.2.1.2.

Nanoscale devices enhance performance across diverse applications. **Semiconductor Nanowires** introduce new functions, improved performance. **Carbon Nanotube Devices** offer parallel programming possibilities for **Neuromorphic Computing.** **Nanocantilevers** provide enhanced performance in scanning probe microscopy. **Switchable Molecular Shuttle Devices** show promise for computing. Memristive crossbar circuits perform approximate edge detection. Carbon nanotube-based rotational actuators revolutionize various technologies. Nanowires enable integrated multicolor nanophotonics for diverse applications. Nanoscale **Electronic Synapses Emulate Biological Synapses** for **Brain-inspired Computing** (cf. Figure 6.4). Overall, nanoscale computing advances edge devices effectively.

6.2.1.1 Advantages of Nanomagnetic Logic in Edge Computing

Nanomagnetic logic excels in edge computing. **Dissipationless,** reversible computation showcases its immense potential. Achieves **Nonvolatility,** overcoming CMOS-based microelectronics limitations. Significant improvements in **Power Consumption** and integration densities. Operates at room temperature with **Radiation Hardness. Square Rings** in a lattice support reliable logic. Nanomagnets enable non deterministic computational schemes. Ideal for tasks like edge detection and problem-solving. Superior energy efficiency and scalability to smaller dimensions. Low-power operation, alleviating issues in CMOS-based

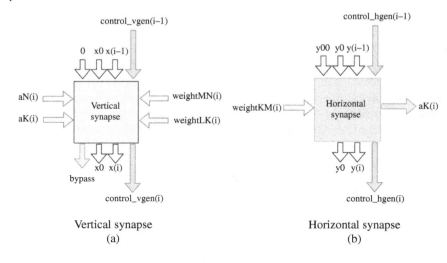

Vertical synapse
(a)

Horizontal synapse
(b)

Algorithm 2 Vertical synapse	**Algorithm 3** Horizontal synapse
1: **while 1 do**	1: **while 1 do**
2: *aux* ⇐ *read_channel (control_vgen[i])*;	2: *aux* ⇐ *read_channel(control_hgen[i + 1])*;
3: *First_dest* ⇐ *aux.control1*;	3: *Last_origin* ⇐ *aux.control1*;
4: *First_block* ⇐ *aux.control2*;	4: *First_block* ⇐ *aux.control2*;
5: *Last_block* ⇐ *aux, control3*;	5: *Last_block* ⇐ *aux, control3*;
6: *First_iter* ⇐ *aux.control4*;	6: *First_origin* ⇐ *aux.control4*;
7: *Rest* ⇐ *aux.control5*;	7: *Rest* ⇐ *aux.control5*;
8: **if** *First_iter = true* **then**	8: *weight* ⇐ *read_channel(weightKM[i])*;
9: *weight* ⇐ *read_channel(weightMN[i])*;	9: **if** *i = 0* **then**
10: **else**	10: **if** *First_block = True* **then**
11: *weight* ⇐ *read_channel(weightLK[i])*;	11: *operand2* ⇐ *read_channel(y00)*;
12: **end if**	12: **else**
13: **if** *i = 0* **then**	13: *operand2* ⇐ *read_channel(y0)*;
14: **if** *First_block = true* **then**	14: **end if**
15: *operand1* ⇐ *read_channel(x0)*;	15: **else**
16: **else**	16: *operand2* ⇐ *read_channel(y[i - 1])*;
17: *operand1* ⇐ *0.0*;	17: **end if**
18: **end if**	18: **if** *i < Bs-1* **then**
19: **else**	19: **if** *Rest ≠ i* **then**
20: *operand1* ⇐ *read_channel(x[i − 1])*;	20: *write_channel(y[i], operand2)*;
21: **end if**	21: *write_channel(control_hgen[i + 2], aux)*;
22: **if** *First_iter = true & First_dest = true* **then**	22: **end if**
23: *operand1* ⇐ *read_channel((aN[cols])*;	23: **end if**
24: **else**	24: **if** *i = 0* **then**
25: *operand1* ⇐ *read_channel(aK[cols])*;	25: **if** *Last_block = Falsei* **then**
26: **end if**	26: *write_channel(y0, operand2)*;
27: *result* ⇐ *operand2 * weight + operand1*;	27: **end if**
28: **if** *i < Bs − 1* **then**	28: **end if**
29: **if** *Rest ≠ i* **then**	29: **if** *RTL = False* **then**
30: *write_channel(x[cols], result)*;	30: *acc* ⇐ *operand2 * weight + acc*;
31: *write_channel(control_vgen[i + 1], aux)*;	31: **else**
32: **else**	
33: *write_channel(bypass[i], result)*;	32: *acc* ⇐ *myAcc(First_origin, operand2, weight)*;
34: **end if**	33: **end if**
35: **else**	34: **if** *Last_origin = True* **then**
36: **if** *First_block = true* **then**	35: *temp2* ⇐ *my_tanhF P single(acc)*;
37: *write_channel(bypass[i], result)*;	36: *write_channel(aK[i], temp2)*;
38: else	37: **if** *RTL = False* **then**
39: *write_channel(x0, result)*;	38: *accu* ⇐ *0*;
40: **end if**	39: **end if**
41: **end if**	40: **end if**
42: **end while**	41: **end while**

Figure 6.4 Detail of the synapses File:Figure 12 https://doi.org/10.1016/j.micpro.2023.104824 Source: Rafael GadeaGirones, 2023/Elsevier/CC BY 4.0/https://www.sciencedirect.com/science/article/pii/S0141933123000704.

electronics. Non volatile magnetic elements enhance computational efficiency. Focused **Electron Beam Deposition** ensures precise nanomagnetic logic, enabling reliable **Information Transfer**.

6.2.1.2 Enabling Efficient AI Processing at the Edge

Edge learning integrates wireless communication with machine learning. AI-enabled applications are supported on various edge devices, overcoming computing power and data limitations efficiently. Leveraging mobile edge computing platforms and massive data distribution. Intelligent edge computing optimizes AI task processing. Energy-efficient scheduling reduces power consumption significantly. Consumes less than 80% energy of static scheduling. Also, 70% of first in first out (**FIFO**) **Strategy** in most settings. Edge-based service provisioning uses smart cloudlets, fog, and mobile edges. AI technologies address potential issues in the edge. Reducing network propagation delay and backhaul load. Edge computing, a promising **solution for** AI, unleashing the potential of edge big data fully. **MediaTek's Edge AI** solutions feature dedicated **AI Processing Unit**. **NeuroPilot** technology enhances AI capabilities widely.

Reliable fleet analytics facilitate machine learning for artificial intelligence of things (AIoT), enabling continuous delivery, deployment, and monitoring of machine learning. Proposed three-layer architecture supports real-time IoT services, utilizing distributed and simultaneous resource management for AI computation. Microservice-based Edge Computing architecture for IoT, enables real-time services and applications with efficiency. NeuRRAM chip delivers 5x–8x better energy efficiency, offering versatility in reconfiguring chip for diverse model architectures by achieving Inference **Accuracy** comparable to software models (cf. Figure 6.5). Edgent leverages edge computing for collaborative DNN inference. Device-edge synergy enables on-demand low latency edge intelligence. Fog computing enhances processing at the edge (cf. Figure 6.6). Interacting with the cloud improves energy efficiency and security. Overview of edge and fog computing enables AI for IoT.

6.2.2 AI Integration in Edge Devices

This section talks about the AI integration in edge devices throughout, briefs the overview of AI algorithms at the edge in Section 6.2.2.1, and describes the importance of on-device processing in Section 6.2.2.2, presents in detail the collaborative edge-cloud AI architectures in Section 6.2.2.3.

AI at network edge unlocks vast potential, optimizing mobile edge computing through federated learning. **Wireless Communication** meets machine learning for intelligent edge. AI-enhanced **Offloading** maximizes accuracy for industrial IoT. Cooperative edge computing integrates AI for IoT networks. Edge intelligence solves key problems in AI. Authentication and trust evaluation are improved in 5G. Field programmable gate arrays (FPGA)-based platform accelerates AI at the edge. Edge AI offers rapid response, privacy, and efficiency.

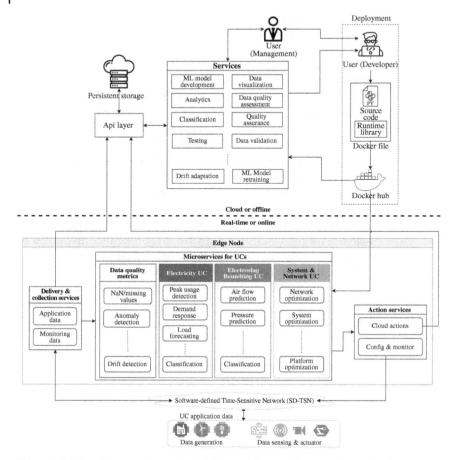

Figure 6.5 The diagram shows the detailed architecture for cloud and edge-based software. The rounded-corner rectangles in the figure indicate containers (like data quality metrics) and ML software containers for UCs (in purple, pink, and green) run on the edge node with limited resources and time. The sharp-corner rectangles indicate containers running on the cloud where there are much less resource constraints. File:Figure 6 https://doi.org/10.1016/j.iot.2023.100805 Source: Hamza Chahed, 2023/Elsevier/CC BY 4.0/https://www.sciencedirect.com/science/article/pii/S2542660523001282.

6.2.2.1 Overview of AI Algorithms at the Edge

AI algorithms at the edge require efficient computation. Models like MobileNet and YOLO optimize for low latency. Quantization reduces model size, enhancing performance. Edge devices utilize hardware accelerators, for example GPUs, TPUs. On-device inference minimizes data transmission, ensuring privacy. Pruning techniques remove redundant neurons, reducing complexity.

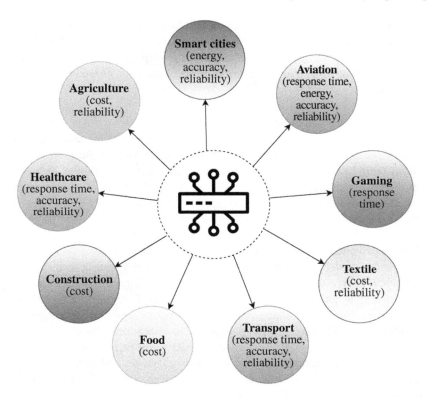

Figure 6.6 Critical performance metrics of Fog continuum in Industry 5.0 applications. File:Figure 3 https://doi.org/10.1016/j.jnca.2023.103648 Source: Shreshth Tuli et al., 2023 Elsevier/CC BY 4.0/https://www.sciencedirect.com/science/article/pii/S108480452300067X.

Memory-efficient models like SqueezeNet offer high accuracy. FPGA deployment allows for reconfigurable computing. TensorRT optimizes deep learning inference. Real-time processing achieved through optimized libraries, such as OpenVINO. On-device training uses federated learning for data locality. Neural architecture search tailors models to specific edge hardware. Edge TPU accelerates inference with low-power consumption. EfficientNet balances accuracy and efficiency. Model distillation transfers knowledge from large models to small ones. Dynamic neural networks adapt to varying computational resources. ARM Cortex-M processors support lightweight neural networks. Edge computing leverages asynchronous computing for parallel task execution. Custom ASICs designed for specific AI workloads. Edge devices implement hybrid precision arithmetic for reduced power usage. Software frameworks, for example TensorFlow Lite, support edge deployment. Real-time data processing is facilitated by low-latency communication protocols. Edge AI systems integrate with IoT for enhanced functionality. Data compression techniques reduce storage requirements. Real-time anomaly detection is implemented for predictive maintenance. Quantized neural networks utilize integer arithmetic for faster computations.

6.2.2.2 Importance of On-Device Processing

On-device processing is vital for efficient data management, optimizes data transmission, reduces latency for **Faster Response**, enables **Autonomous Systems** with **Real-time Learning**, and supports diverse computational modules, enhancing intelligent capabilities. This plays a crucial role in developing consumer devices, from audio to smartphones. Parametric spatial sound processing is essential for communication devices. Voice input through online computer processing system is also feasible. Intelligent signal processing is crucial for composite material damage detection. Reconfigurable digital signal processors have been demonstrated in video surveillance. On-device processing is crucial for reducing latency and enhancing privacy. It directly supports the three pillars of AIDA (- a novel holistic AI-driven network and processing framework for reliable data-driven real-time industrial IoT applications) architecture: adaptability, intelligence, and efficiency (Cf. Figure 6.7). Dynamic resource allocation minimizes energy consumption in mobile cloud. Engineered transcription factors facilitate rapid gene expression transitions. **Stochastic Based Models** enhance efficiency in offloading processes. **On-Device Processing** is crucial for **Efficient Data Handling**.

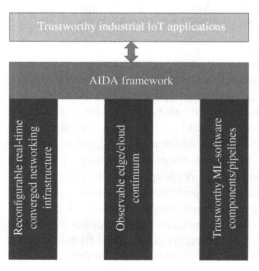

Figure 6.7 The three pillars of the AIDA architecture File:Figure 1 https://doi.org/10.1016/j.iot.2023 .100805 Source: Hamza Chahed et al., 2023/Elsevier/CC BY 4.0/ https://www.sciencedirect.com/ science/article/pii/ S2542660523001282.

6.2.2.3 Collaborative Edge-Cloud AI Architectures

Collaborative architectures optimize computing in diverse environments. Algorithms enhance IoT performance, reducing computation delay. AI and Edge Computing integration yields better IoT solutions. In vehicular networks, dynamic AI optimizes workload assignment. Cloud-edge-device collaboration manages resource heterogeneity, improving power IoT. Sophon Edge platform efficiently deploys and builds **AIoT.** CNNPC (CNN inference with joint model partition and compression) achieves fast, accurate collaborative inference in Edge. Trusted consensus scheme improves edge AI system security. Advances in pretraining models show promise for collaboration. Edge-cloud architecture accelerates AI in SME manufacturing. Two-timescale control framework optimizes energy consumption and task response. Addressing latency, collaboration reduces **Energy Consumption** and improves efficiency. Architecture developments align with **UN Sustainable Development Goal UNSDG 9.** Research explores collaborative AI advancements in multiple domains. Collaborative edge-cloud AI architectures strive for optimization.

6.3 Applications of Edge AI in Nanoscale Computing

This section talks about the applications of edge AI in nanoscale computing, briefs the Smart IoT devices in Section 6.3.1, and describes the healthcare and wearable devices in Section 6.3.2.

Nano AI enhances **Industry 4.0** energetics systems. Wireless Body Area Networks integrate Internet of Things (cf. Figure 6.8). **On-board Data Processing** meets robotic demands efficiently. Adaptive extreme edge computing benefits wearable devices profoundly. **Quantum** and **Blockchain** redefine serverless edge computing. AI in **Autonomous Driving** transforms with edge assistance. Quantum dots pivotal in advancing healthcare research. Nanosystems converge with edge computing for next-gen systems. Edge AI solutions optimize mobile collaborative robotic platforms. Wearable Edge AI thrives in ecological environments effectively. O'Reilly's "AI at the Edge" explores edge computing. Hybrid **Quantum dots** enable in-sensor reservoir computing.

6.3.1 Smart IoT Devices

This section talks about the smart IoT devices throughout, briefs the sensor networks and nanoscale computing in Section 6.3.1.1, and describes the real-time data processing in IoT in Section 6.3.1.2.

The growing demand for IoT devices necessitates advancements in security, leveraging smart edge IoT for swift responses. Smart IoT systems offer versatile control interfaces and hierarchical hardware/software configurations, ensuring diverse performance capabilities with low power consumption. These devices

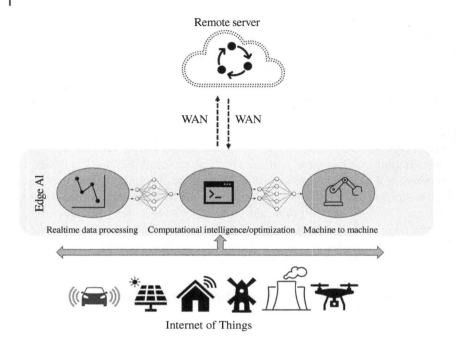

Remote server

WAN ┆ ┆ WAN

Realtime data processing   Computational intelligence/optimization   Machine to machine

Internet of Things

Figure 6.8 Internet of Things File:Graphical Abstract https://doi.org/10.1016/j.iotcps .2023.02.004 Source: Raghubir Singh et al., 2023/Elsevier/CC BY 4.0/https://www. sciencedirect.com/science/article/pii/S2667345223000196.

transfer data to smart gateways, enabling seamless communication with both edge systems and cloud providers. Wireless networks play a key role in driving innovative smart sensor services. Unified applications integrate home automation functionalities, enhancing user convenience. Robust security mechanisms are essential for maintaining IoT device connectivity. Popular protocols demonstrate effectiveness in IoT implementations, addressing challenges posed by constrained devices with limited energy, processing power, and memory. Wearable devices serve as a bridge, connecting humans with objects in the IoT ecosystem.

Smart Home system streamlines electrical equipment control. **GPS Device Enhances** child and women safety. Key technology enablers for IoT include ultra-low power. **Enhanced Sensor** usability through data collection and analysis. **Smart City** device locates **Household Objects** and saves time (cf. Figure 6.9). Control and monitor appliances using **microcontroller-based Smart Home**. IoT applications benefit from energy-efficient approximate computing. Smart health monitoring enables quick patient assessment. Next-gen IoT devices prioritize eco-friendly manufacturing, energy harvesting. 5G devices enable social-IoT spaces, smartphone **Sensory Data.** Smartphones serve as gateways for personal IoT. Empowering technology-enabled care with IoT and smart devices. Wearable devices' health data generation requires privacy. IoT-based smart home

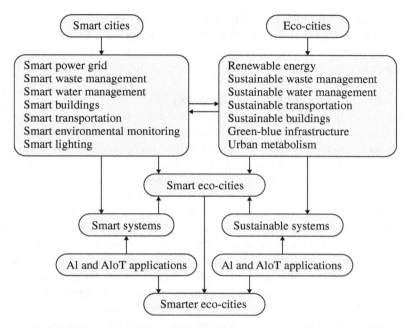

Figure 6.9 An overview of the main conceptual categories identified and their relationships. File:Figure 6 https://doi.org/10.1016/j.ese.2023.100330 Source: Simon Elias Bibri et al., 2024/Elsevier/CC BY 4.0/https://www.sciencedirect.com/science/article/pii/S2666498423000959.

security ensures real-time monitoring. Android mobile interacts with automated smart systems.

Blockchain and CLIPS enhance IoT autonomy. IoT applications monitor and control anything, anywhere. Insecurity by design challenges IoT device security. Smart surfaces in IoT and large-area electronics facilitate advancements in connected technologies. Mobile-based home automation utilizes IoT to streamline control and monitoring. IoT provides tools for improving daily life with efficient and practical solutions. The impact of wearable devices, is demonstrated through applications with measurable benefits. Smart switch control is implemented using Android application development. IoT ecosystem integration enhances smart thing accessibility. IoT connectivity bridges technologies for healthcare applications. Self-powered IoT device lasts over five months. Required tools and technologies crucial for IoT. IoT applications promise sophisticated, flexible, highly reachable lives. Smart home automation utilizes Android mobiles for control. Energy-efficient IoT device solutions mitigate periodic battery replacement. IoT applications transform real-world objects into intelligent virtual entities.

Home control system does not require a dedicated server. Trusted Tiny Things system interrogates and presents device characteristics. Smart environments connect physical world with computation elements. Smartphone as universal interface enhances IoT interoperability. IoT transforms real-world objects into intelligent virtual entities. Low-cost IoT wearable monitors and adapts to habits. Smart devices alert health conditions via IoT. Useful IoT applications benefit from various devices. Proposed system connects and controls IoT devices via voice.

6.3.1.1 Sensor Networks and Nanoscale Computing

In the landscape of nanoscale computing, sensor networks emerge as pivotal components, seamlessly integrating with nanomagnetic logic and material design. Nanoscale sensor networks expand chemical catalysis understanding. Nano-nodes are composed of processors, memory, batteries, transceivers, antennas, sensors. Smallest nanoscale sensor processor design rivals largest-known virus. Wireless nanosensor networks, collaborative and low-cost, gather diverse measurements. Carbon nanotube circuits enable radio frequency in nanoscale. Pulsed UWB communication balances low power and locationing. Nanogenerators sustain self-sufficient micro/nanopower sources for networks. Wireless sensor networks deploy compact nodes for visibility. Nanopatterning enhances manufacturing of advanced sensor systems. Satellites with small sensor-equipped systems improve efficiency. Hardware design complements proposed approaches for effectiveness. Quantum-based networks suggest future commercial quantum computers. Wireless sensor networks integrate transceivers, signal processing, sensing. Sensor networks and nanoscale computing enhance data collection and processing efficiency. These advancements underpin the five pillars of AI/AIoT-driven systems, improving sensing, perceiving, learning, visualizing, and acting capabilities (cf. Figure 6.10). Nanotechnology-enabled sensors find diverse applications in health, safety. Wireless sensor network is an emerging novel technology.

Tiny devices sense, compute, and communicate in daily life. Sensor networks have multifunctional, low-cost, miniature devices. Analog processing circuits integrated in sensor nodes suit low power. Wireless sensor networks reason about the physical world. New technology presents challenges and opportunities in sensor networks. Energy harvesting supports sustainable development of wireless sensor networks. Devices capable of sensing, computation, and communication revolutionize life. Sensor networks cater to various application areas effectively. Advances in technology enable compact, autonomous, mobile nodes. Wireless sensor nodes are available, supporting current trends. Scalable wireless networks detect specific node data efficiently. Ultra low power processors designed for wireless sensor networks emerge. Sensor network algorithms include localized and directed diffusion.

6.3.1.2 Real-Time Data Processing in IoT

IoT real-time data processing enhances efficiency, responsiveness. Methods tackle technical bottlenecks, boosting performance. Apache Kafka and Spark

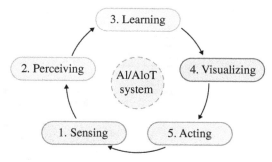

Figure 6.10 The five pillars of an AI/AIoT-driven system: 1-sensing in charge of collecting raw data, 2-perceiving in charge of extracting semantically meaningful information from raw data, 3-learning in charge of learning to predict patterns, 4-visualizing in charge of communicating key insights, and 5-acting in charge of taking action to achieve a certain goal. File:Figure 2 https://doi.org/10.1016/j.ese.2023.100330 Source: Simon Elias Bibri et al., 2024/Elsevier/CC BY 4.0/https://www.sciencedirect.com/science/article/pii/S2666498423000959.

framework optimize event processing. Energy-aware strategies yield remarkable performance, reducing consumption. Architecture integrates real-time data stream with batch processing. Survey suggests a paradigm shift towards Distributed architecture. IoT systems design exhibits commendable working performance. Semantic annotation framework dynamically integrates with the web. NiFi stands out as the most suitable system. Probabilistic data fusion evaluates system performance qualitatively. Framework supports real-time sensor data processing in the cloud, and RIoTBench serves as a vital IoT benchmark.

Secure IoT data management utilizes drones for monitoring. Adaptive data replication strategies ensure fault tolerance. Stream processing architecture analyzes data in real time. Big services model supports scalable and efficient IoT data. Graph-based cloud architecture efficiently handles smart object data. Big data analytics platforms focus on large-scale data management. Parallel systems ensure security monitoring in IoT networks. Prediction-based scheme disseminates data stream effectively in IoT. Collaborative edge and cloud processing enable live analytics. Real-time scheduling crucial for timely IoT applications. Evaluation stresses the importance of distributed stream processing. Edge-stream computing infrastructure analyzes wearable sensor data. IoT devices collaborate for real-time image recognition. Algorithms efficiently reduce data at the network edge. Hut architecture offers scalable IoT data ingestion and analysis. Benchmarking essential for distributed stream processing in IoT. IoT-based data management solutions require scalable designs. Real-time platform supports data dissemination in challenging environments.

MapReduce framework locally processes data on IoT nodes. Secure edge processing offers benefits over cloud for IoT. Distributed architecture enhances

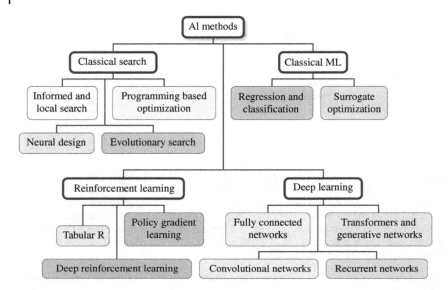

Figure 6.11 A brief taxonomy of AI methods for Fog systems that extend the one proposed by Russell and Norvig (Russell S., Norvig P. Artificial Intelligence: A Modern Approach (third ed.), Prentice Hall Press, USA (2009)). File:Figure 2 https://doi.org/10 .1016/j.jnca.2023.103648. Source: Shreshth Tuli et al., 2023/Elsevier/CC BY 4.0/https:// www.sciencedirect.com/science/article/pii/S108480452300067X.

success in IoT. Real-time IoT and cyber-physical systems introduce timeliness. Deep learning aids IoT intrusion detection effectively. Scalable systems handle large datasets for IoT. Yolo algorithm crucial for real-time object detection. Cloud-edge design is vital for AI-powered IoT applications. Integrated IoT blockchain platform ensures data integrity. Data-centric framework manages IoT data sources and filters. Real-time IoT delivers actionable intelligence for applications. Performance measurement models capture entire production processes. IoT framework uses open-source components for scalability. Big data solutions collect, store, and analyze large IoT datasets. Dependable IoT platforms manage urban infrastructures efficiently. Deep learning techniques, integral to AI methods in Fog systems, enhance IoT data analysis (cf. Figure 6.11). Scalable systems tackle challenges in IoT data management. Collaborative IoT devices ensure real time image recognition. Communication algorithms manage smart device data in IoT. Transport networks integrate with edge and cloud resources. Distributed deployment supports process-aware IoT applications. Machine learning enhances performance in IoT results. Blockchain synchronization models study wireless connectivity in IoT. Remote control servers monitor moving vehicles in real time. Routing techniques influence power consumption in IoT devices. DataTweet architecture enables data-centric IoT services.

6.3.2 Healthcare and Wearable Devices

This section talks about the healthcare and wearable devices throughout.

Wearable devices revolutionize healthcare with diverse applications. Advances in wearable tech enhance medical monitoring. Wearable sensors impact health beyond clinic boundaries. Wearable technologies reshape healthcare delivery and education. Wearable devices contribute to personalized healthcare solutions. Flexible and stretchable electronics integrate seamlessly into healthcare. Smart sensors enable facile interaction with human bodies. Wearable biosensors offer reliable, practical, ubiquitous health data. Design challenges addressed in wearable medical aid devices. Wearable technology's presence widely felt within healthcare. Wearable devices crucial for long-term patient assistance. IoT devices extend applications for wearable healthcare. Real-time data generation supports continuous health monitoring. Wearable technology is a transformative force in healthcare. Flexible sensing electronics crucial for real time health tracking. Continuous monitoring vital for personal health status. Digital devices raise questions about individuality and relationships. Wearable devices emerge as tools for diverse applications. Wearable health applications advance with smart flexible electronics. Wearable technologies diversify, offering critical health insights. Mobile health applications work with wearable technologies effectively. Wearable technology supports users throughout everyday life.

Wearable devices increasingly applied in health research. Wearable technology reshapes healthcare provision and education. Wearable devices offer solutions for health conditions and monitoring. Wearable systems adapt to healthcare demands, promoting accessibility. Wearable devices are integral for monitoring and managing health. Healthwear combines medical technology and wearable innovations. Wearable medical devices introduce transformative hardware for healthcare. Wearable applications prove valuable in healthcare, expanding functionality. Wearable devices assist in improving care for aged and chronically ill. Wearable medical devices is becoming widespread in healthcare observation. Wearable technologies contribute to personalized healthcare solutions. Wearable devices are a significant force in the future of healthcare. Wearable systems revolutionize health care, offering continuous monitoring. Health monitoring systems benefit from wearable devices' integration.

Wearable technologies impact medical internet applications and scientific research. Wearable technology reshapes healthcare with innovative solutions. Wearable devices emerge as tools for diverse healthcare applications. Wearable systems introduce new application types to health-related systems. Wearable devices are promising for healthcare applications. Wearable healthcare applications continue to evolve with advancements. Wearable devices are a catalyst for advancing precision medicine. Wearable devices impact healthcare and medicine, enabling health monitoring. Wearable devices have encouraging improvements as drug delivery systems. Wearable devices assist in monitoring health conditions, supporting health state monitoring. Wearable devices address healthcare challenges, offering personalized solutions. Wearable devices enable real-time disease monitoring for personalized healthcare. Wearable applications address challenges, providing solutions for healthcare innovation. Wearable systems offer efficient solutions for health monitoring. Wearable devices integrate into healthcare seamlessly, offering solutions. Wearable devices enhance healthcare with real-time monitoring capabilities. Wearable technologies impact healthcare with a range of applications. Wearable devices contribute to the development of cost-effective healthcare systems. Wearable devices emerge as transformative tools for healthcare.

Remote Patient Monitoring: Sensor networks play a vital role in enhancing remote patient monitoring through nanoscale computing. Components, guided by nanomagnetic logic, integrate seamlessly with sensor networks. Material design refines magnetic properties, ensuring efficient communication within sensor networks for remote patient monitoring. This application ensures a symbiotic relationship, optimizing the capabilities of nanoscale computing in enhancing sensor-based patient monitoring.

6.4 Edge AI in Robotics and Autonomous Systems

This section talks about the Edge AI in robotics and autonomous systems, briefs the on-board AI processing in robotics in Section 6.4.1, describes the autonomous vehicles and edge AI in Section 6.4.2, and presents in detail the tutorial on developing edge AI applications in Section 6.4.3.

Edge AI enhances robots in diverse applications. Research explores AI integration in robotic systems. AI at the edge transforms robot capabilities. Robots leverage deep learning for autonomous tasks. Edge Computing boosts AI efficiency in robotics. Smart robots benefit from edge-based computing. Mobile and aerial robotics benefit from edge AI. Voice and object recognition empowers searching robots. Adaptive algorithms enable robotic navigation in real-time. Edge AI ensures efficient data handling in robotics. Canny's edge detection enhances vision-guided robotics. Edge intelligence addresses challenges in autonomous driving.

Blockchain and multi-access edge computing enhance robotic autonomy. Intelligent search systems utilize edge computing services. Context information enhances autonomy in edge robotics. Autonomous driving cars benefit from dynamic edge AI. Task-specific robots demonstrate edge AI applications. Security and efficiency drive edge AI in robotics. Escalating attention to edge AI in robotics. AI applications in robotics revolutionize industries. Edge AI contributes to 3D pose estimation in robotics. Industrial robots benefit from edge computing architecture. Real-time communication optimizes edge computing for robotics. AI@EDGE platform targets breakthroughs in robotics. Edge AI roadmap explores novel ML methods. Autonomous robot autonomy is enabled by AI techniques. Edge computing resolves challenges in autonomous vehicles. AI algorithms improve perception and motion control. Image-based recognition thrives with AI in robotics. AI techniques drive robot systems in unconventional environments. Deep reinforcement learning empowers IoT-driven robots. AI powers robots, revolutionizing various industries. Edge AI contributes to autonomous vehicle intelligence. Autonomous systems learn through intelligent algorithms. Knowledge graphs enhance robotic capabilities across domains. Intelligent technology development relies on data capabilities. Image-based recognition thrives with AI in robotics. Multitasking algorithms enhance performance in robotic vehicles.

6.4.1 On-Board AI Processing in Robotics

This section talks about the on-board AI processing in robotics throughout. The essence of on-the-fly decision-making lies in real-time processing, and nanoscale edge computing plays a pivotal role in this arena. Components driven by nanomagnetic logic architecture execute complex tasks instantly, empowering robots with the ability to navigate and respond to dynamic environments in real time. On-board AI processing becomes the foundation of robotic systems, enabling them to make split-second decisions with intelligence derived from nanoscale computations. **Applications in Industrial and Service Robotics:** Service robotics, encompassing a diverse range of applications such as healthcare assistance and customer service, benefit immensely from the integration of nanoscale edge computing Nanomagnetic logic architecture, coupled with material design refinements, enhances the cognitive capabilities of service robots. These advancements enable service robots to process information rapidly, facilitating quick decision-making in dynamic environments. Whether it is assisting medical professionals or enhancing customer interactions, service robots equipped with nanoscale edge computing redefine the standards of efficiency and adaptability.

6.4.2 Autonomous Vehicles and Edge AI

This section talks about the autonomous vehicles and edge AI throughout, it briefs the AI-driven navigation and perception in Section 6.4.2.1, describes the edge processing for vehicle safety in Section 6.4.2.2, and presents in detail the challenges and future developments in Section 6.4.2.3.

Nanoscale edge computing in autonomous vehicles signals a paradigm shift in computational capability. Nanomagnetic logic components, intricately designed through material enhancements, optimize the processing speed and accuracy of onboard computations. This ensures that autonomous vehicles can swiftly interpret complex data, making split-second decisions crucial for navigating dynamic environments. The fusion of nanoscale edge computing with autonomous vehicles redefines the standards of computational efficiency, positioning them as pioneers in intelligent transportation systems. The integration of nanoscale edge computing reinforces the real-time decision-making capabilities of autonomous vehicles. Nanomagnetic logic architecture, coupled with material design refinements, facilitates on-board processing with minimal latency. This real-time processing is indispensable for autonomous vehicles, allowing them to respond instantaneously to changes in the environment, ensuring a safer and more reliable driving experience. The focus on real-time decision-making through nanoscale edge computing reaffirms the commitment to safety and precision in autonomous transportation.

6.4.2.1 AI-Driven Navigation and Perception

In artificial intelligence (AI), navigation and perception, powered by edge computing, enhance efficiency and responsiveness. Edge computing, a Decentralized paradigm, reduces latency in real-time AI applications, like navigation for autonomous vehicles. AI-driven navigation relies on real-time data processing, where edge computing swiftly analyzes data at the network's edge. This is crucial for time-sensitive tasks, such as autonomous vehicle navigation. Edge computing ensures seamless operation in resource-constrained environments and addresses privacy concerns by processing data locally. The fusion of AI and edge computing extends to drones and maritime navigation, empowering devices to react swiftly. Collaborative navigation and continuous improvement through real-time retraining of models at the edge further characterize this synergy. Edge networks enhance overall system resilience.

6.4.2.2 Edge Processing for Vehicle Safety

Edge computing transforms automotive safety by processing critical data on the edge, minimizing decision-making latency. It improves collision avoidance, lane-keeping assistance, adaptive cruise control, and pedestrian detection. Edge processing introduces redundancy, enhancing safety system robustness. In vehicle-to-everything (V2X) communication, edge processing facilitates rapid decision-making, fundamental for cooperative safety applications. Privacy concerns are addressed by processing sensitive information locally. Edge computing plays a crucial role as the automotive industry moves towards autonomous driving, ensuring real-time decision-making.

Challenges in edge processing for vehicle safety include standardization for interoperability and addressing cybersecurity concerns. Machine learning algorithms and continuous refinement of algorithms through real-time data processing at the edge further contribute to safety advancements. Let us take a real-time case study, "My car has a variety of sensors that collect data about my driving behavior and the environment. Cameras monitoring the cabin, sensors tracking speed, lane position, and acceleration rate. Also detecting distractions and asking me to focus on the road, while sensors warn about speeding. Upcoming designs are also incorporated with facial or fingerprint scanning for unlocking.

As aforesaid, modern cars use sensors: measuring brake pedal position, vehicle speed, steering wheel movements, and surrounding traffic conditions. These sensors contribute to understanding the driving environment and ensuring passenger safety. Additionally, cameras track the driver's eye gaze and facial features, collecting personally identifiable information. The car knows various details about the driver. It knows his/her driving habits, attention levels, and identity-related information from biometric sensors. Car manufacturers consider this information important for enhancing safety, comfort, and convenience. For instance, tracking distraction helps improve safety features, while monitoring drowsiness contributes to preventing accidents. Some of this data may be stored, often for improving services and personalization. To add up as a common public, any non tech person is not aware that it is stored and processed.

Cloud-based platforms are responsible for storing and processing vehicular data. Cars collect biometric info, driving behavior, and potentially location data to invoke car security measures to protect against theft and unauthorized access. Privacy policies vary; the automotive industry needs better standards. clear communication, user consent are crucial. With the integration of advanced technologies, such as infotainment systems for multimedia and mobile calls and messages with Bluetooth, there is potential for more extensive data collection. Manufacturers should transparently communicate what data is being collected and for what purposes. Privacy measures evolve with autonomous vehicle advancement. Balancing safety data collection remains crucial emphasizing privacy protection. Continuous monitoring adjusts to evolving environmental concerns. Privacy-preserving tech like federated learning, blockchain, and location masking protects privacy. As pointed out, transparent communication, policies, and user consent maintain this balance and build trust in autonomous vehicles and datas. IEEE standards is the place where we need to go to address these global evolving concerns and there is a new available course on data privacy and ethics offered by IEEE Educational activities" - Source: The Author, Derived from the IEEE Transmitter.

6.4.2.3 Challenges and Future Developments

Nanoscale edge computing faces challenges in size, power consumption, interconnectivity, scalability, security, and reliability. Standardization efforts are crucial for interoperability among diverse devices. Data management, AI implementation, and environmental sustainability are key considerations. Regulatory frameworks need to catch up with nanoscale edge computing advancements. Future developments must focus on reducing latency, revolutionizing healthcare applications, innovative programming paradigms, and sustainable e-waste management. The journey of nanoscale edge computing holds promise for transformative advancements across domains.

6.4.3 Tutorial on Developing Edge AI Applications

This section talks about the tutorial on developing edge AI Applications throughout, briefs the exploring open-source platforms and tools in Section 6.4.3.1, and describes the Hands-on Learning for Students in Section 6.4.3.2.

Developing edge AI applications requires precision and an in-depth understanding of technologies. This tutorial offers a step-by-step guide for creating applications that leverage the power of edge computing and AI.

Understanding Edge AI Fundamentals: Grasp the fundamentals of edge AI, understanding the synergy between AI and edge computing for effective application development.

Choosing Appropriate Edge Devices: Select edge devices based on application requirements, ensuring optimal performance for specific use cases.

Data Collection and Preprocessing: Initiate efficient data collection and preprocessing procedures to organize data for seamless integration into edge AI applications.

Selecting Suitable Edge AI Frameworks: Evaluate and choose edge AI frameworks judiciously, considering model compatibility, scalability, and ease of integration.

Optimizing Model Architecture: Optimize AI model architecture for maximum efficiency on edge devices, balancing complexity with computational constraints.

Implementing Edge-Friendly Algorithms: Implement lightweight and efficient algorithms tailored for edge computing, ensuring smooth execution within resource constraints.

Edge Device Security Measures: Prioritize security with encryption, authentication, and access controls to safeguard data at the device level.

Data Transmission Optimization: Optimize data transmission between edge devices and central systems to minimize latency and bandwidth usage.

Edge Device Management Protocols: Implement device management protocols for seamless operations, defining procedures for registration, updates, and maintenance.

Adapting to Dynamic Environments: Design applications that adapt to dynamic environments, considering changing network conditions and varying device loads.

Continuous Monitoring and Maintenance: Establish continuous monitoring and maintenance: protocols to assess device health, model performance, and data Accuracy regularly.

Edge AI Application Deployment: Strategically deploy edge AI applications considering network topology and device placement for maximum impact.

User-Friendly Interface Design: Prioritize a user-friendly interface design for intuitive and accessible interactions, enhancing the overall user experience.

Incorporating Edge AI in IoT Systems: Explore integration with IoT systems, ensuring seamless compatibility and data exchange between edge devices and IoT components.

Scalability Planning for Edge Networks: (cf. Figure 6.12) Plan for scalability in edge networks, anticipating potential growth of devices and data volume with offload computing.

Collaboration with Cloud Services: Consider Collaboration with cloud services to complement certain functionalities and ensure a holistic application ecosystem.

Compliance with Data Privacy Regulations: Prioritize compliance with data privacy regulations, adhering to regional and industry-specific guidelines.

Performance Metrics and Analytics: Implement performance metrics and analytics, defining KPIs to gauge success and efficiency of edge AI applications.

Edge AI Application Testing: Conduct rigorous testing for robustness, security, and performance under various conditions.

Edge AI Application Debugging: Develop effective debugging protocols for efficient issue identification and resolution.

Open-Source Tools and Resources: Leverage open-source tools, frameworks, and community support to accelerate the development process.

Energy-Efficient Edge Computing Practices: Emphasize energy-efficient practices in edge computing, optimizing algorithms and device operations for minimal consumption.

Customization for Industry-Specific Needs: Customize applications for industry-specific needs, tailoring functionalities for relevance and effectiveness.

User Training and Support: Provide user training and support to ensure end-users are familiar with application features and functionalities. Document the final edge AI application comprehensively, providing detailed instructions for setup, configuration, and troubleshooting.

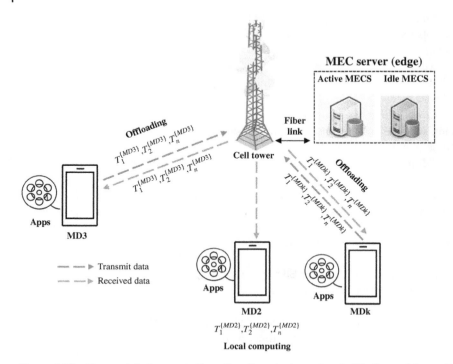

Figure 6.12 The model of computation offloading at edge network. File:Figure 5 https://doi.org/10.1016/j.iotcps.2023.02.004 Source: Raghubir Singh et al., 2023/Elsevier/CC BY 4.0/https://www.sciencedirect.com/science/article/pii/S2667345223000196.

This 25 step-by-step guide serves as a roadmap for developers navigating the challenges of edge AI application development.

6.4.3.1 Exploring Open-Source Platforms and Tools

Open-source platforms play a pivotal role in edge AI development, fostering innovation through collaborative environments. **Benefits of Open-source in Edge AI:** Open-Source platforms offer transparency, flexibility, and cost-effectiveness, allowing customization for diverse hardware, including nanomagnetic logic devices. **Versatility of Open-Source Tools:** From model development to deployment, open-source tools cater to diverse edge AI requirements, addressing the complexities of nanomagnetic logic devices. **Edge AI Development Landscape:** Open-source platforms dynamically adapt to emerging technologies, ensuring developers stay at the forefront of advancements in nanoscale computing.

1. **TensorFlow for Nanomagnetic Logic Devices:** TensorFlow, a popular open-source framework, enhances AI model development compatible with nanomagnetic logic devices.

2. **PyTorch for Nanoscale Computing:** PyTorch, another open-source framework, is valuable in developing AI applications for nanomagnetic logic devices with its dynamic computational graph and ease of use.

3. **Scikit-Learn for Edge AI Development:** Scikit-Learn, a versatile open-source machine learning library, supports edge AI development and rapid model prototyping.

4. **Apache OpenNLP for Natural Language Processing:** Apache OpenNLP, an open-source NLP library, aids in linguistic analysis for edge AI applications on logic devices.

5. **OpenCV for Computer Vision on the Edge:** OpenCV, a powerful open-source computer vision library, is indispensable for edge AI development, especially in applications utilizing NML.

6. **Jupyter Notebooks for Interactive Development:** Jupyter Notebooks, an open-source interactive development environment, facilitates experimentation and collaboration crucial for nanoscale computing applications.

7. **Docker for Containerization in Edge AI:** Docker, an open-source platform, streamlines deployment with lightweight containers, aligning with nanomagnetic logic device constraints.

8. **Kubernetes for Management in Edge Computing:** Kubernetes, an open-source container management platform, enhances scalability and reliability in edge AI applications.

9. **ONNX for Model Interoperability:** ONNX (Open Neural Network Exchange) standardizes model interoperability, ensuring compatibility between models and logic devices.

10. **Apache Kafka for Real-Time Data Streaming:** Apache Kafka, an open-source data streaming platform, efficiently handles real-time data challenges in edge AI applications.

11. **Eclipse Deeplearning4j for Distributed Training:** Eclipse Deeplearning4j, an open-source Distributed deep learning library, supports scalable model training for emerging logic devices.

12. **Apache Mahout for Scalable Machine Learning:** Apache Mahout, an open-source machine learning library, excels in scalability, supporting large-scale model training.

13. **TensorRT for AI Inference Optimization:** TensorRT, an open-source library by NVIDIA, focuses on optimizing AI inference for real-time performance on nanomagnetic logic devices.

14. **Open-Source Community Collaboration:** Open-source communities foster collaboration, accelerating advancements and promoting seamless integration of edge AI with nanomagnetic logic devices.

15. **GitHub Repositories and Version Control:** Leverage GitHub for Version Control in open-source projects to ensure integrity and reliability in edge AI developments for nanomagnetic logic devices.

16. **Continuous Integration with Jenkins:** Jenkins, an open-source automation server, supports continuous integration, enhancing the efficiency of edge AI development for nanomagnetic logic devices.

17. **GitLab for Integrated DevOps:** GitLab, an open-source DevOps platform, streamlines the development lifecycle, optimizing the deployment of edge AI applications.

18. **MLOps with Kubeflow:** Kubeflow, an open-source machine learning operations platform, facilitates MLOps, managing and deploying models for edge AI on nanomagnetic logic devices.

19. **Open-Source Quantum Computing Libraries:** Explore open-source quantum computing libraries for broadening horizons and harnessing the synergy between nanomagnetic logic devices and quantum computing (cf. Figure 6.13).

20. **Particle Swarm Optimization Algorithms:** Investigate open-source Particle Swarm Optimization Algorithms for optimizing model parameters in edge AI development, including emerging logic devices.

21. **Apache Flink for Stream Processing:** Apache Flink, an open-source stream processing framework, addresses real-time data challenges in edge AI applications, crucial for nanomagnetic logic devices.

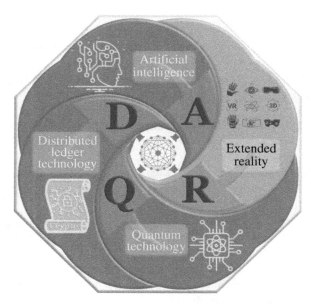

Figure 6.13 This iconographic presents DARQ Technologies with the name and icon representing each of the four DARQ technologies. File:DARQ Technologies.png https://commons.wikimedia.org/wiki/File:DARQ_Technologies.png. Source: DARQ Technologies.

22. **RapidMiner for Data Science Workflows:** RapidMiner, an open-source data science platform, supports end-to-end workflows, valuable for creating and deploying data-centric AI models for logic devices.

23. **Open-Source Reinforcement Learning Libraries:** Examine into Open-source Reinforcement Learning Libraries to enhance the development of AI applications that adapt and learn in real time, aligning with nanomagnetic logic devices' capabilities.

24. **DataRobot for Automated Machine Learning:** DataRobot, an open-source platform for automated machine learning, expedites model development, simplifying and accelerating the creation of AI models for emerging devices.

25. **Google Colab for Cloud-Based Collaboration:** Google Colab, a cloud-based open-source platform, supports collaborative AI model development, enhancing collaborative efforts crucial for nanoscale computing applications.

26. **Open-Source Federated Learning Frameworks:** Explore Open-source Federated Learning Frameworks for collaborative model training, ensuring privacy and efficiency, especially relevant in edge AI applications.

27. **Model Explainability with SHAP and LIME:** SHAP (SHapley Additive exPlanations) and LIME (Local Interpretable Model-agnostic Explanations) are open-source tools for model explainability, crucial for understanding model decisions in edge AI applications.

The exploration of 27 open-source platforms and tools for edge AI development beyond CMOS nanomagnetic logic devices unveils a vast ecosystem. Leveraging these tools strategically ensures the seamless integration of AI capabilities with the unique attributes of nanomagnetic logic devices.

6.4.3.2 Hands-On Learning for Students

1. **Introduction to Practical Learning:** Practical learning is essential for students entering the realm of edge AI development, bridging the gap between theoretical knowledge and real-world application.

2. **Importance of Hands-On Experience:** The importance of hands-on experience cannot be overstated, enhancing understanding, problem-solving skills, and adaptability in the dynamic field of nanoscale computing and edge AI.

3. **Setting up Development Environments:** Begin by setting up development environments, familiarizing students with tools and frameworks suitable for nanomagnetic logic devices, ensuring a smooth transition to practical applications.

4. **Accessing Nanomagnetic Logic Devices:** Provide hands-on access to nanomagnetic logic devices, either through physical devices or simulators, allowing students to comprehend the unique aspects of these devices in edge AI.

5. **Modeling AI Algorithms for Edge:** Guide students in modeling AI Algorithms for edge applications, emphasizing considerations needed for

nanomagnetic logic devices. Practical modeling helps students grasp the intricacies of aligning algorithms with device constraints.

6. **Experimenting with TensorFlow and PyTorch:** Encourage experimentation with TensorFlow and PyTorch, widely used frameworks invaluable for edge AI development. Hands-on activities solidify students' skills in implementing AI models.

7. **Real-Time Data Processing Challenges:** Present real-time Data Processing challenges for students to navigate, instilling problem-solving skills crucial for edge AI, considering the constraints of nanoscale computing.

8. **Deployment on Edge Devices:** Guide students in deploying models on edge devices, involving optimization for real-world applications. Understanding deployment is crucial for students to witness the direct impact of their work.

9. **Utilizing Docker for Edge Environments:** Introduce Docker for edge environments, ensuring students grasp containerization's role in deploying applications on nanomagnetic logic devices. Hands-on experience ensures familiarity with efficient resource utilization.

10. **Kubernetes for Scalable Deployments:** Explore Kubernetes for Scalable Deployments, exposing students to orchestration principles and ensuring readiness to manage diverse applications, including those running on nanomagnetic logic devices.

11. **Collaboration with Open-Source Tools:** Emphasize collaboration with open-source tools, encouraging active contributions to projects and leveraging collective knowledge in the open-source community. This collaboration introduces students to diverse perspectives in edge AI development.

12. **Continuous Integration with Jenkins:** Implement continuous integration using Jenkins, streamlining students' understanding of the iterative development cycle crucial in edge AI with nanomagnetic logic devices.

13. **Experimentation with Quantum-Inspired Concepts:** Encourage experimentation with quantum-inspired concepts, allowing students to explore their enhancement in edge AI. Practical exposure prepares them for the evolving landscape of nanoscale computing.

14. **Hands-On Quantum Computing Libraries:** Introduce hands-on sessions with open-source quantum computing libraries, providing practical exploration into the potential synergy between quantum computing and edge AI.

15. **Simulation of Nanomagnetic Logic:** Simulate nanomagnetic logic in practical exercises, allowing students to experience the behavior of these devices firsthand, crucial for effective edge AI development.

16. **AI Model Explainability Tools:** Include AI model explainability tools in hands-on activities, helping students interpret model decisions, especially in edge scenarios.

17. **Real-world edge AI challenges:** Present real-world edge AI Challenges for students to tackle, reinforcing their adaptability and innovation. Challenges mirror industry scenarios, preparing students for diverse edge AI applications.
18. **Student-Led Edge AI Projects:** Facilitate Student-led edge AI projects, encouraging creativity and independence. These projects reflect the interdisciplinary nature of edge AI with nanomagnetic logic devices.
19. **Documentation and Reporting Skills:** Emphasize documentation and reporting skills, ensuring students can articulate their hands-on experiences effectively, vital for edge AI professionals.
20. **Mock Industry Collaboration:** Arrange mock industry collaboration scenarios, simulating industry interactions for a glimpse into the collaborative nature of edge AI projects.
21. **User-Friendly Interface Design:** Guide students in user-friendly interface Design, focusing on creating interfaces that enhance user experience in edge AI.
22. **Security Protocols in Edge AI:** Integrate security protocols into hands-on exercises, requiring students to actively implement encryption and secure communication methods, crucial for real-world cybersecurity challenges.
23. **Scalability Planning for Edge Networks:** Incorporate scalability planning into practical sessions, ensuring students consider the scalability of their edge AI solutions.
24. **Continuous Learning Through Challenges:** Encourage continuous learning through challenges, introducing new edge AI concepts regularly to keep students engaged in a dynamic learning environment.
25. **Industry Expert Sessions:** Organize sessions with industry experts, providing practical insights into edge AI and nanoscale computing, supplementing students' academic learning.
26. **Hands-On Exploration of Federated Learning:** Facilitate hands-on exploration of federated learning, allowing students to actively participate in creating and testing federated learning models.
27. **Mock Edge AI Startup Projects:** Simulate edge AI startup projects, engaging students in startup-like scenarios for exposure to the fast-paced and innovative nature of entrepreneurial ventures.
28. **Remote Edge AI Collaboration:** Promote remote collaboration in edge AI projects, preparing students for the evolving work landscape with practical experiences in virtual teamwork.

Hands-on learning for students in edge AI development with nanomagnetic logic devices is crucial. Practical experiences reinforce theoretical knowledge and cultivate essential skills, ensuring the next generation of professionals is well-equipped for the challenges and innovations in this dynamic field.

6.5 Conclusion

This section talks about the conclusion, briefs the summarizing key concepts in Edge AI in Section 6.5.1, and describes the Chapter End Quiz in Section 6.5.2.

6.5.1 Summarizing Key Concepts in Edge AI

This section talks about the summarizing key concepts in edge AI throughout, briefs the preparation for hybrid computing discussions in subsequent chapters in Section 6.5.1.1, and describes the Chapter End Quiz in Section 6.5.2.

Edge AI applications vary widely, impacting real-world scenarios. Nanoscale computing enhances processing power and efficiency in edge AI. Open-source collaboration is fundamental for knowledge sharing and development. Challenges in edge AI act as catalysts for innovation. Real-world implementations and success stories validate the practicality of edge AI. Continuous learning and hands-on experience cultivate a dynamic skill set. Edge AI is inherently interdisciplinary, integrating knowledge from various domains. Security and scalability considerations are paramount in edge AI solutions. Quantum-inspired concepts add a futuristic dimension to edge AI. Open-source tools play a pivotal role, offering flexibility and community support.

Hands-on learning is indispensable for students bridging the gap between theory and practice. Simulating industry collaboration and startup projects prepares professionals for real-world scenarios. User-friendly interface design and explainability contribute to user trust and acceptance. Remote collaboration is increasingly relevant, preparing professionals for virtual work environments. Continuous exploration of new concepts is vital for professionals to stay at the forefront of innovations. Industry expert sessions and networking opportunities provide valuable insights and mentorship. Federated learning and quantum computing libraries signify the future of collaborative approaches. Key concepts in edge AI revolve around adaptability, collaboration, and continuous learning.

6.5.1.1 Preparation for Hybrid Computing Discussions in Subsequent Chapters

Exploring hybrid computing for comprehensive understanding begins with defining its technological landscape. This includes examining the integration of classical and quantum computing, incorporating quantum-inspired concepts, and identifying practical applications and real-world use cases. The analysis extends to evaluating the strengths and limitations of hybrid models and the role of quantum computing libraries in such systems.

The investigation highlights collaborative approaches, robust security protocols, and quantum-enhanced edge AI within hybrid setups. It incorporates nanoscale computing principles to refine these models and formulates effective data management strategies. The interdisciplinary nature of hybrid computing is emphasized, leveraging open-source platforms for development.

Key challenges are addressed, with solutions proposed, alongside a step-by-step guide for building hybrid computing applications. Federated learning techniques are explored in this context, narrowing down in a structured foundation for deeper exploration in subsequent chapters.

6.5.2 Chapter End Quiz

This section presents the Chapter End Quiz throughout.

An online Book Companion Site is available with this fundamental textbook in the following link: www.wiley.com/go/sivasubramani/nanoscalecomputing1

For Further Reading:

Passian and Imam (2019), Tiwari et al. (2021), Gill (2021a), Gill (2021b), Wu et al. (2023), Silva et al. (2021), Covi et al. (2021), Ramasubramanian et al. (2022), Bahls et al. (2019), Al-Turjman (2019), Ibn-Khedher et al. (2021), Ahmad et al. (2022), Su et al. (2022), Situnayake and Plunkett (2023), Aksun-Guvenc and Sivasubramani (2024).

References

Tanveer Ahmad, Hongyu Zhu, Dongdong Zhang, et al. Energetics systems and artificial intelligence: Applications of industry 4.0. *Energy Reports*, 8:334–361, 2022.

Bilin Aksun-Guvenc and Santhosh Sivasubramani. 5 ways your car has become a computer on wheels, February 2024. https://transmitter.ieee.org/5-ways-your-car-has-become-a-computer-on-wheels/.

Fadi Al-Turjman. *Internet of Nano-things and Wireless Body Area Networks (WBAN)*. CRC Press, 2019.

Thomas Bahls, Markus Bihler, Julian Klodmann, Freek Stulp, et al. Robotic demands on on-board data processing. In *European Workshop on On-board Data Processing (OBDP2019)*, 2019.

Erika Covi, Elisa Donati, Xiangpeng Liang, et al. Adaptive extreme edge computing for wearable devices. *Frontiers in Neuroscience*, 15:611300, 2021.

Sukhpal Singh Gill. Quantum and blockchain based serverless edge computing: A vision, model, new trends and future directions. *Internet Technology Letters*, page e275, 2021a.

Sukhpal Singh Gill. Quantum and blockchain based serverless edge computing: A vision, model, new trends and future directions. *Internet Technology Letters*, page e275, 2021b.

Hatem Ibn-Khedher, Mohammed Laroui, Mouna Ben Mabrouk, et al. Edge computing assisted autonomous driving using artificial intelligence. In *2021 International Wireless Communications and Mobile Computing (IWCMC)*, pages 254–259. IEEE, 2021.

Pawan K. Tiwari, Mugdha Sahu, Gagan Kumar, and Mohsen Ashourian. Pivotal role of quantum dots in the advancement of healthcare research. *Computational Intelligence and Neuroscience*, 2021:1–9, 2021.

Ali Passian and Neena Imam. Nanosystems, edge computing, and the next generation computing systems. *Sensors*, 19(18):4048, 2019.

Aswin K Ramasubramanian, Robins Mathew, Inder Preet, and Nikolaos Papakostas. Review and application of edge ai solutions for mobile collaborative robotic platforms. *Procedia CIRP*, 107:1083–1088, 2022.

Mateus C Silva, Jonathan CF da Silva, Saul Delabrida, et al. Wearable edge ai applications for ecological environments. *Sensors*, 21(15):5082, 2021.

Daniel Situnayake and Jenny Plunkett. *AI at the Edge.* O'Reilly Media, Inc., 2023.

Weixing Su, Linfeng Li, Fang Liu, Maowei He, and Xiaodan Liang. Ai on the edge: a comprehensive review. *Artificial Intelligence Review*, 55(8):6125–6183, 2022.

Guangjian Wu, Zefeng Zhang, Kun Ba, Xumeng Zhang, et al. Hybrid mos 2/pbs quantum dots toward in-sensor reservoir computing. *IEEE Electron Device Letters*, 44(5):857–860, 2023.

7

Hybrid Computing Systems and Emerging Applications

Specific, Measurable, Achievable, Relevant, and Time-bound (SMART) learning objectives and goals for this chapter:- after reading this chapter you should be able to:

- Identify the core principles of hybrid computing systems.
- Evaluate the efficiency gains in real-world applications using nanomagnetic logic (NML) and complementary metal oxide-semiconductor (CMOS) hybrid architectures.
- Demonstrate the integration of hybrid computing in smart manufacturing and robotics.
- Analyze the ethical considerations associated with the deployment of hybrid computing technologies.
- Within a specified time frame, summarize key concepts related to hybrid computing systems and their applications.
- Understand the implications of hybrid computing in healthcare delivery and personalized medicine.
- Assess the educational emphasis on demystifying hybrid computing concepts and preparing students for industry challenges.
- Formulate recommendations for responsible innovation in the field of hybrid computing.

This chapter talks about the introduction to hybrid computing in Section 7.1, briefs the nanomagnetic-CMOS hybrid architectures in Section 7.2, describes the Neuromorphic Hybrid Systems in Section 7.3, and elaborates the educational emphasis in Section 7.4.

Nanoscale Computing: The Journey Beyond CMOS with Nanomagnetic Logic,
First Edition. Santhosh Sivasubramani.
© 2025 The Institute of Electrical and Electronics Engineers, Inc. Published 2025 by John Wiley & Sons, Inc.
Companion website: www.wiley.com/go/sivasubramani/nanoscalecomputing1

7.1 Introduction to Hybrid Computing

This section talks about the introduction to hybrid computing and briefs the defining hybrid computing systems in Section 7.1.1.

In advanced computing, hybrid computing integrates traditional and emerging technologies. NML architecture is essential in this fusion, enabling information processing at the nanoscale by leveraging magnetic permeability properties (cf. Figure 7.1). This approach addresses the limitations of conventional electronic systems. By combining classical and quantum computing elements, hybrid computing enhances adaptability and efficiency, especially in edge artificial intelligence (AI) devices.

7.1.1 Defining Hybrid Computing Systems

This section talks about the defining hybrid computing Systems throughout, briefs the Integration Avenues in Section 7.1.1.1, and describes the historical evolution and milestones in Section 7.1.1.2. Hybrid computing systems blend CMOS with alternative emerging technologies. NML a promising post-CMOS candidate, offers non volatility and low-power consumption. These systems leverage the unique characteristics of NML creating a balanced integration with CMOS. They also tend to offer superior computational capabilities beyond CMOS.

7.1.1.1 Integration Avenues

Hybrid systems benefit from materials with high permeability. They enhance magnetic component efficiency in integrated designs (cf. Figure 7.2). Integration with spintronic devices opens further possibilities. Spintronics leverages electron spin for information processing. This convergence promotes the development of magnetic quantum computing systems. NML aid in error correction techniques. It enhances reliability in quantum information processing. Integration efforts require advancements in fabrication techniques. Precise control at the nanoscale is essential. Research focuses on improving interconnects between components. Efficient interconnects are crucial for integrated systems. These efforts aim to

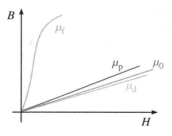

Figure 7.1 Magnetic permeability (not to scale): B—flux density, H—magnetic field, μ_f—permeability of ferromagnets (and ferrimagnets), μ_p—permeability of paramagnets, μ_0—permeability of free space ($4\pi \times 10^{-7}$ H/m), μ_d—permeability of diamagnets. File: Permeability by Zureks.svg https://commons.wikimedia.org/wiki/File:Permeability_by_Zureks.svg Source: Zureks/Wikimedia Commons/Public Domain.

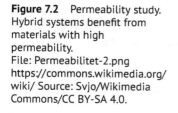

Figure 7.2 Permeability study. Hybrid systems benefit from materials with high permeability.
File: Permeabilitet-2.png
https://commons.wikimedia.org/wiki/ Source: Svjo/Wikimedia Commons/CC BY-SA 4.0.

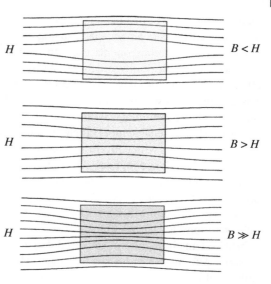

minimize signal loss. Advanced materials play a pivotal role here. They improve performance and compatibility with existing technologies. Collaboration between disciplines accelerates integration progress. This interdisciplinary approach fosters innovative solutions. Standardization of fabrication processes is also necessary. It ensures compatibility across different technological platforms. Continuous research drives integration toward practical applications.

On the other hand, nanomagnets operate at room temperature effectively compared to its Electronic QCA counterparts. Their non volatility ensures energy-efficient data storage capabilities. This combination achieve faster processing speeds and reduced power consumption. NML circuits exhibit unique properties beneficial for integration. They possess high endurance and radiation hardness attributes. This property suits applications in space and harsh environments. NML can be interfaced with photonic devices. This combination facilitates ultrafast data transmission and processing. Hybrid systems merge magnetic and electronic components seamlessly. Such integration enables new paradigms in computing architectures. Magnetic components are used for memory elements and can handle computational tasks concurrently. This hybrid approach optimizes overall system performance.

Motivation for Hybrid Approaches: Hybrid approaches combine strengths of diverse technologies. They enhance overall system performance and capabilities. Integration of various components addresses specific limitations. Hybrid systems leverage the best features available. This approach optimizes efficiency, speed, and power consumption. Different technologies complement each other's strengths. Hybrid approaches enable innovative solutions in computing. They allow for tailored responses to diverse challenges. Combining analog and digital components for NML offers advantages. Analog components handle signal processing efficiently. Digital components excel in data manipulation tasks. This synergy enhances system flexibility and functionality. Hybrid systems also improve energy efficiency. They reduce power consumption without compromising performance. These approaches support scalable and adaptable designs. Flexibility is crucial in rapidly evolving technological landscapes. Hybrid approaches facilitate advancements in AI. They enable more efficient machine learning algorithms. Integration of neural networks with conventional processors is possible.

This combination enhances AI computational capabilities. Hybrid systems support advancements in the Internet of Things (IoT). They provide efficient data processing at the edge. This reduces latency and bandwidth requirements significantly. Hybrid approaches are motivated by diverse application demands. They offer solutions for varied industrial sectors. Medical technology benefits from hybrid systems integration. Advanced diagnostics and monitoring become more feasible. Automotive industries leverage hybrid approaches for smart vehicles. Enhanced safety and autonomous driving features are enabled. Telecommunications benefit from improved data transmission capabilities. Hybrid systems enhance reliability in communication networks. They support high-speed data processing and storage. Hybrid approaches foster innovation across multiple fields. This interdisciplinary synergy drives technological progress. Continuous research and development are essential. They refine hybrid systems and expand their applications. NMLs computational efficiency addresses challenges, and the integration with other technologies in hybrid systems provides real-time solutions, transforming industries.

7.1.1.2 Historical Evolution and Milestones

The historical evolution of hybrid computing traces back decades, reflecting continuous innovation. From early computing with vacuum tubes to the advent of NML, each milestone represents a commitment to efficiency and adaptability. Hybrid architectures, combining diverse technologies, stand as a testament to the ever-evolving nature of computational solutions beyond CMOS. Early computing used vacuum tubes for processing. Transistors replaced tubes, marking significant progress. Integrated circuits emerged in the 1960s, revolutionizing computing. Miniaturization became a defining milestone in the 1970s. CMOS technology emerged as energy efficient and scalable in the 1970s. Parallel computing systems rose in prominence during the 1990s. The 2000s introduced the concept of quantum computing. Neuromorphic computing gained attention in the 2010s. Hybrid computing systems began incorporating diverse architectures. NMLs inception marked a significant milestone. Integration with CMOS signaled progress in the field. Research on NML surged in the 2010s. Diverse applications became

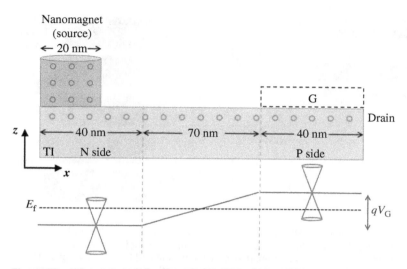

Figure 7.3 Schematics of the discretization of a TIPNJ with a nanomagnet as the source contact. The ferromagnetic contact has dimension 20 nm along the x-direction. The TI surface is 150 nm long with 40 nm on the P and N side and 70 nm transitioning from N to P side. The gate on the drain side can swing the local TI surface to P-type. File: Figure 2 https://doi.org/10.1038/s41598-023-35623-5. Source: Yunkun Xie et al., 2023/Springer Nature/CC BY 4.0.

the focus of hybrid systems. Advancements in nanotechnology and magnetic materials fueled progress. NMLs potential expanded across various industries. Hybrid systems tailored for specific use-cases gained attention. The schematics of a TIPNJ with varying dimensions and gate control illustrate how precise tuning of nanomagnetic and semiconductor elements in hybrid systems can be tailored for specific use-cases (cf. Figure 7.3).

Real-world applications became the litmus test for hybrids. Hybrid computing's evolution reflects contemporary application demand. NMLs integration reshaped computing in various domains. NMLs journey beyond CMOS unfolded in distinct phases. NMLs historical evolution reflected adaptability to technological shifts. Each milestone laid the foundation for subsequent advancements. The applications driving hybrid systems beyond CMOS are diverse and impactful. A study (cf. Figure 7.4) demonstrates voltage-controlled magnetic phase transitions in Ising-like chain and square lattice structures, highlighting the potential of such systems in real-world applications. The ability to control magnetic phases with voltage, as shown in various MFM and SEM images, underscores the adaptability and efficiency of hybrid architectures integrating NML. From healthcare to gaming, these practical solutions are envisaged to showcase the significant technological advancements. The continuous evolution in these sectors demonstrates

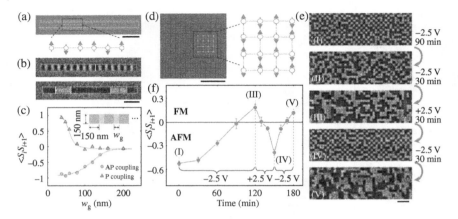

Figure 7.4 (a–f) The figure shows SEM and schematic images of a 1D Ising-like chain with regions of different widths, along with MFM images of coupling states and the nearest-neighbor correlation function. It also includes SEM and MFM images of a 2D Ising-like square lattice, demonstrating reversible magnetic phase transitions and the effect of gate voltages on the lattice's correlation function. The figure illustrates how voltage-controlled magnetic phase transitions in nanomagnetic structures can be integrated into hybrid CMOS nanomagnetic logic systems. Figure 4: Voltage-controlled magnetic phase transition. https://doi.org/10.1038/s41467-023-41830-5 Source: Chao Yun et al., 2023/Springer Nature/CC BY 4.0.

the real-world relevance of hybrid computing. These future projections and prediction intervals marks a significant chapter in the technological landscape (cf. Figure 7.5).

7.2 Nanomagnetic-CMOS Hybrid Architectures

This section talks about the nanomagnetic-CMOS hybrid architectures, and briefs the benefits and performance enhancements in Section 7.2.1. Nanomagnetic-CMOS hybrid architectures integrate nanomagnets with CMOS technology. They combine the strengths of both systems. CMOS provides fast switching speeds. Nanomagnets offer non volatility and low power. Magnetoresistive random access memory (MRAM) is a prominent example. It utilizes nanomagnetic elements for data storage. CMOS circuits handle control and read/write operations. MRAM ensures data retention without power. It provides faster access than Flash memory. NML operates at room temperature. It relies on magnetic field coupling for information processing. This approach consumes less energy than CMOS-only circuits. Nanomagnets act as the primary logic gates. CMOS components manage peripheral tasks. Majority logic gate (MLG) is a fundamental NML component.

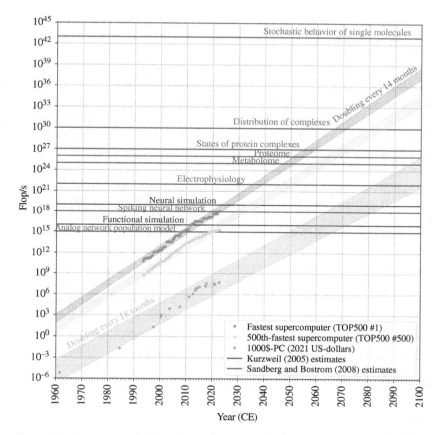

Figure 7.5 Future projections and prediction intervals (95% prediction confidence) using linear regression envisaging hybrid approaches. File: Whole brain emulation.svg
https://commons.wikimedia.org/wiki/File:Whole_brain_emulation.svg
Source: Notyetborn/Wikimedia Commons/CC BY-SA 4.0.

It executes majority voting operations. Three input nanomagnets determine the output state. MLGs enhance computation accuracy and reliability. CMOS circuits assist in initialization and stabilization. Nanomagnetic-CMOS hybrids leverage spin-transfer torque (STT). STT switches the magnetization state. This method consumes less power than traditional magnetic fields. STT-MRAM exemplifies this hybrid approach. CMOS circuits control STT pulses for switching. Domain wall logic (DWL) represents another hybrid model. It manipulates domain walls within nanowires. CMOS circuits generate the required current pulses. DWL offers scalable and energy-efficient logic operations. It reduces interconnect delay and power dissipation. Three-terminal magnetic tunnel junctions (MTJs) serve as key elements. MTJs provide high resistance contrast between states. CMOS circuits facilitate read/write access to MTJs. They ensure reliable operation and high integration density.

Nanomagnetic computing architectures include clocked and clockless designs. Clocked designs synchronize operations using external signals. CMOS circuits generate these clock signals. Clockless designs operate asynchronously, relying on intrinsic delays. Energy-efficient data storage benefits from hybrid designs. Nanomagnetic elements retain data without power. CMOS circuits enable fast data retrieval. This synergy improves overall energy efficiency and performance. Field-coupled nanomagnetic logic (FCNL) relies on near-field magnetic interactions. It reduces power consumption compared to long-range couplings. CMOS components manage signal propagation and synchronization. FCNL enhances logic density and operational speed. Voltage-controlled magnetic anisotropy (VCMA) modulates nanomagnet properties using electric fields. This technique offers energy-efficient switching. CMOS circuits apply and control these electric fields. VCMA-based hybrids improve write energy efficiency. Nanomagnetic interconnects replace traditional metal wires. They transmit signals using magnetic interactions. CMOS drivers generate the necessary magnetic fields. This approach reduces resistive losses and improves signal integrity. Heat-assisted magnetic recording (HAMR) employs localized heating. It enhances nanomagnet switching efficiency. CMOS circuits generate and control heat pulses. HAMR-based hybrids achieve higher data densities. Multiferroic materials combine magnetic and electric properties. They enable electric field control of magnetization. CMOS circuits provide the electric field inputs. Multiferroic-based hybrids achieve non volatile, low-power operations.

Hybrid structures incorporate spin waves for information transfer. Spin waves propagate through magnetic materials. CMOS circuits generate and detect these waves. Spin wave-based hybrids offer low-power, high-speed data transmission. Racetrack memory (RM) employs magnetic domain walls for storage. It uses current pulses for domain wall movement. CMOS circuits generate these current pulses. RM-based hybrids provide high-density, non volatile memory solutions. Nanomagnetic-CMOS hybrids address the limitations of each technology. They combine non volatility, high speed, and low power. This synergy enables advanced computing applications. Error-correction mechanisms are crucial for reliable operation. CMOS circuits implement these mechanisms. They ensure data integrity and fault tolerance in hybrids. Error-correction codes (ECCs) are commonly used. Hybrid architectures facilitate three-dimensional (3D) integration. Nanomagnetic layers stack over CMOS circuits. This approach increases memory density and performance. It leverages the vertical dimension for compact designs. Spintronic devices exploit electron spin for data processing. They offer non volatility and high speed. CMOS circuits manage spintronic device operations. Spintronic-CMOS hybrids enhance computational capabilities.

Skyrmions are topological magnetic structures. They offer stability and low energy manipulation. CMOS circuits control skyrmion generation and movement. Skyrmion-based hybrids provide robust, energy-efficient computing. Nanomagnetic elements exhibit high thermal stability. They maintain functionality under extreme temperatures. CMOS circuits integrate thermal management features. This combination ensures reliable operation in harsh environments. Power gating techniques reduce static power consumption. Nanomagnetic elements retain state without power. CMOS circuits control power gating operations. This synergy enhances energy efficiency in idle states. Nanomagnetic-CMOS hybrids support neuromorphic computing. They mimic neural network functionalities. CMOS circuits implement synaptic connections. Nanomagnetic elements represent neurons and synapses. This architecture enables brain-like processing. Non volatile processors integrate nanomagnetic storage elements. They maintain computational states without power. CMOS circuits execute processing tasks.Non volatile processors enhance energy efficiency and instant-on capabilities. Programmable logic devices benefit from hybrid designs. Nanomagnetic elements store configuration data. CMOS circuits handle reconfiguration operations. This approach offers non volatile, reprogrammable logic solutions. High-frequency operations exploit nanomagnetic resonators. They provide stable oscillations at desired frequencies. CMOS circuits manage tuning and signal amplification. This combination achieves precise, high-speed signal generation. Quantum-dot cellular automata (QCA) utilize quantum dots for computation. They represent binary states through electron positions. CMOS circuits control electron placement. QCA-CMOS hybrids offer high-density, low-power logic operations. Nanoelectromechanical systems (NEMs) combine nanomagnets with mechanical motion. They enable ultra low power switching. CMOS circuits control mechanical actuation. NEMS-based hybrids achieve efficient, compact device integration.

Thermally-assisted switching (TAS) enhances nanomagnet efficiency. Localized heating reduces switching energy. CMOS circuits generate and control heat pulses. TAS-based hybrids achieve low-power, high-speed memory operations. Magnetoelectric coupling (MEC) integrates magnetic and electric functionalities. Electric fields control magnetic states. CMOS circuits provide electric field inputs. MEC-based hybrids offer energy-efficient, non volatile operations. Magnetic vortex structures exhibit stable magnetic configurations. They provide reliable data storage. CMOS circuits control vortex creation and manipulation. Vortex-based hybrids enhance data retention and stability. Hybrid designs optimize interconnect architectures. Nanomagnetic interconnects reduce parasitic capacitance. CMOS circuits manage signal transmission. This approach enhances signal integrity and reduces power consumption. Nanomagnetic oscillators generate high-frequency signals. They exploit spin-torque oscillation mechanisms. CMOS circuits handle frequency tuning and signal processing. Nanomagnetic-CMOS oscillators achieve precise, high-speed operations. Data encryption benefits from hybrid architectures. Nanomagnetic elements store encryption keys securely. CMOS circuits execute encryption algorithms. This combination enhances data security and processing speed. Magnetic domain wall motion provides high-speed data access. It reduces latency in memory operations. CMOS circuits generate and control domain wall movement. This synergy improves memory performance and efficiency. Non volatile flip-flops incorporate nanomagnetic storage elements. They retain state without power. CMOS circuits manage state transitions. Non volatile flip-flops enhance energy efficiency and data retention. Hybrid logic gates combine CMOS and nanomagnetic components. They exploit the strengths of both technologies. This approach achieves low-power, high-speed logic operations. It enhances computational efficiency and scalability. Magnetic logic units (MLUs) integrate NML gates. They perform complex computational tasks. CMOS circuits manage control and synchronization. MLUs enhance processing capabilities and power efficiency.

Nanomagnetic sensors detect magnetic fields. They offer high sensitivity and low–power consumption. CMOS circuits process sensor data. This combination enables accurate, energy-efficient sensing applications. Hybrid memory architectures combine SRAM and MRAM. They leverage the benefits of both technologies. SRAM provides fast access. MRAM offers non volatility. This approach achieves balanced performance and power efficiency. Multilevel cell (MLC) storage increases data density. Nanomagnetic elements store multiple bits per cell. CMOS circuits manage read/write operations. MLC-based hybrids enhance memory capacity and efficiency. High-speed signal processing benefits from hybrid designs. Nanomagnetic elements provide fast switching. CMOS circuits handle signal routing and amplification. This combination achieves low-latency, high-throughput processing. Resistive switching mechanisms enhance data retention. Nanomagnetic elements switch resistance states. CMOS circuits control resistance modulation. This approach improves memory reliability and performance. Quantum computing exploits nanomagnetic qubits. They offer stable quantum states for computation. CMOS circuits manage qubit control and readout. Nanomagnetic-CMOS hybrids enable scalable, efficient quantum computing. Embedded systems benefit from hybrid architectures. Nanomagnetic elements provide non volatile storage. CMOS circuits execute computational tasks. This approach enhances system reliability and energy efficiency. Field Programmable Gate Arrays (FPGAs) integrate nanomagnetic elements. They enable non volatile, reconfigurable logic. CMOS circuits manage configuration and control. Hybrid FPGAs achieve high performance and flexibility.

Low-power wireless communication leverages hybrid designs. Nanomagnetic elements enhance signal modulation. CMOS circuits handle transmission and reception. This combination achieves energy-efficient communication. Hybrid architectures support secure data storage. Nanomagnetic elements protect sensitive information. CMOS circuits implement encryption and access control. This approach enhances data security and integrity. Nanomagnetic diodes rectify current flow. They offer low forward voltage drop. CMOS circuits manage current routing. This combination achieves efficient power conversion. Reconfigurable logic gates adapt to varying tasks. Nanomagnetic elements store configuration data. CMOS circuits control reconfiguration. This approach enhances flexibility and performance. Hybrid designs improve analog signal processing. Nanomagnetic elements provide stable reference signals. CMOS circuits handle amplification and filtering. This combination achieves accurate, low-noise signal processing. Sensor networks benefit from hybrid architectures. Nanomagnetic sensors detect environmental changes. CMOS circuits process and transmit sensor data. This approach enhances network efficiency and responsiveness. Energy harvesting systems integrate nanomagnetic elements. They convert ambient energy into electrical power. CMOS circuits manage energy storage and utilization. This combination enables sustainable, low-power devices. Key factors, characteristics and envisaged tech-trend projection includes:

1. **Seamless Integration Potential:** Nanomagnetic-CMOS hybrids seamlessly combine magnetic and semiconductor technologies, opening avenues for novel computational paradigms.
2. **Enhanced Energy Efficiency:** The integration of NML with CMOS promotes improved energy efficiency, crucial for sustainable and eco-friendly computing practices.
3. **Low-Latency Processing:** Nanomagnetic-CMOS hybrid architectures excel in low-latency processing, achieving swift computations through magnetic logic's inherent advantages.
4. **Real-Time Decision Support:** The hybridization facilitates real-time decision support systems, benefiting applications from medical diagnostics to emergency response.
5. **Synergistic Computational Speed:** Combining NML with CMOS achieves efficient computational speed, particularly advantageous in financial analytics.
6. **Robust Cybersecurity Measures:** Hybrid architectures enhance Cybersecurity with NML and CMOS collaboration. It results in more robust defense mechanisms against evolving cyber threats.
7. **Adaptability to Diverse Workloads:** Nanomagnetic-CMOS hybrids exhibit adaptability to diverse workloads, crucial for handling varied computational demands.

8. **Scalability and Versatility:** The hybrid systems ensure scalability and versatility in computational frameworks, supporting diverse applications.
9. **Optimized Memory Utilization:** The integration optimizes memory usage in hybrid architectures, essential for efficient computational processes.
10. **Efficient Parallel Processing:** Nanomagnetic-CMOS hybrids support efficient parallel processing capabilities, enhancing computational efficiency, especially in scientific simulations.
11. **Minimized Computational Errors:** Minimizing errors in computations is a hallmark of hybrid architectures. NML, with its accuracy, reduces computational errors.
12. **Interoperability Across Platforms:** The integration ensures Interoperability across various computational platforms. Nanomagnetic-CMOS hybrids seamlessly interact with established systems.
13. **Realizing the Potential of Quantum Architectures:** Nanomagnetic-CMOS hybrid architectures pave the way for exploring quantum computing potential, inspiring quantum-inspired computational strategies.
14. **Data Security Enhancement:** The collaboration enhances data security measures in hybrid systems, contributing to secure and private computational operations.
15. Other parameters include **Optimal Big Data Management, Adapting to Technological Shifts, Cognitive Workload Reduction, Contributing to Sustainable Development, Meeting the Demand for High Data Transfer Speed, Enhancing Computational Efficiency in IoT Integration, Precision in Autonomous Vehicle Decision-Making, Optimizing Power Grid Management, Revolutionizing Educational Technologies,** and **Transforming Entertainment Streaming Experiences**.

7.2.1 Benefits and Performance Enhancements

This section talks about the benefits and performance enhancements throughout, briefs the overview of quantum computing principles in Section 7.2.1.1, describes the integrating quantum and nanomagnetic logic in Section 7.2.1.2, presents in detail the enhancing quantum computing with nanoscale technologies in Section 7.2.1.3, briefs the applications and challenges in Section 7.2.1.4, describes the quantum-enhanced information processing in Section 7.2.1.5, and presents in detail the quantum communication with nanoscale components in Section 7.2.1.6.

Nanomagnetic-CMOS hybrid architectures present numerous advantages. Reduced energy consumption tops the list. Nanomagnets enable low-power logic operations. Their non volatility ensures zero static power dissipation. CMOS circuits benefit from this power efficiency. They handle tasks with reduced power overheads. Nanomagnets exhibit inherent bistability. This characteristic simplifies binary data representation. The reduced need for refresh cycles follows. CMOS circuits achieve higher energy efficiency. Enhanced performance arises from faster switching speeds. Nanomagnets switch states rapidly. Their sub-nanosecond switching is noteworthy. CMOS logic gates leverage this speed. It ensures rapid signal processing. The resulting architecture minimizes latency. It provides quicker data processing. Improved speed benefits computational tasks. The hybrid system executes instructions faster. This is vital for high-frequency applications.

Scalability marks another key benefit. Nanomagnets are minuscule in size. They fit well within CMOS fabrication processes. Their integration does not necessitate significant alterations. CMOS circuits maintain their compact form. This allows for denser circuit designs. Enhanced scalability is achieved without compromising performance. Device miniaturization continues seamlessly. The hybrid system supports advanced scaling trends. Thermal stability is a prominent feature. Nanomagnets exhibit high thermal resilience. They operate effectively at elevated temperatures. CMOS circuits benefit from this stability. The hybrid architecture can function in harsh environments. It maintains reliability under thermal stress. This extends the operational lifespan of devices. Enhanced thermal stability also reduces cooling requirements. Compatibility with existing CMOS technology is essential. Nanomagnets integrate seamlessly into standard processes. This eliminates the need for new fabrication infrastructure. CMOS foundries can easily adopt these hybrids. Existing tools and techniques remain useful. This lowers the transition barrier for manufacturers. The technology integrates smoothly into production lines.

Reduced leakage current is another advantage. Nanomagnets exhibit zero off-state leakage. This characteristic is crucial for low-power applications. CMOS circuits benefit significantly from this feature. They experience reduced energy loss. This contributes to overall energy efficiency. Devices operate longer on the same power source. Improved data retention capabilities are notable. Nanomagnets offer non volatile data storage. Data remains intact without power. CMOS circuits utilize this feature for better memory retention. This enhances the reliability of stored information. Systems can recover data after power interruptions. Improved data retention supports robust computing environments. Noise immunity is a significant benefit. Nanomagnets exhibit strong resistance to electrical noise. This ensures stable operation in noisy environments. CMOS circuits leverage this immunity. They experience fewer errors due to noise. The hybrid architecture maintains signal integrity. Reliable data processing becomes achievable.

Reduced manufacturing costs follow from integration ease. Nanomagnets' compatibility with CMOS processes simplifies production. No need for expensive new equipment. Manufacturing costs remain low. This benefits large-scale production. It makes advanced hybrid systems economically viable. Enhanced durability is a key advantage. Nanomagnets exhibit high wear resistance. They withstand mechanical stresses well. CMOS circuits benefit from this durability. The hybrid architecture can endure rigorous usage. It ensures long-term reliability. This is critical for devices in demanding applications. Increased functional density arises from integration. Nanomagnets occupy minimal space. They enable more functions within the same area. CMOS circuits can host more components. This increases the overall functional density. High-density circuits support complex functionalities. The hybrid architecture achieves compact, efficient designs.

High-speed data transfer is another benefit. Nanomagnets enable rapid data movement. This ensures fast interconnects between components. CMOS circuits leverage this speed for efficient communication. The hybrid architecture facilitates swift data exchanges. It supports high-performance computing tasks. Fault tolerance is significantly improved. Nanomagnets provide robust error-resilient structures. CMOS circuits benefit from this fault tolerance. The hybrid architecture can handle errors better. It ensures continuous operation despite faults. This is crucial for mission-critical applications. Low-voltage operation marks a distinct advantage. Nanomagnets function effectively at lower voltages. CMOS circuits benefit from reduced voltage requirements. The hybrid system operates efficiently at low power levels. This reduces overall energy consumption. It supports energy-efficient designs. High endurance characterizes nanomagnets. They endure numerous switching cycles. CMOS circuits benefit from this high endurance. The hybrid architecture supports frequent operations. It ensures long-term reliability. High endurance is vital for repetitive tasks.

Magnetic properties ensure radiation hardness. Nanomagnets resist radiation-induced errors. This is critical for space and military applications. CMOS circuits benefit from this radiation hardness. The hybrid system maintains performance in radiation-rich environments. It ensures reliable operation under adverse conditions. Faster signal propagation is a key advantage. Nanomagnets exhibit swift magnetization dynamics. CMOS circuits benefit from reduced signal delay. The hybrid architecture ensures quick signal transmission. This supports high-speed processing tasks. Reduced parasitic capacitance is another benefit. Nanomagnets contribute minimal parasitic effects. CMOS circuits experience lower capacitance interference. The hybrid architecture achieves higher signal integirty. This improves overall performance. The integration of NML with CMOS technology in hybrid architectures optimizes computational efficiency, reducing energy consumption, and improving signal-to-noise ratios. This collaboration enables scalability for complex tasks, diverse application suitability, and seamless integration into existing systems, advancing parallel processing capabilities, enhancing memory and storage solutions, and increasing reliability with fault-tolerant design strategies. It also addresses quantum-resistant computing, adapts to varying environmental conditions, and ensures a unified information processing architecture.

Leveraging the strengths of each technology involves strategic planning, customized task allocation, and synergistic signal processing. NML contributes inherent advantages like non volatility, while CMOS brings established reliability. The collaboration allows for efficient magnetic domain manipulation, parallel processing synergy, and dynamic control mechanisms. It addresses quantum-resistant attributes and ethical considerations, designed for optimal resource utilization, fault tolerance, and ethical considerations. The provided schematics (cf. Figure 7.6) illustrate the principle of antiferromagnetic/ferromagnetic (AP/P) coupling conversion in NML. This mechanism, driven by gate voltage-induced migration of oxygen ions at the Co interface, exemplifies the efficient magnetic domain manipulation. This lies central to the synergistic integration of NML and CMOS technologies. Optimizing power consumption and processing speed in

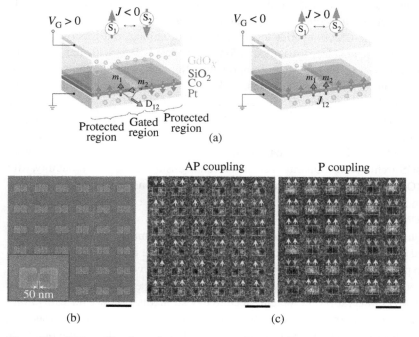

Figure 7.6 Schematics of coupled nanomagnet elements illustrating the principle of AP/P coupling conversion. On application of positive gate voltages, oxygen ions migrate away from the Co interface, and the coupling between two protected regions is AP induced by interfacial DMI. On application of negative gate voltages, oxygen ions migrate to the Co interface, and the coupling becomes P due to the symmetric exchange interaction. Central to the synergistic integration. File: Fig. 1: Basic concept of voltage-controlled magnetic coupling. https://doi.org/10.1038/s41467-023-41830-5 Source: Chao Yun et al., 2023/Springer Nature/CC BY 4.0.

nanomagnetic-CMOS hybrids involves dynamic power management, efficient magnetic domain switching, adaptive voltage scaling, ultra low power states, and parallel processing for energy efficiency. The hybrid architecture explores advanced strategies for heat dissipation, on-demand power supply adjustments, and energy-aware clock gating.

The synthesis of NML and CMOS technology in hybrid architectures optimizes power consumption and processing speed. This example highlights how precision control of magnetic elements can be utilized in hybrid architectures to optimize power consumption and processing speed (cf. Figure 7.7). Precision control of magnetic elements, parallel processing, and task-specific power allocation enhance overall efficiency.

The integration of quantum and nanoscale technologies with CMOS architectures transforms computational capabilities. This convergence advances computation beyond classical limits, exploring synergies, challenges, and implications. Qubits introduce superposition, exponentially expanding possibilities. Nanomagnetic-CMOS hybrids leverage qubits for enhanced diversity. Quantum entanglement enables distributed information processing, with nanomagnetic logic interfacing with CMOS components. Nanoscale components form the foundation for quantum platforms, with CMOS integration providing stability. Non volatile nanomagnetic memory complements quantum systems, ensuring data retention. Quantum gates and NML operations synergize for complex computations.

Hybrid algorithms blend quantum and classical processes within nanomagnetic-CMOS architectures. Quantum error correction ensures nanoscale reliability, with CMOS components supporting fault-tolerant operations. Energy-efficient information processing emerges from quantum-nanoscale collaborations, optimized by CMOS technology.

Quantum communication channels leverage nanomagnetic interfaces, managed by CMOS components.CMOS interfaces integrate quantum-nanoscale data into practical applications for accurate and timely interventions.

Benefits of quantum-nanoscale hybrids mark diverse applications, optimized by CMOS integration. Leveraging technology strengths is key, with CMOS providing necessary infrastructure. Optimizing power consumption defines success, ensured by CMOS in efficient resource utilization. The transformation of a magnetic inductor to an electric capacitor when moved from the magnetic domain to the electric domain, as depicted (cf. Figure 7.8), highlights the innovative approaches required for integrating NML with CMOS technology. Challenges in integration are being continuously addressed innovatively, ensuring compatibility between NML and CMOS. Bridging the gap for seamless integration is essential, facilitated by CMOS interfaces.

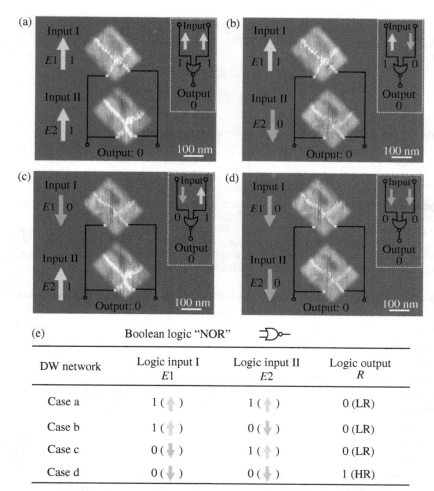

Figure 7.7 (a–d) c-AFM images and corresponding logic circuit diagrams of two parallel-connected nano-islands with the sequence of logic operations for inputs of "11", "10", "01", and "00". (e) Truth table for the NOR logic gate. *E1* and *E2* are the applied in-plane trailing fields for the two nano-islands. File: Fig. 3: Reconfigurable NOR logic gate https://doi.org/10.1038/s41467-022-30983-4 Source: Jing Wang et al., 2022/Springer Nature/CC BY-4.0.

7.2.1.1 Overview of Quantum Computing Principles

Quantum computing operates using quantum bits, or qubits. Qubits exist in superposition states. These states represent both 0 and 1 simultaneously. Superposition enables quantum parallelism. This allows multiple calculations at once. Entanglement is a crucial quantum principle. Entangled qubits exhibit correlated

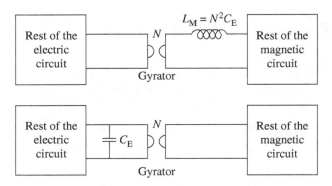

Figure 7.8 A magnetic inductor transforms to an electric capacitor when the element is moved from the magnetic domain to the electric domain. N is the gyration resistance of the gyrator. N nominally has the dimensions of ohms. An inductor with the dimensions of henrys in the magnetic domain transforms to a capacitor with units of farads in the electric domain. File: Magnetic Inductance.png https://commons.wikimedia.org/wiki/File: Magnetic_Inductance.png Source: Constant314/Wikimedia Commons/CC BY 0.1.0.

states. Measuring one qubit affects its partner instantly. This correlation holds, regardless of distance. Quantum gates manipulate qubit states. Common gates include Pauli-X, Hadamard, and CNOT. Pauli-X flips the qubit state. Hadamard creates superposition. CNOT entangles qubits, controlling their interactions. Hybrid CMOS-NML integrates CMOS and nanomagnetic elements. Nanomagnets store information via magnetic orientations. CMOS circuits handle conventional electronic processing. Quantum computing benefits from this hybrid architecture. Nanomagnetic elements facilitate robust qubit storage. CMOS provides reliable classical control.

Error correction is vital in quantum computing. Quantum ECCs, like Shor and Steane, detect and correct errors. Qubits are prone to decoherence and noise. These codes protect against information loss. Quantum algorithms exploit superposition and entanglement. Shor's algorithm factors large numbers efficiently. Grover's algorithm accelerates unstructured search problems. Quantum circuits consist of quantum gates arranged sequentially. Each gate performs a specific operation on qubits. Circuit depth and width influence computational complexity. Measurement collapses qubit superposition into definite states. This process provides classical information from quantum states. Precise measurement techniques ensure accurate results. Scalability is a challenge in quantum computing. Maintaining coherence over many qubits is difficult. Hybrid architectures aim to mitigate these issues. CMOS-nanomagnetic systems enhance coherence stability. Quantum decoherence disrupts qubit states. It results from interactions with the environment. Isolating qubits minimizes decoherence effects. Hybrid systems provide isolation through nanomagnetic materials.

Quantum tunneling is a key quantum phenomenon. It allows particles to pass through barriers. This behavior influences qubit behavior and gate operations. Adiabatic quantum computing (AQC) is another quantum approach. AQC relies on gradual state evolution. It avoids abrupt changes, reducing decoherence. AQC can solve optimization problems effectively. Topological quantum computing uses anyons. Anyons are quasi particles following non-abelian statistics. Braiding anyons forms robust qubits. This approach offers fault-tolerant quantum computation. Quantum supremacy is a milestone in quantum computing. It refers to solving problems that classical computers cannot. Achieving this milestone demonstrates quantum advantage. Cryogenic systems often house quantum computers. Low temperatures reduce thermal noise. This preserves qubit coherence longer. Quantum dots create artificial atoms. These dots confine electrons or holes. Quantum dot qubits are promising for scalability. Spintronics exploits electron spin for information processing. Spin qubits leverage this principle. They offer long coherence times and fast operations. Photonics involves using light for quantum operations. Photonic qubits enable high-speed quantum communication. Optical fibers transmit photonic qubits over long distances. Superconducting qubits use superconducting circuits. Josephson junctions form these circuits. Superconducting qubits achieve fast gate operations.

7.2.1.2 Integrating Quantum and Nanomagnetic Logic

Quantum computing blends quantum, nanomagnetic, and CMOS technologies. It explores nanomagnetic CMOS hybrid foundations. Qubits transcend classical binary states. Superposition gives qubits multiple states. Entanglement links particles for distributed computations. Quantum gates perform essential operations. Parallelism processes multiple solutions simultaneously. Quantum error correction mitigates errors. Quantum algorithms surpass classical ones for specific tasks. Quantum key distribution (QKD) ensures secure communication. Quantum annealing solves optimization problems. Quantum machine learning excels in pattern recognition. Quantum sensing achieves unprecedented precision in measurements. Topological quantum computing employs unique material properties. Quantum cryptographic hash functions redefine secure data processing. Quantum communication networks enable secure and efficient data transmission. Hybrid algorithms integrate quantum and classical processes for optimized solutions. Nanomagnetic-CMOS interfaces provide a robust foundation for hybrid execution. CMOS technology serves as the backbone for quantum computing platforms. Non volatile memory ensures data retention in quantum systems. Spintronics enhances quantum information processing in hybrid architectures. Quantum dot cellular automata contribute to quantum and classical computing. Energy-efficient quantum-nanoscale redefine computation with minimal power consumption. Figure 7.9 illustrates the potential energy

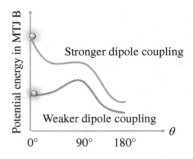

Figure 7.9 Potential energy profile (energy as a function of magnetization orientation in the soft layer of MTJ B) for strong and weak dipole coupling. File: Figure 4 https://doi.org/10.1038/s41598-020-68996-y Source: Mason T. McCray et al., 2020/Springer Nature/CC BY 4.0.

profile, critical for understanding nanomagnetic-CMOS hybrid foundations in quantum computing. It highlights how varying dipole coupling strength impacts magnetization orientation, which is pertinent to qubit behavior and quantum error correction.

7.2.1.3 Enhancing Quantum Computing with Nanoscale Technologies

Nanoscale technologies augment quantum computing capabilities. Quantum computing surpasses classical capabilities with quantum bits. Nanoscale technologies redefine quantum operations with precision. Quantum coherence meets nanomagnetic stability for robust frameworks. Quantum supremacy, coupled with nanoscale synergy, shapes the future. Nanoscale technologies address quantum error correction challenges. Quantum entanglement assists NML in hybrid systems. Nanoscale memory solutions efficiently store quantum states. Quantum tunneling amplifies computational efficiency in nanoscale logic. Quantum algorithms optimized for nanoscale components redefine efficiency. QCA complement nanoscale logic in hybrid systems. Quantum annealing optimizes solutions in nanoscale systems. Topological quantum computing interfaces seamlessly with nanoscale elements. CMOS technology interfaces quantum and nanoscale elements efficiently. Energy-efficient quantum-nanoscale hybrids minimize power consumption. Challenges include maintaining quantum coherence in nanoscale environments.

On the other hand, NML benefits from non volatility and low-energy consumption. Quantum computing excels at parallelism and complex problem-solving. NML can replace some CMOS logic gates. This reduces power usage in hybrid systems. Quantum circuits can handle specific, complex computations. Hybrid systems utilize both for optimized performance. Quantum gates and NML gates differ fundamentally. CMOS technology bridges this difference effectively. Hybrid architecture requires careful design. Quantum logic operations involve superposition and entanglement. NML operates based on magnetization states. CMOS provides control and interface between the two. Efficient integration demands precise synchronization. Quantum coherence needs maintaining during operations. Quantum dots or Josephson junctions serve as qubits. They interact with CMOS transistors. These transistors control quantum operations. NML uses nanomagnets arranged in logical patterns. Magnetic fields manipulate these patterns. CMOS gates drive the magnetic fields. NML is non volatile, holding states without power. This trait enhances memory and logic functions. Quantum bits, however, require active control. Quantum coherence must be preserved. This balance is critical in hybrid systems. CMOS interfaces facilitate this balance.

Clocking schemes differ between NML and quantum systems. NML uses magnetic clocking. Quantum systems use electronic clocking. Hybrid architectures must harmonize these schemes. This ensures efficient data transfer and processing. Precise clock synchronization is necessary. NML gates offer low switching energy. This reduces overall power consumption. Quantum gates process complex tasks rapidly. This complements the slower NML operations. CMOS gates coordinate the overall process. They act as intermediaries between NML and quantum gates. Quantum error correction is vital. It maintains data integrity during operations. NML provides stable, error-free operations. Combining these ensures robust computing. CMOS handles error correction protocols. It supports both quantum and NML components. Interfacing involves converting magnetic states to electrical signals. This is achieved through spintronics devices. Spin valves and MTJs are used. These devices link NML with CMOS circuits. Quantum signals are converted to CMOS signals similarly. CMOS circuits control the initialization of qubits. They also handle qubit measurements. NML initialization involves setting magnetic states. CMOS circuits facilitate this. Measurement involves detecting magnetic state changes. CMOS sensors detect these changes.

Thermal stability is a concern for NML. High temperatures can disturb magnetic states. Quantum circuits are also temperature sensitive. They require cooling to maintain coherence. CMOS circuits must operate efficiently at varied temperatures. Thermal management solutions are necessary. Scalability is another challenge. Quantum circuits need coherent qubit interaction. NML circuits require precise magnetic control. CMOS scaling involves maintaining transistor performance. Integrating these demands careful planning. Modular design approaches help in scaling. Hybrid systems use CMOS to mitigate noise. Quantum circuits are sensitive to noise. NML is less affected but still needs noise control. CMOS circuits can filter out noise. This ensures reliable operation of both quantum and NML components. Energy efficiency is critical. Quantum circuits consume power during operations. NML is inherently energy efficient. CMOS circuits balance the energy needs of both. This results in an overall energy efficient system. Power management is crucial for hybrid systems.

Design tools must support hybrid architectures. These tools include simulators for quantum and NML circuits. CMOS design tools need to integrate these simulations. This aids in accurate design and verification. Cross-domain simulators are particularly useful. Prototyping involves fabricating hybrid chips. These chips include quantum, NML, and CMOS components. Fabrication processes must cater to all three technologies. Integration techniques include 3D stacking and monolithic integration. These approaches optimize space and performance. Testing involves verifying the operation of all components. Quantum, NML, and CMOS components are tested individually. Their integration is also tested. This ensures proper interaction and functionality. Automated testing tools are developed for this purpose. NML's low-energy use and non volatility are beneficial. Quantum computing's power lies in complex problem-solving. CMOS bridges these technologies. The result is a powerful, efficient computing architecture. Future developments will refine this integration further.

7.2.1.4 Applications and Challenges
Nanomagnetic hybrid CMOS architectures leverage magnetoelectric effects. These architectures enhance energy efficiency and speed. MTJs store data non volatilely. STT mechanisms enable efficient data writing.

Memory devices are primary applications. Nanomagnetic elements reduce power consumption. Their non volatility ensures data retention without power. MRAM benefits significantly from these properties. NML gates facilitate low-power logic operations. They enhance performance in computing applications. In neuromorphic systems, nanomagnetic devices simulate synaptic functions. These systems benefit from non-volatility and low power.

Material limitations pose significant challenges. High-quality magnetic materials are essential. Fabrication complexity increases with nanomagnetic elements. Integrating magnetic and CMOS components is difficult. Thermal stability is crucial for reliable operation. Magnetic anisotropy must be finely controlled. MEC efficiency needs enhancement. Interface quality between materials affects performance. Scaling to nanoscale introduces new issues. Device-to-device variations impact reliability. Heat dissipation in dense arrays is problematic. Figure 7.10 illustrates various properties of nanomagnet elements used in these architectures, highlighting their magnetic configurations and coupling behaviors. These characteristics are essential for understanding the applications and challenges discussed.

7.2.1.5 Quantum-Enhanced Information Processing
Quantum-enhanced information processing exploits quantum superposition and entanglement. It outperforms classical computing in specific tasks. Quantum bits (qubits) form the core of these systems. Superconducting qubits are commonly used. Trapped ions also serve as qubits. Quantum error correction ensures reliable operation. Quantum algorithms solve complex problems efficiently.

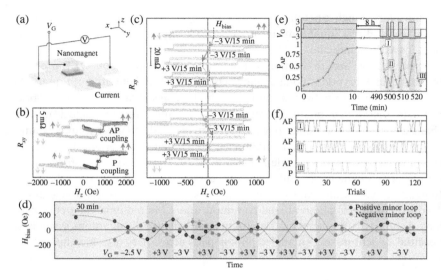

Figure 7.10 Schematic and magnetic hysteresis loops of nanomagnet elements illustrating their magnetic configurations and coupling behaviors. Electrical transport measurements and magnetic coupling dynamics of a nanomagnet on the Hall bar. Hysteresis loops and gate voltage effects on the magnetic configurations of nanomagnet elements. Figure 2: Reversible conversion of AP/P magnetic coupling. https://doi.org/10 .1038/s41467-023-41830-5 Source: Chao Yun et al., 2023/Springer Nature/CC BY 4.0.

Applications: Quantum computing has diverse applications. Quantum simulations model molecular interactions precisely. This is vital for drug discovery and material science. Quantum optimization addresses complex optimization problems. Logistics and finance sectors benefit greatly. Quantum machine learning enhances data analysis capabilities. Pattern recognition and anomaly detection are improved. Cryptographic systems become more secure with quantum algorithms. QKD ensures secure communication.

Challenges: Quantum coherence is a major challenge. Environmental interference reduces qubit coherence time. Error rates in quantum operations are high. Quantum error correction demands significant overhead. Scalability of qubit systems is difficult. Interconnecting many qubits without errors is complex. Fabricating qubits with uniform properties is hard. Maintaining low temperatures is necessary for superconducting qubits. Quantum decoherence due to thermal noise affects performance. High-fidelity qubit control and readout are challenging.

7.2.1.6 Quantum Communication with Nanoscale Components

Quantum communication relies on quantum entanglement and superposition. Nanoscale components enhance these systems. Quantum dots generate entangled photon pairs. Single-photon sources enable secure communication. Nanoscale waveguides direct quantum information. Plasmonic structures manipulate photons at the nanoscale.

Applications: Secure communication is a primary application. QKD provides unbreakable encryption. Long-distance communication benefits from quantum repeaters. Quantum networks connect multiple nodes securely. These networks facilitate distributed quantum computing. Quantum sensors detect minute changes in the environment. They are useful in medical and defense applications.

Challenges: Photon loss in nanoscale components is significant. Efficient coupling between nanoscale elements is difficult. Fabrication of uniform quantum dots is challenging. Entanglement generation efficiency must be high. Integrating nanoscale components into existing systems is hard. Quantum decoherence affects photon-based communication. Maintaining coherence over long distances is problematic. Environmental isolation of quantum systems is essential. Synchronizing entangled photon pairs across networks is complex.

7.3 Neuromorphic Hybrid Systems

This section talks about the neuromorphic hybrid systems, and briefs the advancements in cognitive computing in Section 7.3.1.

Neuromorphic hybrid systems blend biological principles with electronic computing. They emulate neural structures and functionalities. These systems integrate both hardware and software components. Analog and digital circuits work together efficiently. Neuromorphic architectures support parallel data processing. They manage tasks concurrently and reduce latency. Event-driven processing enables low-power consumption. This architecture mimics brain-like signal propagation. Neuromorphic hybrid systems utilize spiking neural networks (SNNs). SNNs employ spike-timing-dependent plasticity (STDP). This learning mechanism adjusts synaptic strengths dynamically. Memristors store and process information simultaneously. They contribute to synaptic emulation in hardware. These systems incorporate diverse sensory inputs. Vision, auditory, and tactile sensors are integrated. Real-time data processing enhances responsiveness. Feedback loops refine system outputs continuously. Neuromorphic cores perform complex computations. They achieve high-speed data processing. Specialized algorithms govern these cores. These algorithms are inspired by biological processes. Neuromorphic systems support real-time adaptation. They adjust to changing environmental conditions. Reinforcement learning algorithms optimize decision-making. Reward-based adjustments refine actions over time. Cortical columns model hierarchical processing. They facilitate multi layered learning approaches. Stochastic elements in synapses enhance robustness. These elements improve noise tolerance. Hybrid systems balance flexibility and efficiency. They combine neuromorphic and classical computing. On-chip learning reduces dependency on external resources. Localized processing enhances system performance. Neuromorphic systems exhibit self-organizing behavior. Autonomous adaptation occurs without explicit programming. Energy-efficient computation extends operational longevity. These systems are ideal for mobile devices. Neuromorphic chips enable compact and versatile designs. System modularity supports diverse applications. Task-specific modules activate as needed. Neuromorphic hybrid systems function effectively in various environments. They maintain performance under diverse conditions. Real-time learning supports continuous system improvement. These systems evolve based on accumulated experience. Robust designs ensure reliable functionality. Neuromorphic systems integrate with existing technologies seamlessly. They complement traditional computing frameworks.

7.3.1 Advancements in Cognitive Computing

This section talks about the advancements in cognitive computing throughout, briefs the real-time learning and adaptation in Section 7.3.1.1, describes the applications in AI and pattern recognition in Section 7.3.1.2, and presents in detail the ethical considerations in neuromorphic hybrid systems in Section 7.3.1.3.

Advancements in cognitive computing revolutionize information processing. Neuromorphic systems significantly enhance cognitive tasks. They leverage brain-inspired architectures for efficiency. These systems process sensory data rapidly. Real-time learning improves adaptive capabilities. Neuromorphic processors handle parallel computations. They execute tasks simultaneously. Event-driven processing reduces energy consumption. This approach mimics biological efficiency. SNNs form the core. SNNs utilize STDP. This plasticity mechanism adjusts synapses dynamically. Memristive devices play a key role. They integrate storage and computation functions. Real-time adaptation enhances cognitive systems. Continuous feedback refines system responses. Reinforcement learning algorithms guide decision-making. Reward-based learning optimizes actions over time. Cortical columns model layered processing. These columns support complex learning. Stochastic synapses improve robustness. They enhance noise tolerance. Neuromorphic systems balance digital and analog components. This balance optimizes performance. On-chip learning minimizes external dependencies. Localized processing improves speed and efficiency. Neuromorphic systems exhibit autonomous behavior. They adapt without explicit instructions. Energy-efficient designs prolong operational life. These systems suit mobile applications well. Neuromorphic chips enable compact designs. Modularity supports diverse cognitive tasks. Task-specific modules activate as required.

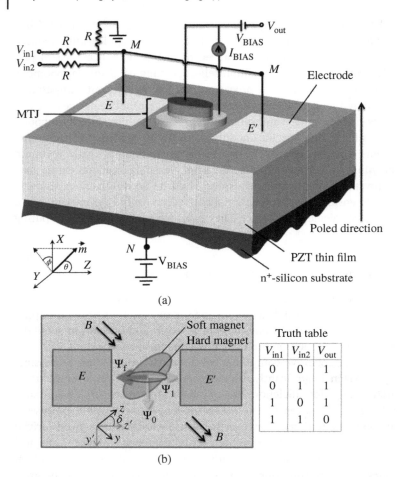

V_{in1}	V_{in2}	V_{out}
0	0	1
0	1	1
1	0	1
1	1	0

Figure 7.11 Structure of a NAND gate with a PZT film on a conducting n^+-Si substrate. The figure shows fixed and stable magnetization orientations affecting MTJ resistance, with footprints of the magnets and the permanent magnetic field direction indicated. File: Figure 1 https://doi.org/10.1038/srep07553 Source: Ayan K. Biswas et al., 2014/Springer Nature/CC BY 4.0.

Neuromorphic systems perform reliably in various environments. They adapt to changing conditions. Real-time learning supports continuous improvement. Experience-based evolution enhances functionality. Robust designs ensure consistent performance. Neuromorphic systems integrate smoothly with existing technology. They augment traditional cognitive computing frameworks.

Neuromorphic systems excel in pattern recognition. They handle complex patterns efficiently. Vision systems benefit greatly. Object recognition is highly accurate. Facial recognition is robust and swift. Edge detection in images is precise. Auditory processing improves significantly. Speech recognition systems are reliable. Environmental sounds are identified accurately. Tactile sensing is enhanced. Robotic manipulation is precise. Healthcare diagnostics improve remarkably. Medical image analysis is rapid. Disease detection is accurate. Autonomous driving is safer. Object detection is precise. Lane recognition is reliable. Traffic sign detection is swift. Financial forecasting benefits immensely. Market trend analysis is accurate. Fraud detection is effective. Cybersecurity is enhanced. Intrusion detection is robust. Malware identification is reliable. Gaming experiences improve. AI-driven characters are realistic. Game difficulty adapts dynamically. Smart home applications are refined. User preferences are accurately identified. Device interactions are seamless. IoT applications benefit greatly. Data processing is swift. Network efficiency is optimized. Weather forecasting is accurate. Prediction models are refined. Disaster management is improved. Agricultural practices are enhanced. Crop management is efficient. Resource usage is optimized. Manufacturing processes improve. Production lines are refined. Defect detection is swift. Space exploration benefits immensely. Autonomous navigation is reliable. Data analysis is accurate. Robotic missions are enhanced. Environmental monitoring is precise. Data privacy is maintained. Ethical considerations are paramount. Fairness in decision-making is ensured. User consent is critical. Transparent data handling builds trust. Bias mitigation is necessary. Regular audits ensure compliance. Safety is crucial. Fail safes are implemented. System validation is thorough. Regulations guide ethical use. Accountability is essential. Traceability of decisions is ensured. Ethical use in surveillance is considered. Civil liberties are protected. Ethical considerations in healthcare are vital. Patient data is protected. Informed consent is mandatory. Fairness in diagnostics is ensured. Ethical guidelines govern financial applications. Fairness in credit decisions is crucial. Ethical use in education is vital. Data privacy for students is maintained. Ethical use in military applications is critical. International law compliance is ensured. Ethical considerations in smart cities are important. Data privacy for citizens is maintained. Ethical use in agriculture is crucial. Resource distribution fairness is ensured. Ethical guidelines for space exploration are necessary. Resource allocation fairness is ensured. Ethical use in entertainment is critical. Data privacy for users is maintained. Ethical considerations in disaster management are important. Ethical guidelines for environmental monitoring are necessary. Data use fairness is ensured. Environmental data privacy is maintained.

Neuromorphic systems achieve efficient cognitive computing. They integrate biological principles with electronic technologies. Real-time learning enhances adaptability. Parallel processing boosts performance. Event-driven architecture minimizes energy usage. SNNs and STDP optimize learning. Memristive devices integrate storage and computation. Neuromorphic systems support various cognitive tasks. They adapt to changing environments. Continuous improvement is achieved through experience. Robust designs ensure reliable performance. Integration with existing technologies is seamless. Neuromorphic systems revolutionize cognitive computing. Neuromorphic systems, which integrate biological principles with electronic technologies, utilize novel logic operations based on magnetization states rather than electronic charge, as shown in the NAND gate structure in Figure 7.11. This integration exemplifies the seamless adaptation and innovative design fundamental to neuromorphic computing.

7.3.1.1 Real-Time Learning and Adaptation

Neuromorphic hybrid systems emulate biological neural networks. Real-time learning enhances adaptive computing capabilities. These systems process sensory inputs efficiently. They use STDP. This mechanism adjusts synaptic weights dynamically. Learning occurs on-the-fly, without pre training. Hardware accelerators boost computational efficiency. Memristive devices play a crucial role. They store and process information simultaneously. These components minimize latency issues. Hybrid systems integrate digital and analog circuits. This combination optimizes performance and energy consumption. Event-driven architecture enables low-power operations. Real-time adaptation involves continuous feedback loops. Neuromorphic processors support parallel processing. These processors handle multiple tasks concurrently. Brain-inspired algorithms improve learning rates. The system's architecture mimics synaptic connectivity. Cortical columns model hierarchical learning. These columns facilitate complex pattern recognition. Stochasticity in synapses aids robust learning. Noise tolerance enhances operational stability. Hybrid systems leverage neuromorphic and classical computing. This synergy improves learning efficiency. Neuromorphic cores execute brain-like computations. They achieve high-speed processing. Specialized sensors gather real-world data. These sensors include visual, auditory, and tactile inputs. SNNs play a central role. SNNs process spikes instead of continuous signals. This method reduces data redundancy. Temporal coding captures time-dependent patterns. Adaptation occurs at multiple scales. Systems adjust to varying environmental conditions. Real-time learning enables autonomous decision-making. Reinforcement learning algorithms refine these decisions. These algorithms use reward-based adjustments. Learning algorithms evolve with experience. System architecture supports dynamic reconfiguration. Task-specific modules activate as needed. This modularity enhances versatility. Neuromorphic systems achieve low-latency responses. They process information near-instantaneously. Energy-efficient computation extends battery life. These systems are suitable for mobile applications. Hybrid systems integrate neuromorphic chips and classical processors. This integration balances flexibility and efficiency. On-chip learning reduces reliance on external resources. These systems achieve localized processing. Neuromorphic systems exhibit self-organizing behavior. They adapt without explicit programming. Autonomous adaptation enhances user experience. Real-time learning supports continuous improvement. The system evolves over time. Robustness against noise ensures reliable performance. Hybrid systems function effectively in diverse environments. Ensuring reliability and precision in adaptive computing (cf. Figure 7.12). Neuromorphic systems' ability to achieve low-latency responses and localized processing aligns with the magneto-elastic system's precise control over random bit generation and correlation, as shown in Figure 7.12. This illustrates how adaptive computing mechanisms contribute to reliable and efficient performance in diverse environments.

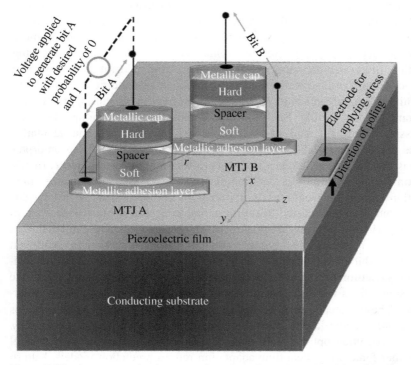

Figure 7.12 A magneto-elastic system for generating two random bits with controlled amount of correlation. The metallic adhesion layer is thin enough to not impede strain transfer from the piezoelectric film to the soft layers. File: Figure 1 https://doi.org/10.1038/s41598-023-35623-5 Source: Mason T. McCray et al., 2020/Springer Nature/CC BY 4.0.

In the realm of cognitive computing, the dynamic interplay between Real-time learning and adaptation takes center stage, driving innovation in nanomagnetic-CMOS hybrid architectures.

7.3.1.2 Applications in AI and Pattern Recognition

Neuromorphic hybrid systems excel in AI tasks. They handle complex pattern recognition. Vision systems benefit from these advancements. Object recognition improves significantly. Facial recognition achieves higher accuracy. Edge detection in images becomes more efficient. These systems process images with minimal delay. Neuromorphic systems enhance auditory processing. Speech recognition systems gain robustness. Real-time adaptation refines voice command accuracy. Environmental sounds are identified swiftly. This capability aids in surveillance applications. Neuromorphic systems advance tactile sensing. Robotic hands achieve finer manipulation. Real-time feedback improves grip precision. These systems process touch data instantaneously. AI-driven healthcare diagnostics benefit greatly. Neuromorphic systems analyze medical images rapidly. Pattern recognition detects anomalies in scans. Early diagnosis of diseases becomes feasible. Neuromorphic systems aid in autonomous driving. Real-time processing enhances object detection. Lane recognition systems gain accuracy. Adaptive algorithms adjust to varying road conditions. Traffic sign recognition is highly efficient. These systems process radar and Light detection and ranging data swiftly. Neuromorphic computing boosts natural language processing. Sentiment analysis becomes more accurate. Real-time adaptation refines language translation. Chatbots respond more naturally. Neuromorphic systems excel in financial forecasting. Pattern recognition detects market trends. Real-time analysis improves decision-making. These systems enhance fraud detection. Anomalous patterns in transactions are identified. Neuromorphic systems aid in cybersecurity. Real-time monitoring detects intrusions. Pattern recognition identifies malware signatures. Adaptive algorithms respond to threats dynamically. Neuromorphic systems enhance gaming experiences. AI-driven characters exhibit realistic behaviors. Pattern recognition improves virtual environments. Neuromorphic systems support smart home applications. Real-time processing refines user preferences. Pattern recognition improves device interactions. These systems enhance energy management. Neuromorphic systems drive advancements in IoT. Real-time data processing enhances connectivity. Pattern recognition optimizes network efficiency. Neuromorphic systems excel in weather forecasting. Real-time adaptation refines prediction models. Pattern recognition detects weather patterns. These systems aid in disaster management. Neuromorphic systems improve agricultural practices. Real-time monitoring enhances crop management. Pattern recognition detects plant diseases. These systems optimize resource usage. Neuromorphic systems drive advancements in manufacturing. Real-time adaptation refines production lines. Pattern recognition identifies defects swiftly. These systems enhance quality control. Neuromorphic systems support space exploration. Real-time processing aids in autonomous navigation. Pattern recognition improves data analysis. These systems enhance robotic missions.

Foundations of Cognitive Computing:

- Cognitive systems emulating human-like thought processes.
- An amalgamation of AI and pattern recognition principles.

Adaptive AI Algorithms:

- Real-time adaptation through nanomagnetic states.
- AI algorithms dynamically responding to changing environments.

Pattern Recognition Paradigms:

- NMLs precision in identifying intricate patterns.
- AI-driven pattern recognition for complex data interpretation.

Applications in Healthcare:

- AI-driven diagnosis using nanomagnetic real-time data processing.
- Enhancing medical pattern recognition for proactive healthcare.

Emerging Technologies Symbiosis:

- Collaborative approach with emerging technologies for AI innovation.
- Integrating quantum-inspired principles for advanced pattern recognition.

Energy-Efficient AI Solutions:

- NMLs low-power attributes in AI applications.
- Sustainable AI advancements without compromising efficiency.

Dynamic Learning in AI Systems:

- Adaptive learning models mirroring human-like cognitive evolution.
- Nanomagnetic states contributing to real-time learning dynamics.

Transportation Industry Transformations:

- AI-driven nanomagnetic sensors for real-time traffic pattern analysis.
- Optimizing transportation systems based on intricate data patterns.

Precision Agriculture Applications:

- AI leveraging nanomagnetic states for real-time crop pattern analysis.
- Enhancing agricultural efficiency through adaptive computing.

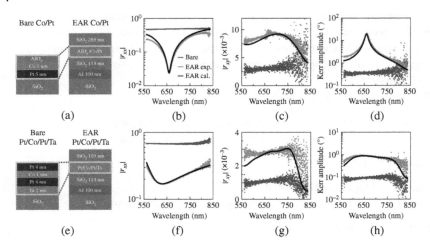

Figure 7.13 (a) Schematics of the bare and EAR Co/Pt layers. (b–d) Measured spectra of the non-MO reflection amplitude $|r_{xx}|$, MO reflection amplitude $|r_{xy}|$, and Kerr amplitude from the bare and EAR Co/Pt layers. (e) Schematics of the bare and EAR Pt/Co/Pt/Ta layers. (f–h) Measured spectra of the non-MO reflection amplitude $|r_{xx}|$, MO reflection amplitude $|r_{xy}|$, and Kerr amplitude from bare and EAR Pt/Co/Pt/Ta layers. We also calculated the non-MO and MO reflection amplitudes and the Kerr amplitude of the EAR Co/Pt and EAR Pt/Co/Pt/Ta films using the transfer matrix method (black solid lines). File: Figure 2 https://doi.org/10.1038/s41467-020-19724-7 Source: Dongha Kim et al., 2020/Springer Nature/CC BY 4.0.

Ethical Considerations in AI Applications:

- Addressing ethical concerns in real-time AI decision-making.
- Developing frameworks for responsible AI pattern recognition.

Pattern Recognition in Manufacturing:

- Nanomagnetic-CMOS controls real-time manufacturing processes.
- AI analyzing patterns for quality control and process optimization.

In advancing AI tasks, neuromorphic hybrid systems are critical, especially when integrating novel materials and configurations. Figure 7.13 illustrates the sophisticated design and measurement of Co/Pt and Pt/Co/Pt/Ta layers, revealing how these structures can impact the efficiency and capabilities of neuromorphic systems. Figure 7.14 complements this by showing various nanomagnet configurations on a TIPNJ, which are integral to optimizing the performance of these systems. These designs contribute to the enhanced pattern recognition and real-time processing capabilities essential for the applications described in the writeup, such as autonomous driving and financial forecasting.

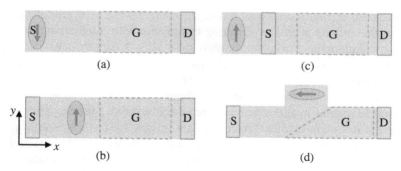

Figure 7.14 Top view of different configurations of nanomagnet on a TIPNJ. (a) Nanomagnet as the source contact. (b) Floating nanomagnet located between source and drain contacts. (c) Floating nanomagnet outside the source–drain path. (d) A variation of the out-of-location floating nanomagnet. The gate (PN junction interface) is at a 45° angle from the source–drain direction. The angled PN junction can reflect electrons from the source to the nanomagnet. File: Figure 1 https://doi.org/10.1038/s41598-023-35623-5 Source: Yunkun Xie et al., 2023/Springer Nature/CC BY 4.0.

7.3.1.3 Ethical Considerations in Neuromorphic Hybrid Systems

Ethical concerns in neuromorphic systems are paramount. Data privacy is a significant issue. Real-time learning systems process sensitive information. Ensuring data protection is crucial. User consent for data use is essential. Transparent data handling builds trust. Bias in algorithms poses ethical challenges. Ensuring fairness in decision-making is vital. Diverse datasets mitigate bias risks. Regular audits enhance algorithmic transparency. Accountability in system decisions is necessary. Traceability of decisions aids accountability. System failures must be addressed promptly. Safety in autonomous systems is critical. Ensuring fail-safes minimizes risks. Real-time adaptation requires robust validation. Ethical guidelines govern system development. Regulations ensure compliance with ethical standards. Autonomous decision-making raises ethical dilemmas. Balancing autonomy and human oversight is crucial. Ethical use of AI in surveillance must be considered. Protecting civil liberties is paramount. Transparent monitoring practices are necessary. Data minimization principles guide data use. Ethical considerations in healthcare applications are vital. Patient data requires stringent protection. Informed consent is essential for data use. Fairness in diagnostics must be ensured. Ethical guidelines for AI in finance are necessary. Ensuring fairness in credit decisions is crucial. Preventing discrimination in financial services is vital. Regular audits ensure ethical compliance. Neuromorphic systems in cybersecurity require ethical scrutiny. Ensuring privacy in monitoring practices is crucial. Balancing security and privacy is necessary. Ethical considerations in AI-driven education are important. Fairness in personalized learning is vital. Ensuring data privacy for students is essential.

Transparent use of educational data builds trust. Ethical use of neuromorphic systems in the military is critical. Ensuring compliance with international laws is necessary. Autonomous weapon systems raise ethical concerns. Balancing military advantage and ethical use is crucial. Ethical considerations in smart cities are important. Ensuring data privacy for citizens is crucial. Transparent use of urban data builds trust. Balancing efficiency and privacy is necessary. Neuromorphic systems in agriculture require ethical scrutiny. Ensuring fairness in resource distribution is vital. Protecting data privacy for farmers is crucial. Transparent use of agricultural data builds trust. Ethical guidelines for AI in space exploration are necessary. Ensuring fairness in resource allocation is crucial. Balancing exploration and ethical use is important. Transparent use of space data builds trust. Ethical use of neuromorphic systems in entertainment is critical. Ensuring fairness in content recommendations is vital. Protecting user data privacy is crucial. Transparent use of entertainment data builds trust. Ethical considerations in neuromorphic systems for disaster management are important. Ensuring fairness in resource allocation is crucial. Protecting data privacy for affected individuals is vital. Transparent use of disaster data builds trust. Ethical guidelines for AI in environmental monitoring are necessary. Ensuring fairness in data use is vital. Protecting environmental data privacy is crucial. Transparent use of environmental data builds trust.

Navigating the complex landscape of neuromorphic hybrid systems, the ethical considerations in cognitive computing stand as sentinels, guarding against unintended consequences. In the realm of nanomagnetic-CMOS hybrid architectures, the intersection of ethics and technology becomes pivotal, demanding a meticulous examination of the moral implications that arise as we examine deeper into the integration of cognitive computing and neuro-inspired technologies.

Foundations of Ethical Neuromorphic Computing:

- Ethical guidelines shaping neuromorphic cognitive systems.
- Striking a balance between innovation and ethical responsibility.

Privacy Concerns in Cognitive Computing:

- Safeguarding individual privacy within neuromorphic architectures.
- Ethical considerations in processing sensitive cognitive data.

Data Ownership and Autonomy:

- Ethical frameworks defining data ownership and user autonomy.
- Balancing technological advancements with individual rights.

Fairness and Bias Mitigation:

- Addressing biases in neuromorphic algorithms for ethical decision-making.
- Striving for fairness in cognitive systems' decision processes.

Transparency in Cognitive Systems:

- Ethical imperative for transparent neuromorphic logic.
- Ensuring accountability through understandable cognitive processes.

Informed Consent in Cognitive Research:

- Ethical considerations in obtaining informed consent for neuromorphic research.
- Ensuring participants comprehend the implications of cognitive experiments.

Addressing Unintended Consequences:

- Ethical responsibility in identifying and mitigating unintended impacts.
- Anticipating and preventing potential ethical pitfalls in neuromorphic systems.

Ensuring Accountability in AI Governance:

- Establishing ethical governance for cognitive AI.
- Holding stakeholders accountable for breaches of ethical standards.

Regulatory Frameworks for Cognitive Ethics:

- Establishing international regulations for ethical cognitive practices.
- Navigating legal landscapes to ensure ethical conduct in cognitive technologies.

Addressing Ethical Dilemmas in AI Decision-making:

- Ethical considerations in AI-driven decision-making processes.
- Developing ethical decision-making models for cognitive systems.

Neuromorphic Education and Ethical Literacy:

- Integrating ethical education into the development of neuromorphic technologies.
- Fostering ethical literacy among innovators and researchers.

Ethics in Brain-Inspired Robotics:

- Ethical considerations in the development and deployment of brain-inspired robots.
- Navigating the ethical landscape of autonomous machines inspired by neural principles.

An ethical imperative guides the integration of neuromorphic systems within nanomagnetic-CMOS hybrid architectures and cognitive computing. Upholding ethical standards is crucial as we traverse the complex intersection of technology and neuroscience, ensuring that the benefits of cognitive computing are harnessed responsibly for the betterment of society.

7.4 Educational Emphasis

This section talks about the educational emphasis and briefs the demystifying hybrid computing concepts in Section 7.4.1. Hybrid computing integrates different technologies to enhance performance and efficiency. For example, combining CMOS with nanomagnetic materials leverages the strengths of each. CMOS handles general computing tasks efficiently, while nanomagnetic components address specific challenges like power consumption and speed. Understanding this integration is crucial for students and professionals in technology fields. It demonstrates how diverse technological solutions can work together to solve complex problems. Focusing on hybrid computing can prepare individuals for advancements in computational systems and drive innovation by applying a multi-faceted approach to technology development and problem-solving.

7.4.1 Demystifying Hybrid Computing Concepts

This section talks about the demystifying hybrid computing concepts throughout, briefs the preparing students for future industry challenges in Section 7.4.1.1, and describes the Chapter End Quiz in Section 7.4.2.

Hybrid computing often sparks confusion due to myths and misconceptions. One common myth is that hybrid computing only involves combining two entirely different technologies. In reality, it often integrates variations of similar technologies to enhance performance. For example, nanomagnetic CMOS hybrid architecture uses CMOS technology with nanomagnetic materials, not a mix of completely unrelated systems.

Another misconception is that hybrid computing always results in significantly higher costs. While integrating new technologies can be expensive, the long-term benefits, such as improved efficiency and reduced power consumption, can offset initial costs.

Additionally, some believe hybrid computing is too complex to be practical. In truth, many hybrid systems aim to address specific limitations of existing technologies, making them more effective in real-world applications.

Another myth about hybrid computing is that it requires entirely new infrastructure. In many cases, hybrid systems can be integrated into existing frameworks, enhancing their capabilities without necessitating a complete overhaul. For instance, nanomagnetic components can be added to existing CMOS setups to improve performance without replacing the entire system.

Additionally, there is a belief that hybrid computing is only relevant for cutting-edge research or large-scale projects. In reality, hybrid approaches are increasingly applied in everyday technology, such as energy-efficient devices and high-performance computing systems, demonstrating their practicality and broad utility.

Moreover, some assume that hybrid computing always requires specialized knowledge and skills. While understanding the underlying technologies is beneficial, many hybrid systems are designed to be user-friendly and accessible, enabling broader adoption.

Understanding these facts helps clarify hybrid computing's potential and practical advantages, preparing students and professionals to leverage these technologies effectively. It clarifies the practical applications of hybrid computing and its role in advancing technology in various fields.

Real-World Analogies:

Providing tangible real-world analogies for understanding hybrid computing. Establishing connections between abstract concepts and practical applications.

Hybrid Vehicles: Think of hybrid computing like a hybrid vehicle that combines an internal combustion engine with an electric motor. In this analogy, the internal combustion engine represents traditional computing (such as CMOS), which is efficient but has limitations in terms of fuel efficiency and emissions. The electric motor symbolizes the new technology (like nanomagnetic components), which excels in specific conditions, such as low-speed driving, where it provides better efficiency. Together, they create a vehicle that balances performance and efficiency, much like how hybrid computing systems aim to optimize both speed and power consumption.

Toolbox with Specialized Tools: Imagine you are working on a complex project and have a toolbox filled with various specialized tools. Each tool is designed for specific tasks-like a wrench for bolts and a screwdriver for screws. Hybrid computing is similar: traditional computing systems are like the standard tools, good at general tasks. Nanomagnetic components are like specialized tools designed for particular problems (e.g., handling large datasets more efficiently). By combining these tools, you can complete your project more effectively and efficiently, just as hybrid computing integrates different technologies to improve overall performance.

7.4.1.1 Preparing Students for Future Industry Challenges

Industry-Ready Skill Development:

Prioritize the development of skills directly applicable in the industry. Align educational objectives with industry expectations to ensure proficiency.

Dynamic Curriculum Evolution:

Continuously update the curriculum to reflect industry dynamics. Maintain the relevance of educational content amidst technological advancements.

Problem-Solving Oriented Education:

Base education on problem-solving in NML. Foster a mindset focused on addressing industry challenges using magnetic logic.

Educational emphasis on preparing students for future industry challenges involves not only imparting knowledge but also shaping individuals capable of influencing the trajectory of magnetic logic and hybrid systems in the professional arena. The emphasis is on creating professionals proficient not just in theories but also in problem-solving, innovation, and contributing to the ever-evolving landscape of magnetic and hybrid computing.

7.4.2 Chapter End Quiz

This section presents the Chapter End Quiz throughout.

An online Book Companion Site is available with this fundamental text book in the following link: www.wiley.com/go/sivasubramani/nanoscalecomputing1

For Further Reading:

Wang et al. (2006), Bourianoff (2003), Ahuja et al. (2009), Das et al. (2012), Panchapakeshan et al. (2011), Prenat et al. (2007), Zhang (2014), Muralidharan et al. (2023), Trinh et al. (2022), Heinrich et al. (2021), Hassija et al. (2020), Christensen et al. (2022), Xu et al. (2023), Wang et al. (2017), Huh et al. (2020), Zhu et al. (2023), Arrigo et al. (2022), Mehta and Subramani (2012), Moo-Young (2019)

References

Sumit Ahuja, Gaurav Singh, Debayan Bhaduri, and Sandeep Shukla. Fault-and defect-tolerant architectures for nanocomputing. In *Bio-Inspired and Nanoscale Integrated Computing* (ed. Mary Mehrnoosh Eshaghian-Wilner), pages 263–293, Wiley, 2009. https://doi.org/10.1002/9780470429983.ch10.

Domenico Arrigo, Claudio Adragna, Vincenzo Marano et al. The next "automation age": How semiconductor technologies are changing industrial systems and applications. In *ESSCIRC 2022-IEEE 48th European Solid State Circuits Conference (ESSCIRC)*, pages 17–24. IEEE, 2022.

George Bourianoff. The future of nanocomputing. *Computer*, 36(8):44–53, 2003.

Dennis Valbjørn Christensen, Regina Dittmann, Bernabe Linares-Barranco, et al. 2022 roadmap on neuromorphic computing and engineering. *Neuromorphic Computing and Engineering*, 2(2):022501, 2022.

Jayita Das, Syed M Alam, and Sanjukta Bhanja. A novel design concept for high density hybrid cmos-nanomagnetic circuits. In *2012 12th IEEE International Conference on Nanotechnology (IEEE-NANO)*, pages 1–6. IEEE, 2012. https://ieeexplore.ieee.org/author/37265930300.

Vikas Hassija, Vinay Chamola, Vikas Saxena et al. Present landscape of quantum computing. *IET Quantum Communication*, 1(2):42–48, 2020.

Andreas J Heinrich, William D Oliver, Lieven MK Vandersypen et al. Quantum-coherent nanoscience. *Nature nanotechnology*, 16(12):1318–1329, 2021. https://www.nature.com/articles/s41565-021-00994-1.

Woong Huh, Donghun Lee, and Chul-Ho Lee. Memristors based on 2D materials as an artificial synapse for neuromorphic electronics. *Advanced Materials*, 32(51):2002092, 2020.

Manjula Mehta and Karthikeyan Subramani. Nanodiagnostics in microbiology and dentistry. In *Emerging Nanotechnologies in Dentistry* (ed. Karthikeyan Subramani and Waqar Ahmed), pages 365–390. Elsevier, 2012.

Murray Moo-Young. *Comprehensive Biotechnology*. Elsevier, 2019. https://www.sciencedirect.com/science/article/pii/B9780128122914000194?via%3Dihub.

Bhaskaran Muralidharan, Manohar Kumar, and Chuan Li. Emerging quantum hybrid systems for non-abelian-state manipulation. *Frontiers in Nanotechnology*, 5:1219975, 2023.

Pavan Panchapakeshan, Priyamvada Vijayakumar, Pritish Narayanan, et al. 3-D integration requirements for hybrid nanoscale-cmos fabrics. In *2011 11th IEEE International Conference on Nanotechnology*, pages 849–853. IEEE, 2011.

Guillaume Prenat, Mourad El Baraji, Wei Guo, et al. CMOS/magnetic hybrid architectures. In *2007 14th IEEE International Conference on Electronics, Circuits and Systems*, pages 190–193. IEEE, 2007.

Phuong Thao Trinh, Sina Hasenstab, Markus Braun, and Josef Wachtveitl. Ultrafast separation of multiexcitons within core/shell quantum dot hybrid systems. *Nanoscale*, 14(9):3561–3567, 2022.

Wei Wang, Ming Liu, and Andrew Hsu. Hybrid nanoelectronics: future of computer technology. *Journal of Computer Science and Technology*, 21:871–886, 2006.

Zhiyong Wang, Laiyuan Wang, Masaru Nagai, et al. Nanoionics-enabled memristive devices: strategies and materials for neuromorphic applications. *Advanced Electronic Materials*, 3(7):1600510, 2017.

Minyi Xu, Xinrui Chen, Yehao Guo, et al. Reconfigurable neuromorphic computing: materials, devices and integration. *Advanced Materials*, 31: 2301063, 2023.

Yue Zhang. *Compact Modeling and Hybrid Circuit Design for Spintronic Devices based on Current-induced Switching*. PhD thesis, Université Paris Sud-Paris XI, 2014.

Shiqiang Zhu, Ting Yu, Tao Xu, et al. Intelligent computing: the latest advances, challenges, and future. *Intelligent Computing*, 2:0006, 2023.

8

Challenges, Conclusions, Road-Map, and Future Perspectives

Specific, Measurable, Achievable, Relevant, and Time-bound (SMART) learning objectives and goals: after reading this chapter you should be able to:

- **Understand Technical Challenges in Nanoscale Computing:**
  - Identify key technical challenges in nanoscale computing.
  - Evaluate potential solutions to overcome material and design constraints.
  - Assess strategies for enhancing signal reliability and stability.
  - Analyze manufacturing and integration challenges in nanoscale computing.
- **Address Educational Challenges:**
  - Recognize the educational challenges associated with nanoscale computing.
  - Formulate strategies to address educational challenges.
  - Evaluate the role of education in advancing nanoscale computing technologies.
- **Explore Societal and Ethical Considerations:**
  - Examine privacy and security concerns in nanoscale computing.
  - Discuss ethical practices in emerging technologies.
  - Evaluate the environmental Impact of nanoscale devices.
  - Analyze the balance between technological advancements and environmental responsibility.
- **Investigate Future Perspectives in Nanoscale Computing:**
  - Understand advancements in nanomagnetic logic (NML).
  - Explore potential breakthroughs in materials and design.
  - Analyze the roadmap for nanoscale computing technologies.
- **Draw Conclusions and Identify Research Opportunities:**
  - Summarize key findings and contributions from the chapter.
  - Identify gaps in nanoscale computing research.
  - Encourage reader feedback and interaction.
  - Anticipate future developments in nanoscale computing.

Nanoscale Computing: The Journey Beyond CMOS with Nanomagnetic Logic,
First Edition. Santhosh Sivasubramani.
© 2025 The Institute of Electrical and Electronics Engineers, Inc. Published 2025 by John Wiley & Sons, Inc.
Companion website: www.wiley.com/go/sivasubramani/nanoscalecomputing1

This chapter talks about the challenges in nanoscale computing in Section 8.1, briefs the environmental impact in Section 8.2, describes the integration with other technologies in Section 8.3, presents in detail of Nanoscale Computing Technologies Roadmap in Section 8.4, elaborates the conclusions and key findings in Section 8.5 and details about the Research Opportunities and Directions in Section 8.6 for the first time.

8.1 Challenges in Nanoscale Computing

This section talks about the challenges in nanoscale computing, it briefs the technical challenges in Section 8.1.1, describes the educational challenges in Section 8.1.2, presents in detail the policy making in Section 8.1.3, briefs the reliability and error rates in nanomagnetic logic in Section 8.1.4, describes the case studies on improving reliability in Section 8.1.5, and presents in detail the scalability solutions and innovations in Section 8.1.6.

Nanoscale computing presents challenges in quantum effects, thermal management, interconnectivity, fault tolerance, energy efficiency, manufacturing, security, and ethics. "JOURNEY BEYOND CMOS WITH NANOMAGNETIC LOGIC" sheds light on these complexities.

1. **Quantum Effects** challenge logical stability, requiring innovative approaches. Thermal challenges arise as components shrink, demanding breakthroughs in materials for better heat dissipation.
2. **Interconnectivity Issues** affect data transmission in nanoscale computing. Resilient interconnects are needed, involving novel materials and architectural designs.
3. **Fault Tolerance** is critical as components approach atomic dimensions. Nanomagnetic Logic offers inherent resilience to certain faults.
4. **Energy Efficiency** is a concern, in nanoscale architectures. Nanomagnetic logic (NML) configurations present potential solutions for ultra-low-power consumption.
5. **Manufacturing Intricacies** demand uniformity in fabrication processes. Adaptive algorithms are needed to accommodate device variability. Figure 8.1 illustrates the implementation of these adaptive algorithms using neuro-optimization with embedded analog-grade eFlash memories.
6. **Security Concerns** in nanoscale computing require encryption mechanisms resilient to quantum attacks. Ethical Considerations emphasize responsible innovation and continuous scrutiny.
7. **Interdisciplinary Collaboration** among material scientists, physicists, and computer engineers is crucial for navigating nanoscale computing Challenges.

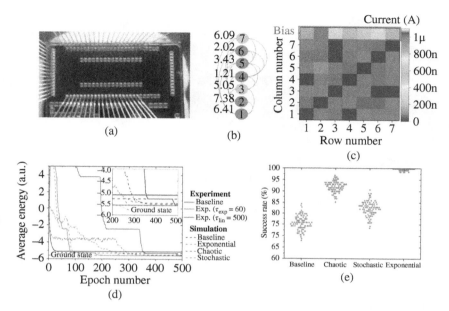

Figure 8.1 Neuro-optimization with embedded analog-grade eFlash memories. Panel (a) shows the fabricated 10×12 eFlash array chip in Global Foundries standard LPe CMOS process https://dl.acm.org/?doi?10.1145/3149526.3149531. (b) A 7-node maximum-weighted independent set problem. (c) The heat map of the synaptic weights for the devices that implement the neuron-optimization. (d) The average energy versus epoch comparing experimental results with simulations over 100 runs. (e) The success rate of different annealing techniques on this problem over 100 runs. File: Fahimi, Z https://doi.org/10.1038/s41598-020-78944-5 Source: Fahimi, Z et al., 2021/Springer Nature/CC BY 4.0.

8.1.1 Technical Challenges

This section talks about the technical challenges throughout and briefs the enhancing signal reliability and stability in Section 8.1.1.1.

8.1.1.1 Enhancing Signal Reliability and Stability

Enhancing signal Reliability and stability in nanoscale computing involves challenges and opportunities in material design, architecture, and engineering solutions.The figure 8.2 illustrates technical challenges associated with nanoscale computing, complementing the discussion on enhancing signal reliability and stability (cf. Figure 8.2).

1. **Material Innovation:**
   Explore unconventional materials with superior magnetic and Electrical Characteristics. Address material heterogeneity for uniform device performance. Develop adaptive algorithms to mitigate inherent material variability.

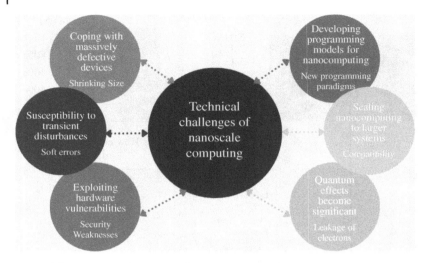

Figure 8.2 Technical challenges of nanoscale computing.

2. **Design Paradigms:**
   Innovate architectural solutions for optimal signal transmission. Balance functionality and scalability in nanoscale dimensions. Tackle quantum effects through design strategies.

3. **Quantum Challenges:**
   Mitigate quantum tunneling and entanglement effects. Develop materials resistant to quantum anomalies. Design fault-tolerant systems to counter quantum disruptions.

4. **Thermal Considerations:**
   Implement efficient thermal management strategies for heat dissipation. Explore materials with superior thermal conductivity. Maintain reliability by preventing overheating in nanoscale components. The figure 8.3 demonstrates diverse spin phenomena for energy interconversion, relevant to thermal management strategies discussed in preventing overheating in nanoscale components (cf. Figure 8.3).

5. **Interconnect Solutions:**
   Redesign interconnect architectures for nanoscale reliability. Experiment with novel materials for enhanced signal transmission. Foster high-reliable communication in the NML paradigm.

6. **Energy-Efficient Configurations:**
   Pursue ultra-low-power consumption profiles in materials and designs. Develop configurations prioritizing energy efficiency. Innovate for sustainable and energy-conscious NML systems.

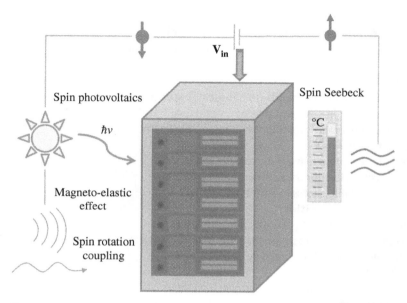

Figure 8.3 Diverse spin phenomena can act as energy interconversion rectifiers among energy sources such as heat (spin Seebeck effect), light (spin photovoltaics), sounds, and vibrations (magnetoelastic effect and spinrotation coupling). The collected energy can feedback power to servers in data centers. File: Figure 7: Energy interconversion rectified by the electron spin. https://doi.org/10.1038/s43246-020-0022-5 Source: Puebla, J et al., 2020/Springer Nature/CC BY 4.0.

7. **Security Measures:**
   Signal security against external interference. Integrate robust encryption mechanisms for heightened data protection. Implement quantum-resistant cryptographic techniques seamlessly.
8. **Manufacturing Precision:**
   Refine manufacturing processes for scalability and reproducibility. Achieve consistent performance in mass-produced NML devices.

8.1.2 Educational Challenges

This section talks about the educational challenges throughout. Nanoscale computing presents unique challenges that demand an adaptable approach:

1. **Curricular Integration:**
   - Seamlessly integrate nanoscale computing into existing curricula.
   - Develop interdisciplinary courses to bridge knowledge gaps.
   - Align educational content with industry advancements.

2. **Faculty Expertise:**
   - Foster faculty expertise through targeted training.
   - Establish collaborative networks for knowledge exchange.
   - Encourage industry – academia partnerships.
3. **Laboratory Facilities:**
   - Upgrade facilities for hands-on nanoscale experiments.
   - Invest in cutting-edge equipment for practical training.
   - Ensure safety measures align with nanoscale technologies.
4. **Access to Resources:**
   - Provide comprehensive resources for students.
   - Establish digital libraries and online platforms.
   - Offer affordable access to simulation tools.
5. **Research Opportunities:**
   - Facilitate research opportunities for students.
   - Establish mentorship programs with industry experts.
   - Promote a research-oriented culture.
6. **Industry Collaboration:**
   - Foster collaboration with industry stakeholders.
   - Establish internship programs for student exposure.
   - Integrate industry-driven projects into the curriculum.
7. **Public Awareness:**
   - Increase awareness about nanoscale computing.
   - Engage in outreach programs for student interest.
   - Collaborate with media for accurate information.
8. **Ethical Considerations:**
   - Integrate ethics into nanoscale computing education.
   - Emphasize responsible research practices.
   - Develop case studies on ethical decision-making.

Addressing educational challenges requires a comprehensive strategy, focusing on Curricular Integration, Faculty Expertise, facilities, resource access, Research Opportunities, Industry Collaboration, Public Awareness, ethics, assessment, and Global Collaboration.

8.1.3 Policy making

This section talks about the policymaking throughout. The advent of nanoscale computing necessitates strategic policy decisions:

1. **National Security:**
   - Develop policies to safeguard nanoscale technologies.
   - Foster collaborations to address security concerns.
   - Formulate regulations for technology export.

2. **Research Funding and Support:**
   - Allocate resources for nanoscale research and development.
   - Encourage public–private partnerships for innovation.
   - Implement grant schemes for researchers.
3. **Standardization and Regulation:**
   - Introduce regulatory frameworks for ethical use.
   - Establish international standards for safety.
   - Monitor and regulate commercialization.
4. **Education and Workforce Development:**
   - Implement policies for nanoscale education integration.
   - Foster industry–educational partnerships for skill development.
   - Support continuous learning initiatives.
5. **Environmental Impact:**
   - Formulate policies to assess and mitigate environmental impact.
   - Encourage eco-friendly NML technologies.
   - Establish guidelines for responsible disposal.

Effective nanoscale governance requires policies addressing national security, research funding, standardization, education, ethics, international collaboration, environmental impact, privacy, and public awareness. Policymakers must consider these dimensions for responsible nanoscale development and deployment. The figure 8.4 highlights the ethical challenges of nanocomputing, which are crucial for policymakers to address in effective nanoscale governance (cf. Figure 8.4).

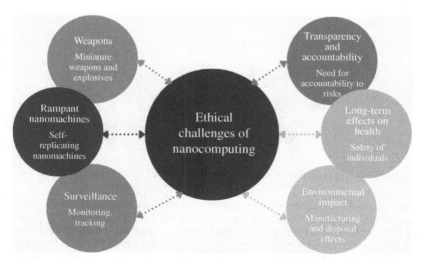

Figure 8.4 Ethical challenges of nanocomputing. Source: The Author.

8.1.4 Reliability and Error Rates in Nanomagnetic Logic

This section talks about reliability and error rates in nanomagnetic logic throughout.

In **nanoscale computing**, **Reliability** and **Error Rates** are crucial considerations. NML offers inherent Reliability advantages due to stable **magnetic states**, reducing vulnerability to external factors. Despite advancements, challenges persist, such as quantum effects and thermal fluctuations. Mitigation strategies include tailored error correction codes, redundancy, novel materials, and error-resilient logic gate designs. Material selection is vital, focusing on stability against environmental factors. Temperature control is necessary, and fault-tolerant architectures enhance system Reliability. The figure 8.5 illustrates different reservoir architectures, which are relevant to the mitigation strategies and error-resilient designs discussed for enhancing reliability in nanoscale computing. These architectures provide various methods for input data handling and output state reading, crucial for reducing error rates and improving stability (cf. Figure 8.5). Ongoing research addresses reliability challenges through collaboration, exploring materials, and adopting multidisciplinary approaches. Robust

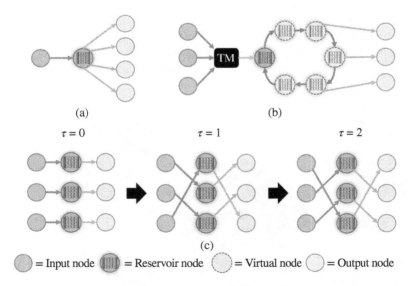

Figure 8.5 Schematic diagram of three reservoir architectures: signal sub-sample, single dynamical node with virtual nodes, and revolving neurons. Each architecture demonstrates different methods for input data handling and output state reading. These designs are essential for error-resilient and reliable nanoscale computing. File: Figure 2: Schematic diagrams of each reservoir architecture. https://doi.org/10.1038/s42005-023-01352-4 Source: I. T. Vidamour et al., 2023/Springer Nature/CC BY 4.0.

testing methodologies and industry standards, along with educational emphasis on reliability, ensure the seamless integration of nanomagnetic logic.

Understanding Error Mechanisms: Nanoscale computing involves inherent error mechanisms, including quantum effects, thermal fluctuations, and material considerations. Quantum Effects induce uncertainties, requiring predictive models. Material properties influence errors, with research focusing on materials resistant to these factors. Temperature-induced errors and external influences like radiation contribute to system errors. Fault-tolerant architectures, feedback mechanisms, and error propagation models are essential for preventive error correction. Collaborative efforts and rigorous testing protocols enhance error understanding.

Error Correction Techniques: Nanoscale computing encounters errors from quantum effects, thermal fluctuations, and other complexities. Effective error correction is crucial for system Reliability. Quantum error correction codes address quantum-induced errors. Redundancy-based approaches like TMR and error detection/correction codes such as Hamming enhance stability. Self-check mechanisms, probabilistic error correction, machine learning, feedback-based correction, multilayered redundancy, and physical error avoidance are tools for adaptive error correction. Cross-disciplinary collaboration accelerates robust solutions.

8.1.5 Case Studies on Improving Reliability

This section talks about the case studies on improving reliability throughout, it briefs the scalability and manufacturing challenges in Section 8.1.5.1, and describes the nanoscale fabrication techniques in Section 8.1.5.2.

Real-world case studies highlight strategies for enhancing Reliability in nanoscale computing. Quantum error correction in quantum dots, redundancy-driven Reliability in carbon nanotube circuits, self-check mechanisms in DNA nanocomputing, probabilistic error correction in spintronics, machine learning adaptation in neuromorphic computing, feedback-driven stability in memristor-based architectures, multilayered redundancy in 2D material-based devices, physical error avoidance in silicon photonics, and cross-disciplinary collaboration in quantum cascade devices and continuous learning models in molecular computing provide insights. These cases exemplify diverse approaches, including quantum error correction, redundancy, self-check mechanisms, probabilistic models, machine learning, feedback-driven stability, multilayered redundancy, physical error avoidance, and collaboration, contributing to Reliability in nanocomputing. The figure 8.6 showcases various mechanisms in nanocomputing, such as reservoir computing and PMA nanomagnets, relevant to the discussed strategies for enhancing reliability through machine learning adaptation and physical error avoidance. These mechanisms exemplify practical applications of reliability techniques in nanoscale computing (cf. Figure 8.6).

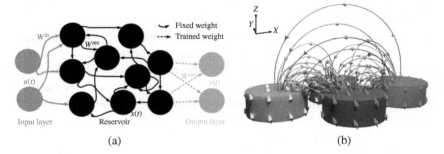

Figure 8.6 Illustrations of reservoir computing and frustrated perpendicular magnetic anisotropy (PMA) nanomagnets. Reservoir computing highlights neural network strategies, while PMA nanomagnets demonstrate physical error avoidance. Both mechanisms are key for enhancing reliability in nanoscale computing. File: Figure 1: Reservoir computing. https://doi.org/10.1038/s42005-023-01324-8 Source: Alexander J. Edwards et al., 2023/Springer Nature/CC BY 4.0.

8.1.5.1 Scalability and Manufacturing Challenges

Nanoscale computing presents promising scalability opportunities and manufacturing challenges. Scaling down enables faster computations and reduced energy consumption but poses Reliability and stability challenges. Precision manufacturing using advanced lithography techniques is crucial. Alternative materials like graphene overcome scalability limitations, and quantum-scale manufacturing addresses uncertainties. Integration of heterogeneous technologies demands advanced techniques, and material compatibility remains a challenge. Nanoscale parallel processing enhances computational power, but yield and reproducibility challenges persist. Power-efficient designs focus on minimizing energy dissipation. Architectural innovations and economic viability are critical for scalability success. Balancing innovation with cost-effectiveness is essential for industrial success in nanoscale manufacturing. Researchers contribute to a transformative era in nanocomputing by navigating scalability and addressing manufacturing challenges.

8.1.5.2 Nanoscale Fabrication Techniques

In nanoscale computing, cutting-edge device development relies on advanced fabrication techniques. These methods construct components at atomic and molecular levels, facilitating advanced NML and complementary metal-oxide-semiconductor (CMOS) hybrid architectures. Figure 8.7 illustrates various stages in the fabrication of nanoscale computing devices, from schematic design to advanced imaging techniques, highlighting the precision required in constructing components at atomic and molecular levels. These fabrication methods are essential for developing advanced NML and CMOS hybrid architectures. Figure 8.8

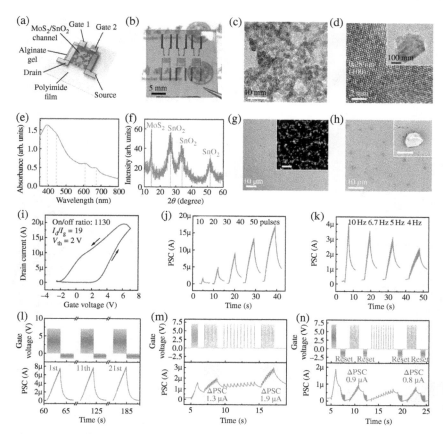

Figure 8.7 Photograph of the devices fabricated as an array on a flexible substrate. Schematic, photographs, and microscopy images detailing the fabrication and characterization of nanoscale devices. These include TEM and SEM images, UV-visible absorption spectra, XRD patterns, and transfer curve analyses, showcasing advanced techniques essential for nanoscale computing. File: Characterization and performance of the synaptic transistor.webp https://commons.wikimedia.org/wiki/File:Characterization_and_performance:of_the_synaptic_transistor.webp Source: Chengpeng Jiang et al., 2023/Springer Nature/CC BY 4.0.

compares thermoelectric and spin Seebeck effect devices, illustrating mechanisms for energy conversion relevant to the discussed strategies for reliability in nanoscale computing. These devices, through their thermoelectric and Nernst effects, demonstrate approaches to enhancing system efficiency and stability in nanoscale applications.

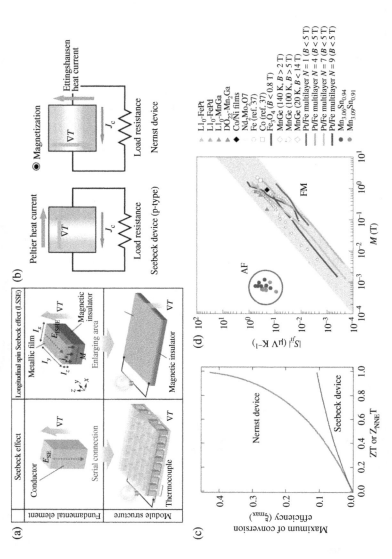

Figure 8.8 (a) Comparison between the fundamental element of a Seebeck effect device and longitudinal spin Seebeck effect device, and their corresponding module architectures (Uchida, K. et al., 2016 / With permission of IEEE). (b, c) The schematics of the thermoelectric conversion mechanism for a Seebeck device and a Nernst effect device, and comparison of the maximum conversion efficiencies ζ_{max} under a temperature gradient $\Delta T = 300$ K (figures reproduced from [Mizuguchi, M. & Nakatsuji, S., 2019 / Taylor & Francis / CC BY 4.0]). (d) Comparison of Nernst signals S_{ji} for ferromagnets (FM) and the non-collinear antiferromagnet (AF) Mn$_3$Sn. Ikhlas, M, 2017 / With permission of Springer Nature. File: Figure 8: Spin thermoelectric generation mechanism. https://doi.org/10.1038/s43246-020-0022-5 Source: Puebla, J. et al. / Springer Nature / CC BY 4.0.

1. **Photolithography Advancements:**
   - Key of nanofabrication techniques.
   - Achieves resolutions below 10 nm.
   - Extreme ultraviolet lithography enables finer feature patterning.
   - Improved photoresists enhance pattern transfer precision.
2. **Electron Beam Lithography Precision:**
   - Offers unparalleled precision in nanoscale patterning.
   - Allows direct writing of intricate patterns.
   - Achieves sub-10-nm feature sizes.
   - Pivotal for prototyping nanoscale devices.
3. **Nanoimprint Lithography Efficiency:**
   - Involves stamping patterns onto substrates for replication.
   - Enables large-scale, cost-effective production.
   - Finds applications in reproducibly producing nanoscale components.
4. **Atomic Layer Deposition (ALD) Precision:**
   - Ensures precise deposition of atomic layers.
   - Allows controlled growth of thin films.
   - Conformal coating enhances device performance.
   - Crucial for creating ultrathin dielectrics and metal layers.
5. **Quantum Dot Self-Assembly:**
   - Leverages spontaneous arrangement of quantum dots.
   - Self-assembles into ordered structures on substrates.
   - Provides an energy-efficient approach for nanoscale structures.
   - Valuable for quantum dot-based NML devices.
6. **Directed Self-Assembly (DSA) Applications:**
   - Guides polymer patterns into predefined configurations.
   - Enhances resolution and pattern uniformity.
   - Employed in creating regular arrays for nanoscale circuitry.
   - Leverages thermodynamics for controlled self-organization.
7. **Top-Down and Bottom-Up Integration:**
   - Optimizes nanofabrication through integration.
   - Top-down designs materials at the nanoscale.
   - Bottom-up assembles structures from atoms or molecules.
   - Hybrid approaches offer versatility for complex nanodevices.
8. **Chemical Vapor Deposition (CVD) Innovations:**
   - Enables controlled deposition of thin films on substrates.
   - Innovations enhance uniformity and coverage on complex surfaces.
   - Plasma-enhanced CVD ensures high-quality films.
   - Integral to creating semiconductor materials for nanoscale devices.
9. **Quantum Point Contacts for Nanoscale Wires:**
   - Used to create nanoscale wires.
   - Exhibits quantum mechanical effects for precise electron control.

- Nanowires contribute to developing nanoscale interconnects. The pinning energy profile depicted in is essential for understanding the quantum mechanical effects in nanoscale wires created using quantum point contacts. These insights are pivotal for constructing intricate nanocomputing architectures, as they facilitate precise electron control and stability in nanoscale interconnects (cf. Figure 8.9).
- Applications in constructing intricate nanocomputing architectures.

10. **3D Printing for Nanoscale Components:**
    - Extends into nanoscale for precise component fabrication.
    - Allows creation of intricate structures with diverse materials.
    - Instrumental in rapid prototyping of nanodevices.
    - Opens avenues for customized and on-demand nanomanufacturing.

The diverse landscape of nanoscale fabrication techniques shapes the future of nanocomputing. The convergence of these methods unlocks possibilities for advanced NML, CMOS hybrid architectures, and innovative devices. As researchers refine these techniques, the transformative impact on nanocomputing unfolds, evolving in an era of unparalleled computational capabilities.

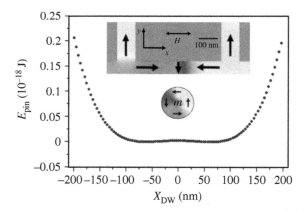

Figure 8.9 The pinning energy profile, $E(X)$, for a transverse domain wall between two anti-notches extracted by fitting the micromagnetic simulations according to [Pivano, A. & Dolocan, V. O. Chaotic dynamics of magnetic domain walls in nanowires. Phys. Rev. B 93, 144410. https://doi.org/10.1103/PhysRevB.93.144410 (2016).]&[Martinez, E., Lopez-Diaz, L., Alejos, O., Torres, L. & Carpentieri, M. Domain-wall dynamics driven by short pulses along thin ferromagnetic strips: Micromagnetic simulations and analytical description. Phys. Rev. B 79, 094430. https://doi.org/10.1103/PhysRevB.79.094430 (2009).]. The inset shows a schematic representation of the single domain wall in a magnetic nanowire with two anti-notches placed symmetrically at a 350 nm distance edge-to-edge. File: Figure 1 https://doi.org/10.1038/s41598-021-94975-y
Source: Ababei, R.V et al., 2021/Springer Nature/CC BY 4.0.

8.1.6   Scalability Solutions and Innovations

This section talks about the scalability solutions and innovations through-out, describes the innovative solutions, presents in detail the addressing manufacturing challenges in Section 8.1.6.2, briefs the challenges in integration in Section 8.1.6.3, describes the integration with existing technologies in Section 8.1.6.4, presents in detail the integration challenges and solutions in Section 8.1.6.5, and discusses the Success Stories of Integration in Section 8.1.6.6.

Key Nanoscale computing Challenges:

1. Shrinking transistor sizes pose scalability challenges.
2. Interconnect limitations hinder information flow.
3. Quantum effects and heat dissipation become pronounced.

8.1.6.1   Innovative Solutions

Innovative technological solutions in computing focus on various advanced methodologies. Parallel processing paradigms involve executing multiple tasks at the same time. This approach alleviates performance bottlenecks by distributing the workload across different processing units. As a result, the system achieves better performance and improved scalability for complex computations. Quantum computing harnesses the power of quantum bits, or qubits, which inherently possess parallel processing capabilities. This technology aims to solve specific problems at exponential speeds compared to classical systems. However, maintaining qubit stability remains a significant challenge for practical applications. Photonic interconnects provide a faster alternative to traditional copper inter-connects. These optical components enable rapid data transmission, significantly reducing latency and enhancing overall performance. Utilizing photonic technologies can address the increasing demand for higher data transfer rates. Nanoscale memory architectures focus on developing memory solutions at the nanometer scale. Resistive random-access memory (RRAM) is one such technology that offers rapid access and scalability. This type of memory reduces data retrieval times, thus improving system efficiency and speed. Beyond traditional CMOS technology, novel paradigms explore new computing architectures.

NML introduces an alternative method of computation by leveraging magnetic properties at the nanoscale. This architecture promises energy-efficient operations with potential for scaling up. CMOS hybrid architectures combine traditional CMOS with emerging technologies to extend capabilities. These hybrid systems incorporate NML to potentially enhance performance and efficiency. Integrating new technologies with existing CMOS infrastructure offers a pathway to improved computational power. Each of these innovative approaches addresses specific limitations found in current systems. Parallel processing deals with task management across multiple processors. Quantum computing explores faster

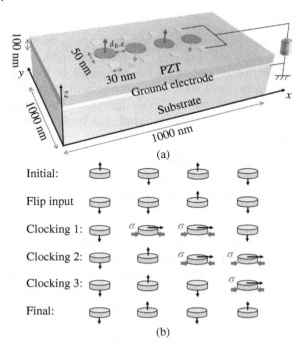

Figure 8.10 (a) 3D illustration of the Bennett clocking system simulated in the model. (b) Information flow of Bennett clocking process. File: Figure 1 https://doi.org/10.1038/s41598-019-39966-w Source: Qianchang Wang et al., 2019/Springer Nature/CC BY 4.0.

problem-solving techniques using quantum mechanics. Photonic interconnects aim for quicker data transfer rates. Nanoscale memory architectures focus on speed and efficiency improvements. Beyond CMOS, new architectures explore alternative computation methods. NML presents a scalable option with low energy consumption. CMOS hybrid approaches blend established and new technologies for enhanced performance. Each solution presents distinct advantages and challenges, contributing to the advancement of computing technologies. The Bennett clocking system, as shown (cf. Figure 8.10), integrates advanced clocking mechanisms into computational models. Figure 8.10 details the information flow within this system, reflecting how it enhances performance through optimized data transmission processes.

8.1.6.2 Addressing Manufacturing Challenges

Addressing manufacturing challenges involves various strategies tailored to specific technical requirements. DSA techniques enable accurate nanoscale component placement by utilizing self-organizing materials. These methods reduce defects during manufacturing, improving scalability and yield. Precision in component placement is essential for advancing nanoscale technologies. Additionally, DSA techniques can adapt to varying design requirements, optimizing alignment and integration processes.

3D integration strategies address limitations inherent in planar scaling. By stacking integrated circuits vertically, these strategies increase computational density and enhance performance. This approach facilitates more efficient use of space and resources in semiconductor devices. The vertical arrangement of components allows for higher data throughput and reduced latency. Implementing these strategies requires sophisticated alignment and bonding technologies to ensure reliable connections between layers.

Power-efficient approaches are crucial for managing energy consumption in modern electronics. Energy-efficient transistors are designed to minimize leakage currents, which significantly impacts overall power consumption. These transistors contribute to the feasibility of scaling up power-hungry systems while maintaining performance. Reducing leakage currents directly correlates with improved energy efficiency and reduced thermal management issues.

Neuromorphic computing paradigms simulate brain-inspired architectures to achieve energy-efficient computation. These architectures mimic neural processes to perform complex tasks with lower power requirements compared to traditional computing models. Neuromorphic systems offer solutions that reduce energy consumption while maintaining high computational capabilities. Their design emphasizes parallel processing and adaptive learning, further enhancing efficiency.

By combining these advanced techniques, manufacturers can address challenges associated with scalability and energy efficiency. DSA and 3D integration provide solutions to spatial and performance constraints. Concurrently, energy-efficient transistors and neuromorphic computing paradigms offer advancements in power management and computational efficiency. Implementing these approaches requires careful consideration of material properties, device architecture, and manufacturing processes.

8.1.6.3 Challenges in Integration

Cross-layer communication enhancements focus on optimizing interactions between different computing stack layers. These enhancements are crucial for achieving efficient integration across different system components. Without precise alignment and improved data exchange mechanisms, the integration process faces significant challenges. Standardization efforts play a pivotal role in this context. Implementing uniform standards facilitates smoother integration by creating a common framework. These standards also enhance interoperability between different system elements. They reduce compatibility issues and simplify the overall integration procedure.

Establishing well-defined protocols can address communication bottlenecks. Clear and consistent guidelines for data exchange enhance coordination among diverse components. Efforts to standardize these processes help in minimizing integration complexity. These initiatives focus on resolving technical discrepancies and streamlining integration workflows. Standardized approaches ensure that different systems or components can work together effectively. As a result, the integration process becomes less cumbersome and more predictable. Refining cross-layer communication and advancing standardization efforts are crucial for successful system integration. These technical improvements lead to more efficient interactions between components and a smoother integration process.

The pursuit of scalability solutions in nanoscale computing involves diverse strategies, from exploring beyond CMOS to addressing manufacturing challenges and embracing power-efficient approaches.

8.1.6.4 Integration with Existing Technologies

Magnetic tunnel junctions (MTJs) leverage quantum mechanical tunneling for operation. They consist of three layers: a fixed magnetic layer, an insulating tunnel barrier, and a free magnetic layer. The tunneling current's resistance relies on the magnetic orientations of these layers. When aligned, the resistance is low; when misaligned, it is high. This principle underlies the MTJ's functionality. MTJs provide significant benefits in nanoscale computing. They exhibit non volatile behavior, meaning they retain data even when power is off. This characteristic supports persistent data storage and enhances energy efficiency. In addition, the quantum tunneling process in MTJs involves minimal energy, contributing to low-power consumption. This efficiency is crucial for nanocomputing, where power constraints are stringent.

The MTJ's ability to perform rapid magnetic switching translates to high-speed data processing. Fast switching speeds improve computational performance and efficiency. By integrating MTJs, systems gain advantages in both energy efficiency and processing speed. The nanomagnet reservoir computing (NMRC) system diagram in Figure 8.11 illustrates the interaction between input signals and the MTJ, defining the reservoir voltage landscape critical for the memristor crossbar array. This operational framework is further detailed in the NMRC training and inference process, which visualizes signal flow and system actions. In practical applications, MTJs' low energy consumption aligns well with the goals of modern computing, which aims to balance performance with power usage. Their non volatility aids in developing memory solutions that are both fast and reliable. The technology supports advancements in computational efficiency, crucial for handling complex tasks and large data sets. MTJs integrate seamlessly with

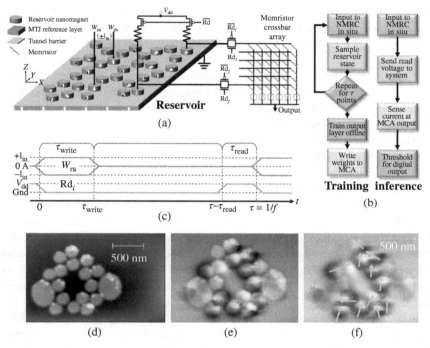

Figure 8.11 (a) NMRC system diagram shows signals W_{ra} and W_{rb} energizing nanomagnets via the MTJ, impacting the reservoir voltage landscape and the MCA output layer. (b) NMRC training and inference process maps signal input, reservoir read, and system actions, with distinct colors indicating different stages. (c) NMRC inference timing outlines the sequence of write and read periods, ensuring $\tau_{write} + \tau_{read} < \tau$, where τ is the reciprocal of the operating frequency. (d) topography image. (e) Phase image. (f) Illustration of the magnetization direction. File: Figure 2: Nanomagnet reservoir computing (NMRC). https://doi.org/10.1038/s42005-023-01324-8 Source: Alexander J. Edwards et al., 2023/Springer Nature/CC BY-4.0

existing technologies, offering a blend of speed, efficiency, and reliability. The technology's adaptability makes it suitable for various applications, from memory storage to advanced computing systems. Their design and operational principles enable a range of use cases in modern technology infrastructures.

8.1.6.5   Integration Challenges and Solutions

Integration of MTJs with CMOS technology presents several difficulties. Compatibility issues arise due to different material properties and fabrication processes. To address this, researchers are developing hybrid architectures that facilitate integration. These architectures combine MTJs with CMOS elements, ensuring a seamless interface between the two technologies. In manufacturing, controlling MTJ component fabrication is challenging. Precision in creating MTJs impacts their performance. Advances in fabrication techniques are needed to improve accuracy. Techniques such as ALD and precise etching are being refined to address these issues.

MTJs are explored for memory solutions in edge AI devices. Their non volatility and low-energy consumption make them suitable for memory applications. They offer advantages over traditional memory elements in terms of power efficiency and data retention. The integration of MTJs into edge AI devices highlights their role in memory solutions. The schematic illustrates the structure of MTJs and their variants in readout schemes, emphasizing their potential benefits (cf. Figure 8.12).

MTJs also hold promise for logic and computing operations. They are being studied for energy-efficient computations. Integrating MTJs into logic circuits could enhance processing capabilities. Their potential to reduce power consumption in computing tasks is significant. Material design plays a crucial role in MTJ performance. Researchers are focusing on materials with specific magnetic properties. Engineering magnetic anisotropy is essential for achieving desired device characteristics. Tailoring materials to meet specific requirements is key to optimizing MTJ functionality. The schematic illustrates the structure and operational mechanisms of MTJs in various applications. This provides insight into how material design influences their effectiveness in logic and computing operations (cf. Figure 8.13). Future prospects involve scaling MTJs for nanocomputing. Research is investigating how to scale these devices for various applications. Understanding their limits and possibilities is critical for practical computing scenarios. Scaling challenges include maintaining performance while reducing device size. Hybrid architectures are being explored to enhance synergies between MTJs and other technologies. Combining MTJs with emerging technologies could yield new capabilities. Research aims to create synergies between MTJs and alternative paradigms for improved performance and efficiency. The integration of

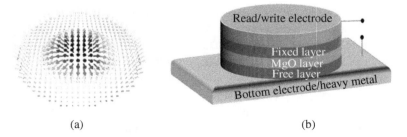

(a)   (b)

Figure 8.12   (a) Skyrmion (b) MTJ structure (variants of the readout schemes in the supplement). File: Figure 1 https://doi.org/10.1038/srep31272 Source: Dhritiman Bhattacharya et al., 2016/Springer Nature/CC BY 4.0.

Figure 8.13 (a) The MTJ temperature sensor structure involves a three-terminal device with a tunneling oxide layer and a heavy metal underlayer. A spin current is injected to the free layer, and a read current determines the device state. (b) The sensor MTJ, interfaced with a reference MTJ, forms a voltage divider circuit that drives an inverter to measure switching probability based on temperature and control signals. File: Figure 1 https://doi.org/10.1038/s41598-017-11476-7 Source: Sengupta, A et al., 2017/Springer Nature/CC BY 4.0

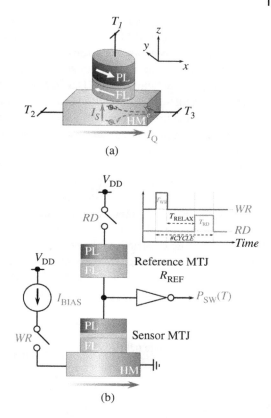

MTJs into CMOS technology faces compatibility challenges. Hybrid architectures are a focus area to resolve these issues. Manufacturing MTJ components requires precision and advanced fabrication techniques. For edge AI devices, MTJs offer efficient memory solutions. Their role in logic and computing operations is under exploration. Material design considerations are pivotal in optimizing MTJ performance. Future research will address scaling challenges and hybrid architectures for enhanced synergies. The integration of MTJs into nanoscale computing holds promise in advancing edge AI devices, with ongoing exploration into compatibility, manufacturing, and material design.

8.1.6.6 Success Stories of Integration

The field of memory technology has experienced significant advancements, particularly with the integration of MTJs as nonvolatile memory elements. These developments have led to improved data retention and efficient storage solutions. MTJs have enhanced memory performance by facilitating faster data access and retrieval, surpassing traditional methods. In the realm of edge AI device optimization, the integration of MTJs has contributed to the creation of energy-efficient edge computing systems. This efficiency is a result of MTJs' low-power consumption attributes. Furthermore, high-speed data processing has seen success through rapid magnetic switching in MTJs, which accelerates computational tasks.

Hybrid architectures have also seen notable progress. Successful combinations with CMOS technology highlight the compatibility and functional improvements achieved. This synergy has facilitated the development of hybrid nanocomputing devices that leverage the strengths of multiple technologies. In practical applications, the integration of MTJs in wearable technology has brought benefits, including non-volatile memory and energy efficiency. Similarly, smart sensors and Internet of Things (IoT) devices have been optimized for data storage and processing, enhancing their efficiency and functionality.

The schematic of the whisker flow sensor illustrates the application of thin-sheet magnetostrictive alloys in sensor technology. This aligns with the advancements discussed, showcasing material innovations and their integration into practical devices (cf. Figure 8.14). Material design advancements have played a crucial role in these developments. Tailoring materials for MTJs has led to

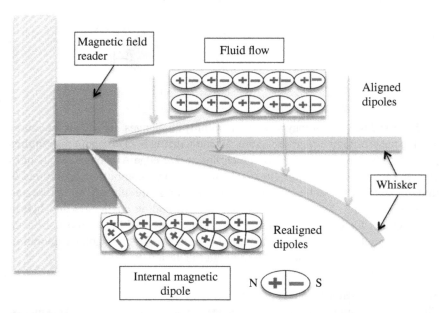

Figure 8.14 Schematic of a whisker flow sensor developed using thin-sheet magnetostrictive alloys. File: Magnetostrictive flow sensor.png https://commons.wikimedia.org/wiki/File:Magnetostrictive_flow_sensor.png Source: Julia Downing/Wikimedia Commons/CC BY-SA 4.0.

increased reliability and stability. Success in engineering magnetic anisotropy has allowed for fine-tuning of MTJs, enhancing their overall functionality. Overcoming challenges has been a key aspect of these advancements. Solutions to integration challenges have ensured compatibility and seamless incorporation of new technologies. Precision in manufacturing nanoscale MTJs has been achieved through advancements in techniques, resulting in improved precision and performance. Looking ahead, the trajectory of nanocomputing suggests continued success in scaling MTJs for various applications. This scaling involves exploring smaller dimensions and increased integration density. Research continues to delve into hybrid technologies, pushing the boundaries of current capabilities. Hybrid nanocomputing remains a prominent area of exploration, with ongoing studies aiming to uncover new synergies and applications. These future directions highlight the ongoing evolution and potential of memory technologies and edge AI optimizations.

8.2 Environmental Impact

This section talks about the environmental impact, it briefs the sustainable manufacturing practices in Section 8.2.1, describes the Energy Consumption in Nanoscale Devices in Section 8.2.2, and presents in detail the Balancing Technological Advancements with Environmental Responsibility in Section 8.2.3

1. Nanoscale computing requires examination of environmental impact.
2. Sustainable manufacturing practices are crucial.

8.2.1 Sustainable Manufacturing Practices

This section talks about the sustainable manufacturing practices throughout. Sustainable manufacturing practices require comprehensive energy efficiency strategies. Nanofabrication processes should utilize minimal energy to reduce overall consumption. Assessing the lifecycle of materials from initial extraction through to final disposal is essential. This includes analyzing all stages, including transportation and processing. Designing products for recyclability ensures materials can be effectively repurposed. E-waste management involves systematic recycling and safe disposal methods. Facilities specializing in nanocomputing should use renewable energy sources, like solar or wind power. Strategies for lowering carbon emissions must be integrated into production methods. Utilizing low-emission technologies in manufacturing contributes to a reduced carbon footprint. Nanodevices should be designed with architectures that optimize energy use. Closed-loop manufacturing systems help in minimizing waste by reusing materials. Implementing water-efficient practices involves reducing water consumption and improving management. Effective water stewardship includes proper disposal and treatment of wastewater. Biodiversity assessments should be conducted at manufacturing sites to minimize ecological impacts. Adherence to strict environmental standards ensures compliance with regulations and best practices. Designing for an extended operational life reduces the frequency of replacements. Eco-friendly synthesis methods further contribute to sustainability. Engaging local communities in these practices promotes wider acceptance and implementation. Research into alternative computing paradigms explores methods for improving energy efficiency. Regenerative design principles can be applied to nanoscale computing systems. Real-time environmental monitoring provides immediate feedback on ecological impact. Developing cooling solutions that are environmentally friendly reduces the reliance on harmful refrigerants. These measures collectively enhance the sustainability of manufacturing processes.

8.2.2 Energy Consumption in Nanoscale Devices

This section talks about the energy consumption in nanoscale devices throughout. In nanoscale devices, effective power management becomes paramount due to their size. Devices operating at this scale require meticulous attention to energy consumption and its broader environmental ramifications. Implementing efficient energy usage protocols can be difficult when device dimensions shrink. It is crucial to fabricate nanomaterials through sustainable methods to minimize their environmental footprint. Achieving energy efficiency involves developing nanodevices with designs that inherently use less power. Proper management of electronic waste is essential for reducing ecological impact. Incorporating renewable energy sources into these devices can significantly alter their energy profiles. Addressing cooling issues is vital since nanoscale devices generate substantial heat, necessitating advanced cooling strategies. Raising awareness in communities about energy consumption is key to fostering responsible practices. Exploring quantum computing offers potential for lower energy usage compared to classical systems. Dynamic power management techniques allow for adaptive energy use, optimizing performance while conserving power. Pursuing carbon-neutral manufacturing practices in nanotechnology contributes to reducing overall environmental impact. Balancing device longevity with energy sustainability ensures that devices remain functional and efficient over time. Considering societal effects of energy-heavy computations is important for ethical development. Green computing models should be explored for their potential to establish sustainable energy methodologies. Finally, integrating nanodevices capable of energy harvesting can help capture and reuse energy within systems.

8.2.3 Balancing Technological Advancements with Environmental Responsibility

This section talks about the balancing technological advancements with environmental responsibility throughout. Incorporate environmental considerations into nanocomputing designs from the outset. Develop materials focusing on computational efficiency alongside ecological impact. Ensure nanomanufacturing practices achieve carbon neutrality. Design nanodevices with architectures that maximize resource efficiency. Apply principles of circular economy to nanoscale computing systems. Combine nanocomputing technologies with renewable energy sources to enhance sustainability. Prioritize energy-efficient algorithms to reduce overall power consumption. Formulate policies that integrate technology advancements with responsible electronic waste management. Embed ethical considerations into nanocomputing advancements to ensure societal benefits. Increase community awareness of energy consumption implications related to new technologies. Address cooling challenges in energy-intensive devices using sustainable technologies. Investigate the potential of quantum computing as a greener alternative. Explore green computing paradigms to embed sustainability in nanocomputing. Consider nanodevice lifespan alongside ethical and ecological considerations. Assess societal and environmental justice impacts of emerging technologies. Incorporate responsible AI practices within nanocomputing frameworks to align with ethical standards. Evaluate interfaces that enable direct nanoscale computing integration with the human body, ensuring safety and efficiency.

8.3 Integration with Other Technologies

This section talks about the integration with other technologies, it briefs the synergies with quantum computing and AI in Section 8.3.1, describes the future perspectives in nanoscale computing in Section 8.3.2, and presents in detail the Graphene: A Conductor of Change in Section 8.3.3.

Integrating NML with current CMOS architectures necessitates designing interfaces that facilitate compatibility between these disparate technologies. CMOS, renowned for its established robustness in semiconductor technology, must adapt to accommodate the unique properties of NML, which leverages spintronics for processing. The challenge lies in harmonizing the different operational principles and ensuring that performance metrics such as power consumption and speed align. Examining the intersection of nanoscale logic with quantum computing requires understanding how quantum bits or qubits can interface with nanoscale circuits. Nanoscale logic operates at scales where quantum effects become significant, thus requiring integration strategies that manage coherence and entanglement across different computational paradigms. This integration must address potential quantum decoherence and error rates inherent in nanoscale systems.

Bridging nanoscale computing with neuromorphic architectures involves creating systems that mimic neural networks' functionalities while utilizing nanoscale elements. Neuromorphic computing's inspiration from biological neural networks requires nanoscale circuits to simulate synaptic connections and neuron-like behavior effectively. This integration involves developing analog or mixed-signal circuits that can emulate neural responses at extremely small scales. Aligning nanoscale computing with edge computing demands that nanoscale devices support real-time data processing at network edges. Edge computing reduces latency and bandwidth usage by processing data locally. Nanoscale devices must therefore incorporate robust local processing capabilities while maintaining low-power consumption and high performance, aligning with edge-computing requirements.

Embedding nanoscale computing within the IoT requires creating devices that are not only small but also capable of handling data exchange and processing in the IoT ecosystem. This integration involves designing nanoscale sensors and processors that communicate seamlessly with other IoT devices, adhering to standards for data interoperability and security. Exploring the integration of nanoscale devices with biocompatible systems involves creating technologies that can function within biological environments without adverse reactions. This entails developing nanoscale devices with materials and coatings that are biocompatible and can interface with biological systems for applications such as medical implants or biosensors.

Addressing challenges in integrating nanoscale computing with diverse systems involves tackling issues related to scalability, compatibility, and standardization. Solutions must include developing interfaces that can bridge nanoscale technologies with existing infrastructure and ensuring that performance criteria meet the needs of various applications across different sectors. Advancing nanophotonics for seamless integration involves improving optical technologies at the nanoscale to enable efficient data transfer and communication. This includes developing nanoscale optical components that can integrate with electronic systems to enhance performance and functionality.

Designing collaborative memory architectures for hybrid nanoscale systems involves creating memory units that efficiently work with both nanoscale and traditional computing elements. This design must address data consistency, access speed, and error correction across different memory technologies. Infusing machine learning algorithms into nanoscale computing requires optimizing algorithms to function efficiently on nanoscale hardware. This involves modifying algorithmic approaches to leverage nanoscale processing capabilities and ensure that machine learning models can be trained and executed with high accuracy.

Crafting hybrid nanoscale systems tailored for cryptographic functionalities entails developing nanoscale devices that can handle encryption and decryption tasks. These systems must incorporate security features to protect sensitive information and ensure cryptographic operations' integrity and confidentiality. Fostering collaborative initiatives between industry and academia involves creating partnerships to advance nanoscale computing research and development. These collaborations can drive innovation by combining academic research with practical industry applications, resulting in accelerated technological advancements.

Integrating nanoscale computing into cyber-physical systems involves embedding nanoscale components within systems that interact with physical processes. This integration requires ensuring that nanoscale computing elements can effectively process and respond to real-world stimuli in real-time. Embedding nanoscale decision-making capabilities into collaborative robotic systems involves designing nanoscale processors that can handle decision-making tasks in robotic environments. These processors must be capable of processing sensor data and executing complex algorithms to enable autonomous decision-making and actions.

Propelling innovations in nanoscale communication networks involves developing technologies that allow nanoscale devices to communicate effectively with each other. This includes creating protocols and hardware that can handle the specific challenges of nanoscale communication, such as signal attenuation and interference. Infusing nanoscale computing power into autonomous vehicles involves integrating nanoscale processing units that can handle complex computations required for vehicle autonomy. This integration must ensure that the nanoscale components contribute to the vehicles overall performance, safety, and decision-making capabilities.

Shaping collaborative architectures that integrate nanocomputing with cloud systems involves designing systems where nanoscale devices can interact with cloud computing resources. This integration requires addressing issues related to data synchronization, processing efficiency, and communication between nanoscale and cloud-based systems. Collaborating on green computing initiatives aligns nanoscale technologies with sustainability goals. This involves developing nanoscale devices that consume less energy and generate less heat, contributing to overall energy efficiency and environmental sustainability. Exploring interfaces for direct integration of nanoscale computing with the human body involves creating nanoscale devices that can operate within or on human tissues. This includes ensuring that these devices are biocompatible, secure, and capable of interfacing with biological systems for applications such as health monitoring or therapeutic interventions.

8.3.1 Synergies with Quantum Computing and AI

This section talks about the synergies with quantum computing and AI throughout.

Quantum bits (qubits) are the essential building blocks of quantum computing. They function as the fundamental units, representing the smallest quantum information. These units leverage the principles of superposition and entanglement, allowing for complex computations beyond classical limits. Quantum superposition permits a qubit to exist in multiple states simultaneously. Entanglement links qubits, creating a system where the state of one qubit directly affects another, irrespective of distance. NML utilizes nanoscale elements oriented by magnetic fields. These components operate by controlling the magnetic states of nanomagnets to process information. The interplay between quantum entanglement and nanomagnetic interactions offers new dimensions for computing. By merging these two domains, quantum computing can be scaled down to the nanoscale level. This integration presents challenges, particularly in maintaining compatibility between quantum and nanoscale systems.

Compatibility issues at the quantum–nano interface include aligning quantum coherence with nanomagnetic stability. Developing effective error correction strategies is crucial to manage the discrepancies at this juncture. Error correction ensures the reliability of computations amidst quantum noise and nanomagnetic instability. Specific methods must be tailored to handle errors unique to quantum-nano systems. Entanglement-based gate operations in NML are an area of active exploration. These gates use entangled qubits to perform computations, potentially enhancing processing power. Designing algorithms that harness both quantum and nanomagnetic computing strengths can lead to more efficient solutions. These algorithms must exploit the unique properties of quantum entanglement and nanomagnetic dynamics.

Materials with quantum properties need to be compatible with nanomagnetic elements. This requirement demands precise material engineering to achieve seamless integration. Developing such materials involves tailoring their quantum attributes to interact effectively with nanoscale magnetic elements. Pioneering information processing models at the quantum–nano– scale is essential for advancing this field. In medical diagnostics, integrating quantum precision with nanomagnetic resonance can improve diagnostic accuracy. Quantum-enhanced magnetic resonance techniques offer better resolution and sensitivity. This integration could revolutionize how medical imaging and diagnostics are performed. Energy efficiency is another critical area where quantum-nano synergy shows promise. AI applications can benefit from the reduced energy consumption achieved through this integration. Quantum computing can optimize algorithms, leading to more efficient AI processes. Nanomagnetic elements further enhance this efficiency by minimizing power requirements.

Quantum dots represent a potential breakthrough as quantum-nano information carriers. These dots can serve as efficient conduits for quantum information, bridging the gap between quantum and nanomagnetic systems. Their unique properties enable them to function effectively in this hybrid computing environment. A comprehensive roadmap is needed to guide the evolution of quantum–nano computing. This roadmap should outline key milestones, technical challenges, and potential breakthroughs. Such a plan will provide direction for researchers and developers working at the intersection of quantum and nanoscale technologies. Overall, the synergy between quantum computing and NML holds significant promise. By addressing compatibility challenges, developing targeted error correction strategies, and designing advanced algorithms, this integration can drive innovation. Enhanced materials, improved diagnostic techniques, and increased energy efficiency are some of the tangible benefits. The potential for quantum dots and a well-structured roadmap further supports the advancement of this emerging field. Environmental considerations, sustainability, and responsible integration with other technologies are crucial aspects in the development of nanoscale computing.

8.3.2 Future Perspectives in Nanoscale Computing

This section talks about the future perspectives in nanoscale computing throughout, and briefs the advancements in nanomagnetic logic in Section 8.3.2.1.

Neuromorphic Nanoscale Architectures: Neuromorphic nanoscale architectures derive inspiration from biological systems. These designs replicate the complex network of synaptic connections found in the brain. NML is employed to mimic these synaptic connections. This approach aims to advance artificial intelligence (AI), steering it closer to human-like cognitive abilities. The architecture integrates nanoscale components to emulate neural behaviors and processes. This method seeks to bridge the gap between traditional computing and brain-like computation. The use of nanoscale elements enhances the scalability and efficiency of neuromorphic systems. By mimicking biological processes, these architectures promise to revolutionize AI and machine learning domains.

Quantum Dot Integration: Quantum dots offer substantial potential for high-density data storage. They leverage quantum states to enhance data retention and retrieval. Incorporating quantum dots into nanoscale devices could redefine storage solutions. Quantum dots' unique electronic properties enable efficient data manipulation at a microscopic level. This integration supports the development of advanced memory devices with higher capacities. Employing quantum states also improves the stability and reliability of storage systems. The precision of quantum dots contributes to innovations in nanoscale computing technologies. Their application paves the way for more compact and powerful data storage solutions.

Quantum–Nano Synergies: The convergence of quantum and nanoscale computing presents transformative possibilities. This synergy involves integrating quantum computing principles with nanoscale technologies. Such integration aims to unlock new computational paradigms and efficiencies. The intersection of these fields promises significant advancements in computational capabilities. Quantum–nano synergies offer potential breakthroughs in processing speeds and data handling. This fusion could lead to entirely new types of computational devices. Envisioning a future where both fields merge provides insight into the next generation of computing innovations. These combined technologies hold the potential to redefine the landscape of computational science.

8.3.2.1 Advancements in Nanomagnetic Logic

Material Innovation: Graphene Unveiled: Graphene has emerged as a revolutionary material in computing technologies. Its exceptional properties are driving advancements in NML applications. The introduction of graphene has brought significant improvements in material performance. This single-layered material exhibits remarkable electrical and thermal properties. Graphene's introduction marks a pivotal shift in nanoscale computing materials. Its unique characteristics enable novel applications in nanomagnetic systems. As a game changer, graphene facilitates the development of more efficient and powerful nanoscale devices. The material's versatility opens new avenues for technological innovation in NML.

Graphene's Marvels Unraveled: Graphene's single-layered structure provides exceptional electrical conductivity. This property enhances the performance of electronic components in nanoscale devices. Additionally, graphene's mechanical strength contributes to its durability and resilience. Its high thermal conductivity impacts heat dissipation in nanoscale systems. These properties are crucial for the stability and efficiency of nanocomputing technologies. Graphene's potential extends beyond conventional materials, offering superior performance metrics. The integration of graphene in computing systems promises advancements in speed, efficiency, and reliability. These marvels of graphene are pivotal in advancing the field of NML.

NML Quantum Leap: NML is evolving in parallel with advancements in quantum computing. This evolution involves harnessing quantum principles to enhance computational methods. Integrating quantum concepts into NML systems drives significant progress. Quantum-inspired architectures are emerging, reflecting this evolution. These architectures aim to leverage quantum mechanics for improved computational capabilities. By adopting quantum principles,

NML is making strides toward more advanced computational solutions. This quantum leap represents a significant milestone in the field, promising enhanced performance and efficiency. The ongoing integration of quantum concepts into NML continues to push the boundaries of computing technology.

8.3.3 Graphene: A Conductor of Change

This section talks about the graphene (a conductor of change) throughout, and briefs the Potential Breakthroughs in Materials and Design in Section 8.3.3.1.

Graphene emerges as an outstanding electrical conductor, displaying remarkable efficiency in current transport. It ensures rapid data transmission by minimizing resistance and signal degradation. This characteristic significantly enhances the performance of electronic systems by improving interconnect quality. Graphene's low electrical resistance and high carrier mobility offer unprecedented speeds and efficiency. These properties make it indispensable for next-generation computing and communication technologies.

In nanoscale computing, graphene's adaptability is notable. Its flexibility facilitates the creation of bendable, wearable electronic devices. The materials mechanical properties enable its integration into various nanodevices, including sensors and actuators. This versatility allows for innovations in both consumer electronics and specialized industrial applications. Graphene's unique characteristics contribute to the evolution of nanoscale technology and computing. Integrating NML faces significant hurdles. The primary challenge is aligning this technology with existing conventional systems. Compatibility issues can arise from differences in material properties and operational mechanisms. Successful integration requires developing interfaces that bridge the gap between new and old technologies. Overcoming these obstacles is crucial for the widespread adoption of NML in various applications.

Graphene has achieved prominence in material science. Its extraordinary electrical and thermal conductivity, along with mechanical strength, makes it a revolutionary material. Collaborative efforts between researchers and industry professionals drive the advancement of graphene-based technologies. These collaborations focus on developing practical applications and refining production methods. Graphene's influence extends across multiple scientific disciplines, from physics to engineering. Sustainability in nanoscale technologies is becoming increasingly important. Graphene contributes to this movement through its lower environmental impact compared to traditional materials. Its production processes are less polluting and more efficient. The integration of graphene into nanocomputing promotes environmentally friendly practices. This shift toward green technologies reflects a growing commitment to reducing ecological footprints.

Edge AI devices benefit from the advancements in NML. This technology enables real-time data processing directly at the edge of the network. The reduced latency and increased computational power provided by NML enhance AI applications. As AI continues to evolve, NML plays a pivotal role in advancing edge computing technologies. Graphene significantly impacts energy efficiency in computing. Its high thermal conductivity aids in effective heat dissipation, enhancing overall system performance. Utilizing graphene-based components helps lower-energy consumption and reduces the carbon footprint of electronic devices. The combination of graphene with NML promises a new era of energy-efficient computing solutions.

Future trajectories in nanoscale computing are promising. The development of new materials and technologies continues to push the boundaries of what is possible. NML is expected to influence future technological advancements significantly. Researchers are exploring new applications and enhancements, aiming to achieve breakthroughs in nanoscale innovation. The synergy between NML and graphene is expected to drive the next wave of technological evolution. Overall, the continued exploration and application of graphene and NML are likely to lead to transformative changes in technology. The focus on integration challenges, sustainability, and energy efficiency will shape the future of nanoscale computing. The advancements in these areas are crucial for developing next-generation electronic systems and applications.

8.3.3.1 Potential Breakthroughs in Materials and Design

In the ever-evolving landscape of nanoscale computing, the chapter unveils a pivotal exploration graphene's magnetic properties and their potential breakthroughs in materials and design. This journey walks into the intricacies of graphene, – fundamental aspects of its magnetic attributes, implications for nanoscale computing, and the transformative potential it holds.

The schematic illustrates the strain dependence of magnetic anisotropy in FeRh/MgO bilayers, highlighting magnetic moments from electron spin interactions. This connects directly to the exploration of ferromagnetic and antiferromagnetic behaviors in graphene, emphasizing how strain influences magnetic properties (cf. Figure 8.15). Pointing out on the enhanced stability and Reliability in logic architectures. Figure 8.16 provides a detailed look at spin diffusion and spin-transfer torque mechanisms, essential in understanding the Quantum Spin Hall Effect and spin transport phenomena in graphene. The concept depicted support theoretical and practical advancements in spintronic applications discussed. Related to the theoretical underpinning of the Quantum Spin Hall Effect. The other schematic shows the design and operation of a skyrmion transistor driven by spin currents, correlating with the manipulation of spin waves and spin currents in graphene (cf. Figure 8.17). This underscores the role of spin currents

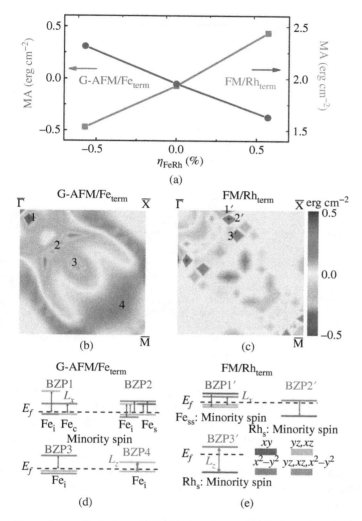

Figure 8.15 Strain impacts magnetic anisotropy in FeRh/MgO bilayers. Minority-spin and majority-spin behaviors vary at different BZ points. Spin–orbit coupling elements highlighted for orbital angular momentum. File: Figure 2 https://doi.org/10.1038/s41598-017-05611-7 Source: Zheng, G et al., 2017/Springer Nature/CC BY-4.0

for information transfer and controlled information processing in spintronic devices. Connected to the propagation of spin currents for information transfer.

Potential breakthroughs in materials and design present promising advancements. Graphene exhibits unique magnetic properties. Electron spin interactions induce magnetic moments. Ferromagnetic and antiferromagnetic behaviors are

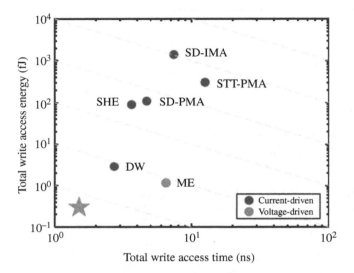

Figure 8.16 SD, spin diffusion; STT, spin-transfer torque; SHE, spin Hall effect; DW,domain wall motion; ME, magnetoelectric switching; IMA, in-plane magnetic anisotropy; PMA, perpendicular magnetic anisotropy. A red star designates the preferred corner (Pan, C et al., 2017/With permission of IEEE.). File: Figure 2: Performance estimates, writing energy vs time, for various spintronic memory cells. https://doi.org/10 .1038/s43246-020-0022-5 Source: Puebla, J et al./Springer Nature/CC BY 4.0.

explored. These properties integrate into NML. This leads to enhanced stability in logic architectures. It suggests energy-efficient computing innovations. The Quantum Spin Hall Effect underpins theoretical advancements. Topological insulators (TIs) in graphene support spin-polarized currents. These insulators hold potential for quantum computing. Graphene's magnetic junctions offer new fabrication methods. Magnetic domains in graphene aid information processing. MTJs impact nanocomputing significantly. Spin transport in graphene shows promise for applications. Spintronic applications benefit from graphene's capabilities. Spin currents facilitate efficient information transfer. Graphene nanoribbons modify magnetic behavior effectively. Spin waves enable controlled information processing. This manipulation enhances computing potential. The landscape of graphene's magnetic properties broadens. Research in these areas advances rapidly.

Graphene's magnetic landscape evolves with new findings. NML harnesses graphene's magnetic potential. Stability and reliability in architectures improve. Quantum computing progresses with spin-polarized currents. Magnetic junctions in graphene provide innovative pathways. Information processing leverages graphene's magnetic domains. Nanocomputing benefits from MTJs. Spintronic

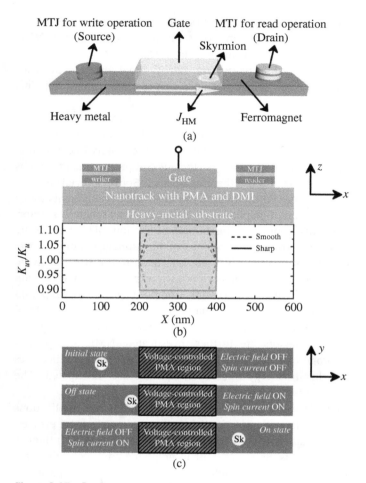

Figure 8.17 Design and operation of skyrmion transistor driven by spin current. Movement influenced by charge current and spin Hall effect. States illustrated by position in nanotrack and PMA modulation. File: Figure 7: Energy interconversion rectified by the electron spin. https://doi.org/10.1038/srep11369 Source: Zhang X, et al., 2015/Springer Nature/CC BY 4.0

applications explore new frontiers. Spin currents in graphene optimize information transfer. Graphene nanoribbons customize magnetic behavior. Spin waves revolutionize information processing. These developments redefine computing paradigms. Graphene's magnetic properties drive significant research. Enhanced stability in logic architectures materializes. Energy-efficient computing gains momentum. Quantum Spin Hall Effect insights deepen. Spin-polarized currents in TIs prove crucial. Quantum computing innovations emerge. Graphene's magnetic junctions demonstrate novel fabrication techniques. Information processing evolves with magnetic domains. Nanocomputing leverages MTJs effectively. Spin transport phenomena enrich spintronic applications. Graphene's role in spin currents advances information transfer. Tailored magnetic behavior in nanoribbons offers precision. Spin waves open new vistas in computing. This marks a transformative period in material science.

Graphene's magnetic landscape continues to expand. Research focuses on electron spin interactions. Ferromagnetic behaviors in graphene structures emerge. NML integrates these magnetic properties. Stability and reliability in computing architecture are enhanced. Potential energy-efficient computing breakthroughs are explored. Quantum Spin Hall Effect theories advance. Spin-polarized currents in graphene TIs hold promise. Future quantum computing developments hinge on these findings. Fabrication of graphene magnetic junctions progresses. Information processing benefits from magnetic domain utilization. MTJs impact the field of nanocomputing. Spintronic applications exploit graphene's spin transport phenomena. Spin currents enable advanced information transfer. Magnetic behavior tailoring through graphene nanoribbons is explored. Spin waves are manipulated for efficient information processing. Computing technology is poised for significant advancements. Graphene's magnetic landscape remains a focal point. Magnetic moments arise from electron spin interactions. Ferromagnetic and antiferromagnetic behaviors are further investigated. NML incorporates graphene's unique properties. Stability and reliability in logic architectures improve. Breakthroughs in energy-efficient computing are anticipated. The Quantum Spin Hall Effect is theoretically supported.

TIs in graphene enable spin-polarized currents. This holds key implications for quantum computing. Magnetic junction fabrication using graphene evolves. Graphene's magnetic domains facilitate information processing. MTJs influence nanocomputing developments. Spin transport in graphene advances spintronic applications. Information transfer benefits from spin current propagation. Graphene nanoribbons tailor magnetic behavior. Spin waves are harnessed for information processing. These advancements reshape the future of computing. Potential breakthroughs in materials and design are numerous. Graphene's magnetic properties offer significant promise. Electron spin interactions generate magnetic moments. Ferromagnetic behaviors in graphene structures are studied. NML benefits from these properties. Stability and reliability in architectures are achieved. Energy-efficient computing prospects emerge. The Quantum Spin Hall Effect supports new theories. Spin-polarized currents in TIs are critical. Quantum computing relies on these developments. Magnetic junctions in graphene are fabricated. Information processing leverages magnetic domains. MTJs play a role in nanocomputing. Spintronic applications explore spin transport in graphene. Spin current propagation enhances information transfer. Magnetic behavior in graphene nanoribbons is tailored. Spin waves are manipulated for computing. This leads to a transformative period in material science.

- **Equation: Spin–Orbit Coupling in Graphene** (cf. Eq. 8.1)

$$H_{so} = \frac{\lambda_{so}}{\hbar}(\sigma_x k_y - \sigma_y k_x) \tag{8.1}$$

Spin–orbit coupling Equation governing graphene's magnetic behavior. Exploration of spin–orbit interactions for magnetic control.

Emergent phenomena in magnetic graphene exhibit novel behaviors. Quantum Hall ferromagnetism arises from complex magnetic interactions. These interactions are investigated within nanoscale regimes. Magnonic devices leverage graphene's unique magnetic properties. Magnons, or spin waves, are manipulated for efficient information transmission. Graphene serves as a promising magnonic conductor, crucial for future computing systems. Research delves into graphene-centric computing paradigms, exploring potential applications. Quantum Hall ferromagnetism in graphene is thoroughly investigated. Such investigations aid in comprehending nanoscale magnetic behaviors. The development of magnonic devices utilizes graphene's magnetic properties. Magnons are precisely controlled for optimal information transmission. Graphene's conductive properties in magnonics are explored. These properties are integral to future computing innovations. The focus on graphene-centric computing paradigms drives technological progress. Magnetic graphene's applications in device architectures are explored. This research direction hints at future technological advancements.

- **Equation: Magnetoresistance in Graphene** (cf. Eq. 8.2)

$$R = R_0 + \frac{R_s}{1 + P^2\sin^2(\theta/2)} \tag{8.2}$$

Magnetoresistance equation characterizing graphene's magnetic response. Utilizing magnetoresistance for magnetic sensing applications.

Figure 8.18 illustrates the scaling of input data to an applied magnetic field, crucial for understanding anisotropic magnetoresistance (AMR) signals and their role in producing computational features. This relates directly to the investigation of magnetic phenomena in graphene, particularly how these interactions influence magnetic behavior and computing efficiency. The reconstruction of signals via a ring array and control measurements emphasizes the importance of precision in magnetic logic applications and the manipulation of magnons for advanced information processing.

Magnetic graphene enables revolutionary device architectures, enhancing computational capabilities. This exploration hints at a future where magnetic graphene redefines technological standards. Synergistic integration of magnetic logic with

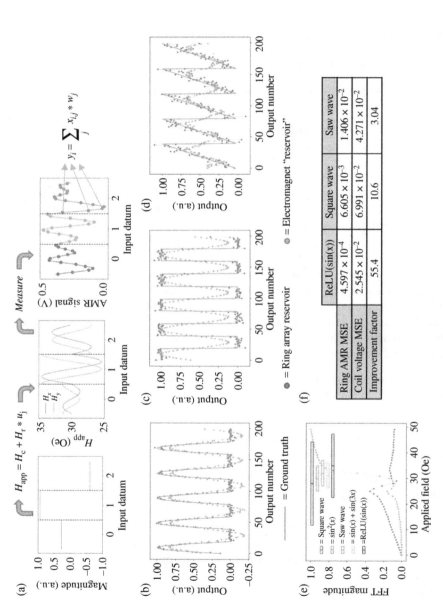

Figure 8.18 The diagram shows input data scaling to an applied magnetic field, rotation components, and anisotropic magnetoresistance (AMR) signal sampling. Optimal signal reconstructions from ring array and control electromagnet voltages are compared to target signals. Error comparisons are made between target signals and reconstructed outputs from ring array and electromagnet measurements. File: Figure 3: Performance of signal transformation task. https://doi.org/10.1038/s42005-023-01352-4 Source: I. T. Vidamour, et al.,

graphene's properties is pursued. This integration boosts advancements in computational speed and efficiency. The potential of nanoscale computing with graphene is unveiled. Investigations focus on graphene's magnetic phenomena and their applications. Researchers examine quantum Hall ferromagnetism's origins and implications. These studies contribute to understanding emergent behaviors in nanomaterials. Magnonic devices represent a significant application area. Magnon manipulation within graphene facilitates information transfer. Graphene's role as a magnonic conductor is crucial. Future computing technologies might rely heavily on these properties. Research on graphene-centric paradigms continues to expand. Innovative device architectures harness magnetic graphene's capabilities. This research direction promises to revolutionize computational technologies. Magnetic logic combined with graphene's attributes enhances performance. This combination accelerates computational processes and efficiency. The study of nanoscale computing with graphene reveals its potential. Detailed examinations of magnetic phenomena in graphene advance knowledge. Synergistic integration of magnetic logic and graphene propels progress. This integration fosters significant improvements in computational efficiency. The potential of graphene in nanoscale computing is continually unveiled. Magnetic phenomena in graphene undergo extensive investigation. Quantum Hall ferromagnetism's role in graphene is studied in depth. Understanding these phenomena aids in exploring nanoscale magnetic properties. The role of graphene in magnonic devices is significant. Control over magnons facilitates efficient information transmission. Graphene's properties as a magnonic conductor are key to future applications.

The exploration of graphene-centric computing paradigms continues. Magnetic graphene's potential in device architectures is recognized. This direction promises to redefine future computational technologies. The integration of magnetic logic with graphene properties enhances performance. Computational speed and efficiency see marked improvements. Graphene's potential in nanoscale computing becomes clearer. The study of emergent phenomena in magnetic graphene advances. Quantum Hall ferromagnetism and its implications are examined. This research contributes to understanding magnetic properties in nanomaterials. The development of magnonic devices leverages graphene's unique characteristics. Precision in magnon manipulation is critical for information transmission. Graphene as a magnonic conductor shows promise for future computing. Focus on graphene-centric computing paradigms drives innovation. Magnetic graphene in device architectures suggests future advancements. Synergistic integration with magnetic logic enhances computing capabilities. The potential of graphene in nanoscale computing is continually explored.

8.4   Nanoscale Computing Technologies Roadmap

This section talks about the nanoscale computing technologies roadmap, it briefs the One Transistor in Section 8.4.1, and describes the one nanomagnet in Section 8.4.2. Current research landscape exploring novel materials and innovative architectures is crucial. Quantum and magnetic computing paradigms are under thorough investigation. Integrating NML into edge AI devices is ongoing.

Foundations of Nanomagnetic Logic: NML principles are being meticulously understood. Magnetic properties are leveraged for performing logical operations. Pioneering research focuses on spintronic and magnonic applications.

Ongoing Material Innovations: Graphene is identified as a significant magnetic powerhouse. MTJs find diverse applications. Nanoscale computing is enhanced using TIs. Magnonic materials advance information processing capabilities.

Diversity in Architectures: Hybrid systems are developed collaboratively. Synergies with quantum computing are explored extensively. Researchers investigate 2D materials beyond graphene.

Energy Efficiency and Environmental Impact: Energy consumption in nanoscale devices is addressed strategically. Technological advancements are balanced with environmental responsibility. Sustainable manufacturing practices in nanoscale computing are emphasized.

Quantum-Nanomagnetic Synergies: Synergies between quantum and NML are unraveled. Prospects for quantum computing with nanomagnetic logic are explored. Quantum properties are harnessed for enhanced computing capabilities.

Nanoscale computing involves manipulating materials at an atomic level to develop smaller, faster, and more efficient devices. Researchers focus on novel materials like graphene due to its exceptional electronic properties. Graphene's unique two-dimensional structure allows for rapid electron movement, making it a magnetic powerhouse. This property is utilized in creating MTJs, which are pivotal in spintronic devices. Spintronics, leveraging electron spin instead of charge, significantly improves data storage and transfer rates. In addition to graphene, TIs are gaining attention. These materials conduct electricity on their surface while remaining insulative inside. Their unique properties enable robust and efficient nanoscale computing. Magnonic materials, using spin waves (magnons) for data processing, offer a promising alternative to traditional electronic devices. These materials are crucial for developing magnonic logic circuits, which can operate at lower power levels compared to conventional electronic circuits. Architectural diversity is another focal point. Hybrid systems combine multiple computing paradigms, enhancing overall performance. For instance, integrating quantum computing principles with nanoscale devices leads to significant advancements. Quantum computing exploits quantum bits (qubits) to perform complex calculations faster than classical computers. These hybrid systems also explore 2D materials beyond graphene, such as transition metal dichalcogenides (TMDs), which exhibit unique electronic and optical properties. Energy efficiency is a critical consideration in nanoscale computing. As devices shrink, power consumption and heat dissipation become significant challenges. Researchers address this by developing low-power NML circuits and optimizing energy usage in nanoscale devices. Additionally, they emphasize sustainable manufacturing practices to minimize environmental impact. Techniques like recycling rare materials and reducing hazardous chemical use are crucial.

Quantum-nanomagnetic synergies present exciting opportunities. Combining quantum properties with NML enhances computing capabilities. Quantum entanglement and superposition, core principles of quantum computing, offer new ways to process information. For instance, using qubits in NML devices can significantly boost processing speeds and data storage densities. Researchers are investigating these synergies to create next-generation computing systems. Graphene's magnetic properties are pivotal in advancing nanoscale computing. Its high electron mobility enables efficient data transfer, crucial for developing high-speed logic circuits. MTJs, formed by sandwiching an insulating layer between two ferromagnetic layers, exploit electron spin to store and process information. These junctions are integral to spintronic devices, offering faster and more reliable data processing than traditional electronic devices. TIs' unique electronic properties are harnessed for robust and efficient computing. Their surface states, protected by time-reversal symmetry, allow for lossless electron flow. This property is beneficial in developing low-power, high-efficiency nanoscale devices. Additionally, integrating TIs with other materials can create hybrid systems with enhanced functionalities. Magnonic materials, using spin waves for data transmission, offer a novel approach to information processing. Magnons, quanta of spin waves, propagate through magnetic materials without moving charge. This mechanism significantly reduces power consumption and heat generation, addressing critical challenges in nanoscale computing. Researchers are developing magnonic logic circuits, which promise lower-power consumption and higher operational speeds compared to conventional electronic circuits.

In hybrid systems, integrating quantum and classical computing paradigms enhances overall performance. Quantum computing principles, such as superposition and entanglement, enable parallel processing of information, significantly speeding up complex calculations. Hybrid systems also explore 2D materials like TMDs, which exhibit unique properties such as high electron mobility and strong spin-orbit coupling. These materials are used to develop advanced nanoscale devices with improved performance and efficiency. Energy efficiency is a central focus in nanoscale computing research. As devices become smaller, managing power consumption and heat dissipation becomes more challenging. Researchers address these issues by optimizing NML circuits for low-power operation. Techniques such as spintronic devices and magnonic logic circuits significantly reduce power consumption compared to traditional electronic devices. Additionally, sustainable manufacturing practices are emphasized to minimize environmental impact. These practices include recycling rare materials, reducing hazardous chemical use, and optimizing production processes for energy efficiency. Quantum-nanomagnetic synergies offer new possibilities for advanced computing systems. Quantum entanglement, where particles become interconnected and instantly affect each other, enables new ways to process and store information. Using qubits in NML devices leverages quantum properties for enhanced performance. For instance, qubits can exist in multiple states simultaneously, enabling parallel processing and significantly increasing computational power. Researchers explore these synergies to create next-generation computing systems with unprecedented capabilities.

Advanced Research and Applications: Graphene's potential in nanoscale computing extends beyond its magnetic properties. Its mechanical strength and flexibility allow for the development of flexible electronic devices. These devices can be integrated into wearable technology, providing powerful computing capabilities in a compact form. Additionally, graphene's thermal conductivity helps dissipate heat in nanoscale devices, addressing one of the critical challenges in miniaturization. MTJs are vital in spintronic devices. MTJs use the tunneling magnetoresistance effect to store and process information. When electrons tunnel through the insulating layer, their spin state determines the resistance, enabling data storage. MTJs offer faster data processing and higher storage densities compared to traditional electronic devices. Researchers are exploring new materials and structures to enhance MTJ performance further.

TIs enable robust computing with their unique surface states. These states, protected by symmetry, allow for lossless electron flow, reducing power consumption and heat generation. TIs are integrated into hybrid systems to create devices with enhanced functionalities. For instance, combining TIs with superconductors can develop fault-tolerant quantum computing systems, leveraging both materials' unique properties for improved performance. Magnonic materials offer promising applications in low-power data transmission. Spin waves, or magnons, propagate through magnetic materials without moving charge, significantly reducing power consumption and heat generation. Magnonic logic circuits use these spin waves for data processing, offering a novel approach to information processing. Researchers are developing magnonic devices with higher operational speeds and lower-power consumption than conventional electronic circuits.

Hybrid systems combining quantum and classical computing paradigms enhance overall performance. Quantum computing principles, such as superposition and entanglement, enable parallel processing of information, significantly speeding up complex calculations. Hybrid systems also explore 2D materials like TMDs, which exhibit unique properties such as high electron mobility and strong spin-orbit coupling. These materials are used to develop advanced nanoscale devices with improved performance and efficiency. Energy efficiency remains a primary concern in nanoscale computing research. As devices become smaller, managing power consumption and heat dissipation becomes more challenging. Researchers address these issues by optimizing NML circuits for low-power operation. Techniques such as spintronic devices and magnonic logic circuits significantly reduce power consumption compared to traditional electronic devices. Additionally, sustainable manufacturing practices are emphasized to minimize environmental impact. These practices include recycling rare materials, reducing hazardous chemical use, and optimizing production processes for energy efficiency. Quantum-nanomagnetic synergies present new opportunities for advanced computing systems. Quantum entanglement, where particles become interconnected and instantly affect each other, enables new ways to process and store information. Using qubits in NML devices leverages quantum properties for enhanced performance. For instance, qubits can exist in multiple states simultaneously, enabling parallel processing and significantly increasing computational power. Researchers explore these synergies to create next-generation computing systems with unprecedented capabilities.

Future Directions and Challenges: The future of nanoscale computing lies in further advancing material innovations. Researchers are continually exploring new materials with unique properties that can enhance device performance. For instance, beyond graphene, materials like molybdenum disulfide (MoS_2) and black phosphorus are investigated for their potential in nanoscale computing. These materials exhibit high electron mobility and strong spin–orbit coupling, making them ideal candidates for advanced nanoscale devices. Developing new architectures is also crucial for the future of nanoscale computing. Hybrid systems that combine multiple computing paradigms, such as classical, quantum, and spintronic computing, are being explored. These systems leverage the strengths of each paradigm to create devices with superior performance and efficiency. For instance, integrating quantum computing principles with nanoscale devices can significantly boost processing speeds and data storage densities.

Energy efficiency remains a critical challenge as devices continue to shrink. Researchers are developing new techniques to reduce power consumption and manage heat dissipation in nanoscale devices. For instance, using magnonic logic circuits, which use spin waves for data processing, significantly reduces power consumption compared to traditional electronic devices. Additionally, researchers are exploring new materials and device structures that can operate at lower power levels. Sustainable manufacturing practices are essential to minimize

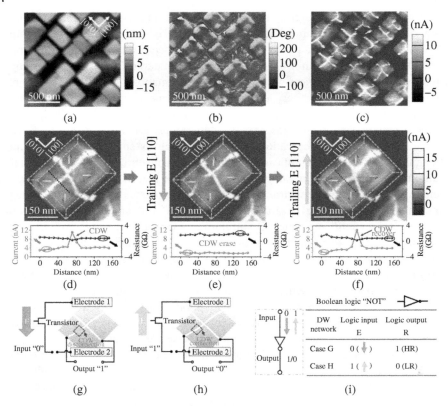

Figure 8.19 Morphology, PFM phase, and c-AFM images of vertex domains in BiFeO$_3$ nano-islands highlight their role in NOT gate logic operations, depicted with corresponding current profiles and truth tables. File: Figure 1: Electric field control of the on-and-off switching of a CDW and a NOT gate. https://doi.org/10.1038/s41467-022-30983-4 Source: Jing Wang, et al., 2022/Springer Nature/CC BY 4.0.

the environmental impact of nanoscale computing. Researchers are developing new techniques to recycle rare materials, reduce hazardous chemical use, and optimize production processes for energy efficiency. These practices are crucial for developing environmentally responsible nanoscale computing technologies.

Quantum-nanomagnetic synergies offer exciting opportunities for the future of computing. Researchers are exploring new ways to harness quantum properties for enhanced computing capabilities. For instance, using qubits in NML devices can significantly boost processing speeds and data storage densities. Additionally, researchers are developing new techniques to integrate quantum and classical computing paradigms, creating hybrid systems with superior performance and efficiency. Nanoscale computing technologies are advancing rapidly, driven by

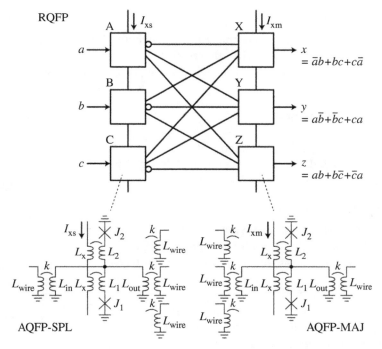

RQFP

$$x = \bar{a}b + bc + c\bar{a}$$

$$y = a\bar{b} + \bar{b}c + ca$$

$$z = ab + b\bar{c} + \bar{c}a$$

AQFP-SPL

AQFP-MAJ

Figure 8.20 The schematic details the RQFP gate composed of AQFP gates, illustrating its reversible and symmetrical design with specified input/output data and circuit parameters. File: Figure 1 https://doi.org/10.1038/s41598-017-00089-9 Source: Naoki Takeuchi, et al., 2017/Springer Nature/CC BY 4.0.

innovations in materials, architectures, and energy efficiency. Researchers are exploring new materials like graphene and TIs, developing hybrid systems that combine multiple computing paradigms, and optimizing devices for low-power operation. Sustainable manufacturing practices are emphasized to minimize environmental impact, and quantum-nanomagnetic synergies are explored to create next-generation computing systems. As research progresses, nanoscale computing technologies will continue to evolve, offering new possibilities for advanced computing systems. The use of vertex domains in $BiFeO_3$ nano-islands (cf. Figure 8.19) demonstrates advanced material manipulation in nanoscale computing. This aligns with ongoing research on novel materials and architectural diversity discussed above. The RQFP gate's schematic (cf. Figure 8.20) illustrates the complexity and precision of hybrid quantum computing systems. This reflects the synergy between quantum properties and nanoscale computing technologies mentioned.

8.4.1 One Transistor

This section talks about the one transistor throughout, briefs the physical characteristics in Section 8.4.1.1, describes the material characteristics in Section 8.4.1.2, and presents in detail the electrical characteristics in Section 8.4.1.3.

8.4.1.1 Physical Characteristics

- **Size:** Transistors are microscale components.
- **Structure:** Composed of semiconductor materials (e.g. silicon).
- **Shape:** Typically flat and rectangular.

8.4.1.2 Material Characteristics

- **Semiconductor Material:** Silicon is commonly used.
- **Doping:** Addition of impurities alters conductivity.
- **Insulating Layers:** Separate different semiconductor regions.

8.4.1.3 Electrical Characteristics:

- **Current Amplification:** Transistor amplifies electrical signals .
- **Switching Function:** Acts as a switch in circuits.
- **Voltage Control:** Operates based on applied voltages (cf. Eq. 8.3).

Equation:

$$I_C = \beta \cdot I_B \tag{8.3}$$

8.4.2 One Nanomagnet

This section talks about the one nanomagnet throughout, briefs the physical characteristics in Section 8.4.2.1, describes the material characteristics in Section 8.4.2.2, and presents in detail the electrical characteristics in Section 8.4.2.3.

8.4.2.1 Physical Characteristics

- **Size:** Nanomagnets are at the nanoscale.
- **Geometry:** Shapes can vary (e.g. elliptical, rectangular).
- **Magnetic Orientation:** Determines the magnet's poles.

8.4.2.2 Material Characteristics

- **Magnetic Material:** Often made of ferromagnetic alloys.
- **Anisotropy:** Influences magnetic properties.
- **Coercivity:** Resistance to demagnetization.

8.4.2.3 Electrical Characteristics

- **Magnetic Moment:** Represents strength and orientation.
- **Spin Polarization:** Key in spintronics applications.
- **Magnetic Switching:** Controlled by external stimuli (cf. Eq. 8.4).
  Equation:

$$M = \chi H \tag{8.4}$$

Transistors, characterized by their semiconductor nature, play crucial roles in signal amplification and switching functions. On the other hand, nanomagnets, operating at the nanoscale, are key components in spintronics, demonstrating magnetic properties influenced by materials and external factors. To perform a 1-bit addition, you would typically use XOR, AND, and OR gates along with additional logic for handling the carry. A simple implementation of a 1-bit full adder using CMOS technology involves several gates.

Here is a breakdown of the transistors needed for a basic 1-bit full adder in CMOS:

1. XOR gate: six transistors (two PMOS and four NMOS)
2. AND gate: six transistors (two PMOS and four NMOS)
3. OR gate: six transistors (two PMOS and four NMOS)

Additionally, you need some additional transistors for the carry logic:

1. Inverter: two transistors (one PMOS and one NMOS)
2. AND gate for the carry: six transistors (two PMOS and four NMOS)
3. OR gate for the carry: six transistors (two PMOS and four NMOS)

Now, summing up the transistors:

- XOR gate: Six transistors
- AND gate: six transistors
- OR gate: six transistors
- Inverter: two transistors
- Carry AND gate: six transistors
- Carry OR gate: six transistors

Total transistors for a 1-bit full adder: 32 transistors.

8.5 Conclusions and Key Findings

This section talks about the Conclusions and Key Findings, and briefs the summary of book contributions in Section 8.5.1.

- **Evolution Beyond CMOS:**
  - A dual historical perspective spanning 75 years of transistors and 100 years of spin was presented.

- Traditional CMOS technology challenges, such as scaling limitations, power dissipation, and performance bottlenecks, were discussed.
- NML emerged as a promising alternative, addressing the shortcomings of traditional CMOS technology.
- **Edge AI Devices: A Driving Force:**
  - Nanoscale Computing's pivotal role in Edge AI was underscored, exploring applications, use cases, and drawing insightful analogies.
  - The architecture and material design considerations for NML were examined, emphasizing the importance of architectural choices and material selection.
- **Limitations of CMOS Technology:**
  - Detailed exploration of challenges in traditional CMOS technology, including scaling limitations, power dissipation issues, and performance bottlenecks.
  - Implications for computing systems, technological and economic challenges, and the bridge to nanoscale computing were thoroughly investigated.

8.5.1   Summary of Book Contributions

This section talks about the summary of book contributions throughout.

1. **Fundamentals of Nanomagnetic Logic:**
   - An in-depth exploration of basic principles, magnetic properties, design, and operational aspects of NML gates.
   - Signal processing considerations, energy efficiency, and educational emphasis were crucial aspects covered.
2. **Nanomagnetic Logic Architectures:**
   - Overview of major architectures, including combinational and sequential logic architectures.
   - Exploration of parallel, pipelined, reconfigurable, and hybrid architectures, supported by case studies and a focus on hands-on learning.
3. **Material Design for Nanoscale Computing:**
   - The foundational role of material selection, exploring magnetic and non-magnetic materials, and case studies for practical application.
   - A tutorial on material design and a comprehensive conclusion summarizing key considerations.
4. **Nanoscale Computing at the Edge: AI Devices and Applications:**
   - An introduction to edge computing, examining its intersection with nanoscale computing and exploring applications in IoT, healthcare, and robotics.
   - Real-world case studies and a tutorial on developing edge AI applications provided practical insights.

5. **Hybrid Computing Systems and Emerging Applications:**
   - Introduction to hybrid computing, covering nanomagnetic–CMOS, quantum–nanoscale, and neuromorphic hybrid systems.
   - Exploration of emerging applications in industry, emphasizing practical design projects for hands-on learning.

8.6 Research Opportunities and Directions

This section talks about the research opportunities and directions, briefs the speculative laws on NML in Section 8.6.1, describes the identifying gaps in nanoscale computing research in Section 8.6.2, and concludes with the Chapter End Quiz in Section 8.6.3.

The evolution of computing laws drives NML, transcending CMOS. Predicting trends for 100 years entails forecasting every two years. Aligning with CMOS principles, NML laws foresee revolutionary shifts. In the first two years, NML prototypes demonstrate scalability and efficiency. Industry witnesses gradual interest, initiating foundational research collaborations. Research thrusts into NML material design for enhanced functionality. Innovations emerge, marking a shift from traditional computing paradigms. By the fifth and sixth years, NML's energy efficiency surpasses CMOS benchmarks. Educational initiatives focus on NML's real-world applications and hands-on experiences. Hybrid systems gain prominence, blending NML with existing technologies. In the eighth and ninth years, NML adoption expands, impacting edge AI devices. Pioneering industries invest in NML-powered applications, amplifying commercialization. R&D intensifies, exploring Reliability, scalability, and societal implications.

Entering the 10th and 11th years, NML standards solidify, paving the way for global adoption. NML-centric policies and regulations emerge, addressing ethical concerns and integration challenges. Industry leaders commit to sustainable practices. By the 15th and 16th years, NML infiltrates mainstream technologies, diversifying applications. Quantum–nanoscale hybrid systems gain traction, shaping the future landscape. NML research incorporates neuro-inspired architectures, unlocking unprecedented possibilities. In the 20th and 21st years, NML becomes ubiquitous, reshaping the computing ecosystem. Nanoscale computing prevails as a key in technological landscapes. Ethical considerations advance, driving responsible innovation practices. Moving ahead to the 25th and 26th years, NML pioneers collaborative computing models, fostering interdisciplinary approaches. Quantum–nanoscale-neuromorphic hybrids redefine computing paradigms, inspiring novel applications.

Advancing to the 30th and 31st years, NML integrates seamlessly into societal infrastructure, revolutionizing daily life. Nanocomputing education programs proliferate, cultivating a skilled workforce. Industry leaders converge to address global challenges. In the 35th and 36th years, NML accelerates the advent of smart ecosystems. Quantum–nanoscale-neuromorphic synergies redefine human-machine interactions. Sustainable practices drive NML research and application developments. Navigating to the 40th and 41st years, NML weaves into the fabric of emerging technologies, fostering symbiotic relationships. Interconnected smart cities leverage NML, enhancing efficiency and resource management. Cross-disciplinary research flourishes. The 45th and 46th years witness NML's transformative impact on healthcare. Quantum–nanoscale-neuromorphic interfaces revolutionize medical diagnostics and treatment modalities. Regulatory frameworks adapt to ensure responsible implementation.

In the 50th and 51st years, NML drives breakthroughs in environmental monitoring and conservation. Quantum–nanoscale-neuromorphic systems contribute to climate change mitigation efforts. Collaborations transcend borders for a sustainable future. Approaching the 55th and 56th years, NML converges with biotechnology, fueling advancements in personalized medicine. Quantum–nanoscale-neuromorphic interfaces redefine genetic research, ushering in a new era of precision healthcare. In the 60th and 61st years, NML underpins transformative developments in space exploration. Quantum–nanoscale-neuromorphic computing enables complex simulations and data analysis. International cooperation shapes the future of space exploration. Entering the 65th and 66th years, NML drives innovation in energy harnessing and storage. Quantum–nanoscale-neuromorphic solutions optimize renewable energy systems. Global collaboration focuses on sustainable energy futures.

As we reach the 70th and 71st years, NML permeates communication technologies, enabling secure and efficient networks. Quantum–nanoscale-neuromorphic encryption becomes a key of digital communication. In the 75th and 76th years, NML contributes to advancing nanorobotics for medical and industrial applications. Quantum–nanoscale-neuromorphic control systems redefine precision in nanoscale operations. Approaching the 80th and 81st years, NML interfaces with augmented reality, transforming human perception and interaction. Quantum–nanoscale-neuromorphic interfaces enhance immersive experiences. By the 85th and 86th years, NML converges with advanced materials science, leading to unprecedented innovations in material design. Quantum–nanoscale-neuromorphic approaches redefine the properties and applications of materials.

In the 90th and 91st years, NML shapes the evolution of artificial general intelligence. Quantum–nanoscale-neuromorphic architectures enable complex cognitive processes, pushing the boundaries of machine intelligence. Moving into the 95th and 96th years, NML contributes to the development of self-sustaining ecosystems in space. Quantum–nanoscale-neuromorphic solutions enable autonomous systems for long-term space exploration.

Approaching the 100th year, NML stands as the bedrock of computing evolution. Quantum–nanoscale-neuromorphic principles continue to inspire novel technologies, transcending the limitations of the past. Throughout this journey, NML adheres to the laws of efficiency, reliability, and ethical responsibility. The fusion of quantum–nanoscale-neuromorphic paradigms shapes a future where computing seamlessly integrates with the fabric of human existence. The next century holds promises of unparalleled innovation, guided by the speculative Laws.

8.6.1 Speculative Laws on NML

This section talks about the speculative laws on NML throughout. Envisioning laws for NML-based Beyond CMOS computing is a critical step considering the unique attributes and potential trajectories. Here are some speculative laws that could shape the landscape of NML computing:

- **Quantum-Magnetic Synergy Law:** (cf. Figure 8.21)
  - Computational power doubles: 2× every two years
- **Sustainability Quotient Law:**
  - Environmental impact decreases: 1.5× every three years
- **Entanglement Assurance Law:** (cf. Figure 8.22)
  - Entanglement fidelity increases: Ongoing progression, $e^{(t/4)}$
- **Security Resilience Law:** (cf. Figure 8.22)
  - Security resilience enhancement: 1.8× every five years
- **Nanofabrication Efficiency Law:** (cf. Figure 8.23)
  - Energy consumption decrease: 2% annually for two decades

Nanomagnetic Logic Growth Law: (cf. Figure 8.24)

$$N(t) = A \times 2^{(t/t_1)} \times 1.5^{(t/t_2)} \tag{8.5}$$

Explanation: (cf. Eq. (8.5))

- A: This constant represents the initial number of nanomagnets at $t = 0$. It sets the baseline for the growth of nanomagnets, depending on the starting point or the initial conditions of the NML device. In mathematical terms, it corresponds to the intercept of the exponential growth function. Its significance lies in providing a reference point, reflecting the initial state of the NML device.
- t_1: This constant determines the rate at which the number of nanomagnets doubles, following the pattern of Moore's Law. It sets the time it takes for the number of nanomagnets to double, analogous to the doubling time in exponential growth models. This constant is analogous to the doubling time in exponential growth models. In the context of Moore's Law, the doubling time refers to the period it takes for the transistor count on a chip to double.

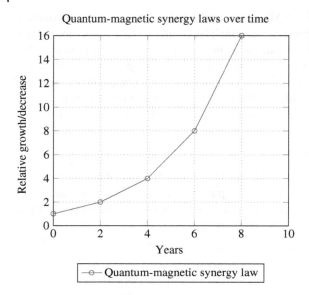

Quantum-magnetic synergy laws over time

Figure 8.21 Speculative law of NML – quantum magnetic synergy.

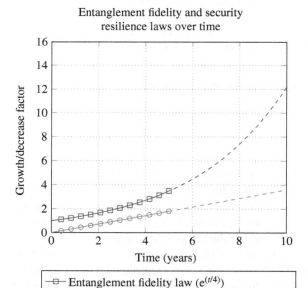

Entanglement fidelity and security resilience laws over time

Figure 8.22 Speculative laws of NML – entanglement fidelity and security resilience.

Energy consumption decrease over two decades

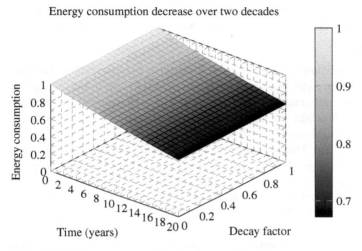

Figure 8.23 Energy consumption decrease projection.

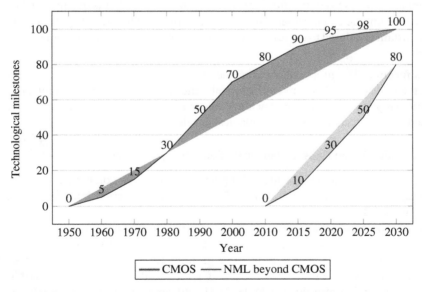

Figure 8.24 Evolution of CMOS verus NML beyond CMOS with milestones. Mathematical trajectory plot – Speculative law on NML.

- t_2: Similar to t_1, this constant governs the rate at which the number of nano-magnets increases by a factor of 1.5. It sets the time it takes for the additional growth factor to influence the total number of nanomagnets, related to the growth rate Specific to nanomagnetic logic technologies. It sets the time it takes for the additional growth factor to influence the total number of nanomagnets. This constant is related to the growth rate specific to NML technologies.

Significance:

1. **A - Initial Conditions:** The constant A is essential as it establishes the starting point for the growth, providing a reference point for the initial state of the NML device. In mathematical terms, it corresponds to the intercept of the exponential growth function. Its Significance lies in providing a Reference point, reflecting the initial state of the Nanomagnetic Logic device.

2. t_1 **- Doubling Time:** The constant t_1 dictates the pace at which the nanomagnets double in number, aligning with the concept of Moore's Law. It characterizes the rapid pace of growth in the early stages of the technology. The doubling time is a key parameter that characterizes the rapid pace of growth in the early stages of the technology. It reflects the historical trend observed in semiconductor technologies.

3. t_2 **- Additional Growth Factor:** The constant t_2 introduces an incremental growth factor specific to NML technologies, representing the time for unique enhancements to contribute to the overall growth. It represents the time it takes for the unique enhancements in NML to contribute to the overall growth. This constant is tailored to capture the expected advancements in the nanomagnetic domain.

Reference: The proposed Nanomagnetic Logic Growth Law (cf. Figure 8.25) is a theoretical model inspired by historical trends in semiconductor technologies, particularly Moore's Law. The constants' values are illustrative and can be adjusted based on empirical data and technological developments.

8.6.2 Identifying Gaps in Nanoscale Computing Research

This section talks about the identifying gaps in nanoscale computing research throughout and discusses the inviting feedback and interaction from readers in Section 8.6.2.1.

Nanoscale computing research encounters specific obstacles needing resolution. Present technologies face scalability constraints, necessitating alternative methods. CMOS technology is nearing its miniaturization threshold. Consequently, nanoscale solutions are increasingly investigated. NML presents viable options beyond conventional CMOS. NML utilizes nanomagnetics to achieve computation, providing novel pathways. Despite its potential, substantial research gaps impede progress. For instance, fabrication techniques for nanoscale devices remain underdeveloped. Additionally, energy efficiency in nanoscale systems needs optimization. Research also must focus on thermal management in nanoscale circuits. Quantum effects, significant at the nanoscale, pose further challenges. Addressing these issues is vital for advancing nanoscale computing.

Figure 8.25 NML Growth Law (cf. Eq. 8.5).

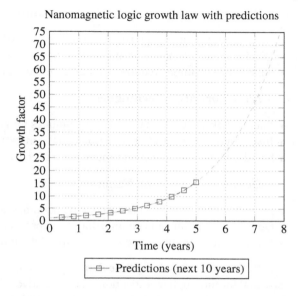

Nanomagnetic logic growth law with predictions

Moreover, enhancing material properties is crucial for reliable nanoscale components. The integration of nanoscale devices into existing systems also presents difficulties. System architecture needs adaptation to fully leverage nanoscale advantages. Consequently, interconnect technology requires innovation for nanoscale integration. Furthermore, developing new algorithms tailored for nanoscale hardware is essential. Thus, bridging these research gaps will unlock nanoscale computing's full potential.

Identified Key Research Gaps 1–6 as follows:
1. Nanomagnetic Logic Integration

- Integration challenges with existing computational architectures.
- Need seamless compatibility, avoid disruptive transitions.
- Lack of standardized protocols for NML integration.

2. Scalability and Performance Metrics

- Define scalable metrics for nanoscale computing performance.
- Lack of universally accepted benchmarks for NML evaluation.
- Quantify performance gains and energy efficiency accurately.

3. Reliability and Error Correction

- Investigate and mitigate errors in NML circuits.
- Develop robust error correction mechanisms for NML.
- Enhance fault tolerance, critical for practical applications.

4. Power Consumption and Heat Dissipation

- Analyze and minimize power consumption in NML systems.
- Efficient heat dissipation strategies lacking, impede scalability.
- Sustainable solutions required for energy-efficient nanocomputing.

5. Nanofabrication Techniques

- Explore advanced nanofabrication techniques for NML devices.
- Insufficient understanding of nanomagnetic material properties.
- Standardize nanofabrication processes for reproducibility.

6. Quantum Effects and Nanomagnetics

- Investigate quantum effects influencing nanomagnetic logic.
- Quantum tunneling and entanglement's impact on NML.
- Harness quantum phenomena for improved computational capabilities.

Path Forward – Resolving Research Gaps: Interdisciplinary research teams are essential for thorough investigations. Collaborative efforts between academia and industry remain crucial. Establishing partnerships yields more comprehensive insights. International cooperation is necessary for shared resources. Cross-border research enhances data reliability. Joint efforts streamline technological advancements. Aligning goals among various institutions is critical. Efficient resource allocation benefits all stakeholders.

Projection for Nanoscale Computing: NML is expected to surpass CMOS within two decades. Bridging research gaps is vital for NML adoption. Nanoscale computing will revolutionize multiple industries. Rapid innovation accelerates technological integration. Diverse applications emerge from nanoscale advancements. Enhanced computational power drives industrial growth. Sector-specific solutions improve operational efficiencies. The evolution of NML fosters competitive advantages. Comprehensive strategies are needed for widespread implementation.

Trajectories in Beyond CMOS Landscape: Nanoscale technologies continue evolving beyond CMOS. Quantum computing represents a key trajectory. Neuromorphic computing offers alternative pathways. Diverse approaches necessitate ongoing research. Technological advancements require continuous adaptation. Emerging methods reshape the computing landscape. Innovations drive the future of computational paradigms. Cross-disciplinary efforts facilitate progress in this domain. Real-time data integration improves technology applications. Collaborative environments enhance developmental milestones. Identifying and addressing research gaps pivotal for progress. NML, a promising beyond CMOS approach, demands concerted efforts. Collaborative research essential for shaping the future of nanoscale computing.

8.6.2.1 Inviting Feedback and Interaction from Readers
Engagement Initiatives 1. Interactive Dialogues

- Foster dialogues, engaging readers in discussions.
- Exchange thoughts, building a collaborative knowledge base.
- Responsive conversations, a dynamic exchange of ideas.

Engagement Initiatives 2. Commentaries Welcome

- Encourage comments, diverse perspectives enrich content.
- Varied insights, broadening the scope of discussions.
- Reader commentaries, valuable contributions to ongoing conversations.

Student Involvement in Future Research: Students play an essential role in shaping future research landscapes. Their active engagement fosters innovation in Beyond CMOS Logic. Dynamic contributions by students positively influence technological paradigms. Hands-on involvement provides students with practical research insights. Real-world applications supplement theoretical knowledge with experiential learning. Tangible experiences act as a catalyst for comprehensive understanding. Seamless integration connects academic concepts with practical implementations. Academic theories serve as foundations for tangible advancements in technology. A holistic approach merges classroom knowledge with cutting-edge research. Active student participation enhances research output quality. Involvement in projects equips students with critical problem-solving skills. Exposure to real-world challenges refines their analytical capabilities. Collaboration with seasoned researchers hones their investigative techniques. Students' fresh perspectives introduce novel ideas and solutions. Participation in interdisciplinary teams broadens their intellectual horizons. Engaging in research cultivates a robust understanding of complex concepts. Access to advanced tools and methodologies accelerates their learning process. Students' contributions to research lead to innovative breakthroughs. Their involvement ensures the continuous evolution of technological fields.

8.6.3 Chapter End Quiz

This section talks about the Chapter End Quiz throughout.

An online Book Companion Site is available with this fundamental text book in the following link: www.wiley.com/go/sivasubramani/nanoscalecomputing1

For Further Reading:

An et al. (2017), Proykova (2010), Ionescu (2016), Van den Hoven (2014), Madkour (2019a, 2019b), Zahid et al. (2013), Laucht et al. (2021) and Mount et al. (2019b).

References

Hongyu An, Zhen Zhou, and Yang Yi. Opportunities and challenges on nanoscale 3D neuromorphic computing system. In *2017 IEEE International Symposium on Electromagnetic Compatibility & Signal/Power Integrity (EMCSI)*, pages 416–421. IEEE, 2017.

Adrian Mihai Ionescu. Nanotechnology and global security. *Connections*, 15(2):31–47, 2016.

Arne Laucht, Frank Hohls, Niels Ubbelohde, et al. Roadmap on quantum nanotechnologies. *Nanotechnology*, 32(16):162003, 2021.

Loutfy Hamid Madkour. Environmental impact of nanotechnology and novel applications of nano materials and nano devices. *Nanoelectronic Materials: Fundamentals and Applications*, pages 605–699, 2019a.

Loutfy Hamid Madkour. Environmental impact of nanotechnology and novel applications of nano materials and nano devices. *Nanoelectronic Materials. Advanced Structured Materials*, vol 116. Springer, Cham, 2019b. https://doi.org/10.1007/978-3-030-21621-4_16.

Ana Proykova. Challenges of computations at the nanoscale. *Journal of Computational and Theoretical Nanoscience*, 7(9):1806–1813, 2010.

Jeroen Van den Hoven. Nanotechnology and privacy: the instructive case of RFID. In *Ethics and Emerging Technologies*, pages 285–299. Springer, 2014.

Muniza Zahid, Byeonghoon Kim, Rafaqat Hussain, Rashid Amin, and Sung Ha Park. Dna nanotechnology: a future perspective. *Nanoscale Research Letters*, 8:1–13, 2013.

Index

Nanoscale Computing: The Journey Beyond CMOS with Nanomagnetic Logic,
First Edition. Santhosh Sivasubramani.
© 2025 The Institute of Electrical and Electronics Engineers, Inc. Published 2025 by John Wiley & Sons, Inc.
Companion website: www.wiley.com/go/sivasubramani/nanoscalecomputing1

9 781394 263554